Maximilian Lackner,
Árpád B. Palotás, and
Franz Winter

Combustion

Maximilian Lackner, Árpád B. Palotás, and Franz Winter

Combustion

From Basics to Applications

Verlag GmbH & Co. KGaA

The Authors

Maximilian Lackner
Vienna University of Technology
Institute of Chemical Engineering
Getreidemarkt 9 /166
1060 Vienna
Austria

Árpád B. Palotás
University of Miskolc
Institute of Chemical Engineering
3515 Miskolc-Egyetemvaros
Hungary

Franz Winter
Vienna University of Technology
Institute of Chemical Engineering
Getreidemarkt 9 /166
1060 Vienna
Austria

All books published by **Wiley-VCH** are carefully produced. Nevertheless, authors, editors, and publisher do not warrant the information contained in these books, including this book, to be free of errors. Readers are advised to keep in mind that statements, data, illustrations, procedural details or other items may inadvertently be inaccurate.

Library of Congress Card No.: applied for

British Library Cataloguing-in-Publication Data
A catalogue record for this book is available from the British Library.

Bibliographic information published by the Deutsche Nationalbibliothek
The Deutsche Nationalbibliothek lists this publication in the Deutsche Nationalbibliografie; detailed bibliographic data are available on the Internet at <http://dnb.d-nb.de>.

© 2013 Wiley-VCH Verlag GmbH & Co. KGaA, Boschstr. 12, 69469 Weinheim, Germany

Cover Design   Formgeber, Eppelheim

Typesetting   Thomson Digital, Noida, India

Printing and Binding   Markono Print Media Pte Ltd, Singapore

Printed in Singapore
Printed on acid-free paper

Print ISBN (Hardcover): 978-3-527-33376-9
Print ISBN (Softcover): 978-3-527-33351-6
ePDF ISBN: 978-3-527-66721-5
ePub ISBN: 978-3-527-66720-8
mobi ISBN: 978-3-527- 66719-2
oBook ISBN: 978-3-527-66718-5

Contents

Foreword

Combustion is a fascinating process which has been quite instrumental in civilization and industrialization. From the hearth fire and cooking stove, via techniques for ore smelting, glass blowing and porcelain making, to steam engines, cars and power plants, combustion has accompanied the history of humankind. Now, in a situation where global warming and air quality deterioration are associated with combustion-generated carbon dioxide and pollutants, it is at the same time important to provide access to affordable but clean and sustainable energy. Combustion as mature technology still dominates today's power generation and transportation, and it is used in a number of important industrial processes and applications. Many professions, trades and businesses are linked to combustion, and many people world-wide depend on their work for car and aircraft manufacturers, for the petroleum, cement or steel industry or even as safety engineers and firefighters. This global situation is not likely to be changed rapidly, in spite of considerable effort to replace fossil by renewable energy, regarding the increase in world population and the desire to raise living standards and productivity accordingly. Combustion thus shows its Janus face today with promises of high-density fuels for an energy-hungry and mobile society on the one hand and threatening pictures of smog-polluted megacities without blue skies on the other.

It is thus time for the present volume as a summary and introduction to combustion fundamentals and applications for the more generally interested reader, including students and practitioners. With a fundament in physics and chemistry, modern concepts of combustion are presented in the necessary detail for a broad overview, without an excess of detail, in coherent and comprehensible fashion. The book provides a clear structure with seven Chapters, starting with some historical facts and interesting details in Chapter 1. The second Chapter introduces fuels with respect to their important properties and physic-chemical characteristics, accompanied by useful tables and literature. With Chapter 3, the fundamental principles of combustion are provided in an instructive form with some illustrative and facile calculation examples. The reader is introduced to the concepts of stoichiometry, the conservation equations and transport processes as well as to the basics of chemical reaction mechanisms and ignition processes. Pollutants are characterized in Chapter 4, which mainly gives some classifications and describes the main sources for specific emissions. For a more in-depth understanding of the different categories of pollutants and their chemical formation and destruction pathways, readers are referred to relevant original literature. The very important aspect of carbon dioxide formation from combustion processes and concepts for carbon dioxide management are presented in Chapter 5. The next Chapter is very instructive regarding the typical technical environments in which combustion is encountered, and it explains many interesting features of combustion devices including those found in heating, power generation, transportation, and in certain industries. As a modern concept, combustion and gasification are seen as somewhat related subjects, with a short access also to gasification strategies. The book concludes with important safety aspects, especially also regarding industrial-scale applications.

In a timely manner, the book offers an overview on an introductory level, and in this respect, it will be useful to a broad community. Certainly, huge tomes could be written for each subject treated in these Chapters, and substantial reviews and literature exists on individual facets of combustion – as for example, conventional and bio-derived fuels, combustion kinetics or specific combustion systems and applications. For a field that is as complex as that of combustion, a guided tour – as in the present case – is helpful to not lose orientation! I wish that you, the reader, may find combustion, not only at a candle-light dinner or for a barbecue, a fascinating object for study, in spite or because of the many challenges presented by its use. I also hope that by understanding the fundamental principles and limitations of combustion better, the community might find suitable replacement strategies for the systems in use today to contribute to a more efficient and cleaner energy use in the near future.

Bielefeld, April 2013 *Prof. Dr. Katharina Kohse-Höinghaus*

Preface

Combustion, the source of comfort and fear, warmth as well as devastation, has always fascinated mankind. It has been and still is one of the most important and most widely used technologies. In 2010, the authors published the "Handbook of Combustion" [1], a five-volume reference work that was very well received by the scientific community. Soon the idea was born to distill the knowledge from the approximately 3200 pages into a digestible textbook for students.

This book is designed to be a compilation of up-to-date knowledge in the field of combustion in a way that even a reader from a different field of expertise can understand the basic principles and applications. The purpose of this textbook is to provide an introduction to combustion science and technology, spanning from fundamentals to practical applications. It deliberately does not dwell too much on the details, although the book aims at providing the necessary knowledge for those wishing to move further into the various sub-disciplines, such as energy efficiency, oxyfuel combustion, gasification, pollutant reduction, or combustion diagnostics.

This book is written not only for undergraduate and graduate students of chemistry, chemical engineering, materials science, engineering and related disciplines, but also for practitioners in the field.

Topics covered are:

- History of combustion
- Fuels
- Combustion principles
- Environmental impacts
- Measurement methods
- Applications
- Safety issues

Each chapter can be studied independently. For further reading, web resources are suggested at the end of each chapter.

The authors are proud to present this textbook and hope that it will serve many technicians, scientists and engineers throughout their studies and careers.

Vienna, June 2013   *M. Lackner, Á. B. Palotás, F. Winter*

Reference

1 Lackner, M., Winter, F., and Agarwal, A.K. (eds) (2010) *Handbook of Combustion*, Wiley-VCH Verlag GmbH, Weinheim, 978-3527324491.

1
History of Combustion

1.1
Introduction

Combustion can be considered the oldest technology of mankind, and one of our most important discoveries/inventions. There are numerous legends around fire, for example, how it was handed to mankind by Prometheus. In Greece and India, there are stories and myths about eternally burning fires, often in relation to religious or supernatural phenomena. Legend has it that the Oracle of Delphi is located at the site of just such a fire [1]. Our ancestors used fire for various purposes, and today it accounts for the major share of primary energy production, approximately 85%. The science of combustion has a long history. Fire was one of the four elements in alchemy, and combustion processes were used for many transformations. Figure 1.1 depicts an alchemist's laboratory from $\sim$ 1600, where combustion plays a pivotal role.

Fire has been used by man for a long time for various purposes, such as cooking, metal production and warfare. However, as combustion phenomena are complex, significant advances in the understanding of combustion theory were only made in the last decades by a close collaboration between experimentalists and theoreticians.

An early observation made by the Flemish alchemist *Johann Baptista van Helmont* (1580–1644) was that a burning material results in smoke and a flame. From this, he concluded that combustion involved the escape of a "wild spirit" (*spiritus silvestre*) from the burning substance. Van Helmont's theory was further developed by the German alchemist *Johann Becher* (1635–1682) and his student *Georg Ernst Stahl* (1660–1734) into the phlogiston theory, according to which all combustible materials contain a special substance, the so-called phlogiston, which is released during combustion. This theory stayed in place for two centuries and was strongly defended by *Joseph Priestley* (1733–1804), who, with *Carl Wilhelm Scheele*, is also credited with the discovery of oxygen. Figure 1.2 shows Priestley's laboratory.

A classic textbook on combustion history is *Michael Faraday*'s *"The Chemical History of a Candle"* [2]. Other names associated with the development of combustion technology are *James Watt* (1736–1819), the inventor of the steam engine (more precisely, he made improvements to the Newcomen steam engine), plus *Rudolf*

Combustion: From Basics to Applications, First Edition. Maximilian Lackner, Árpád B. Palotás, and Franz Winter.
© 2013 Wiley-VCH Verlag GmbH & Co. KGaA. Published 2013 by Wiley-VCH Verlag GmbH & Co. KGaA.

Figure 1.1 Alchemist's laboratory, around 1600. Source: Deutsches Museum, Munich, Germany.

Diesel (1858–1913) and *Nikolaus Otto* (1832–1891), the inventors of their homonymous engines.

During the *Industrial Revolution* (~1750–1850), combustion of fossil fuels began to be used on a large scale for energy production, raw materials, the manufacture of various goods, mainly from steel, and transportation, for example, via steam engines, see Figure 1.3.

The black smoke coming out of factory chimneys during the Industrial Revolution (compare Figure 1.3) was seen as a sign of progress and prosperity.

In archeology and physical anthropology, one uses the so-called three-age system for the periodization of human prehistory into three consecutive time periods,

Figure 1.2 Joseph Priestley's laboratory, around 1773. Source: Deutsches Museum, Munich, Germany.

Figure 1.3 Industrial Revolution (Vintage engraving from 1878 of the Shipyards and shipping of the Clyde, Scotland).

which are named for their respective tool-making technologies: Stone Age, Bronze Age and Iron Age. The latter two are intimately associated with combustion. One can draw an analogy to energy and speak about the "coal age", "nuclear age" and "renewables age", while some observers are talking about a "second coal age" with the renewed interest in energy production from coal. Historically, the coal age – also termed the carboniferous period – is the time when coal was formed: 360 to 290 million years ago. Without doubt, mankind has lived in a "combustion age" for the last centuries.

1.2
Timetable

The following section lists key milestones in man's "taming" of fire, compiled from [3–9].

The process started 1–2 million years ago with natural fires, for example, triggered by a flash of lightning, that man could grab, and progressed to primitive ways of creating fire, for example, by friction (Figure 1.4) or flint stones (Figure 1.5) almost 10 000 years ago, followed by a rapid evolution of technologies:

First 500 000 years and beyond . . .

1–2 million years BC:	Man discovers fire

Before our time . . .

$\sim$ 7000 BC:	Man uses flint stones
$\sim$3000 BC:	Egyptians invent candles, made from beeswax

Figure 1.4 Making fire (Image from the Dutch stone age theme park Archeon).

~1000 BC:	Chinese use natural gas for lamps [1]
500 BC:	Greeks describe combustion with the 4 elements earth, water, air and fire
450 BC:	Herodotus describes oil pits near Babylon

Approaching the middle ages . . .

~100 AD:	Chinese invent gunpowder and fireworks
1242:	Roger Bacon, an English friar, publishes gunpowder formula [9]

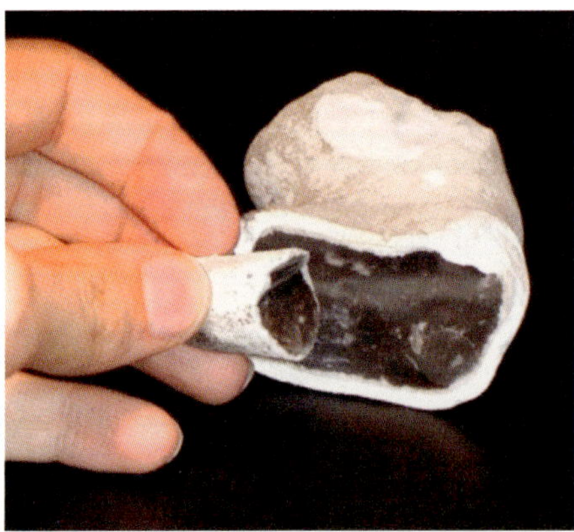

Figure 1.5 Flintstone, found in the Cliffs of Dover/UK.

fifteenth century:	Candles are used for street lighting
500 years ago . . .	
1556:	Georgius Agricola publishes *"De re metallica"*, a book cataloging the state of the art in mining, refining, and smelting of metals.
400 years ago . . .	
1627:	First recorded use of black powder for rock blasting in Hungary [9]
1650:	Otto von Guericke demonstrates that a candle does not burn in a vacuum
1678:	Abbé Hautefeuille describes an engine for raising water, powered by burning gun powder
1698:	Thomas Savery builds a steam-powered water pump for removing water from mines
~1700:	Christiaan Huygens comes up with the idea of an internal combustion engine
300 years ago . . .	
1712:	Thomas Newcomen builds a piston-and-cylinder steam-powered water pump for use in mines (Figure 1.6). It is the first practical device to harness the power of steam to produce mechanical work [10]
1772:	Carl Wilhelm Scheele carries out experiments to split air and discovers oxygen (Figure 1.7).
1778:	Alessandro Volta discovers the analogy between the ignition of combustible gases and fen fire (ghost light) in swamps (Figure 1.8).

Figure 1.6 Newcomen engine [10].

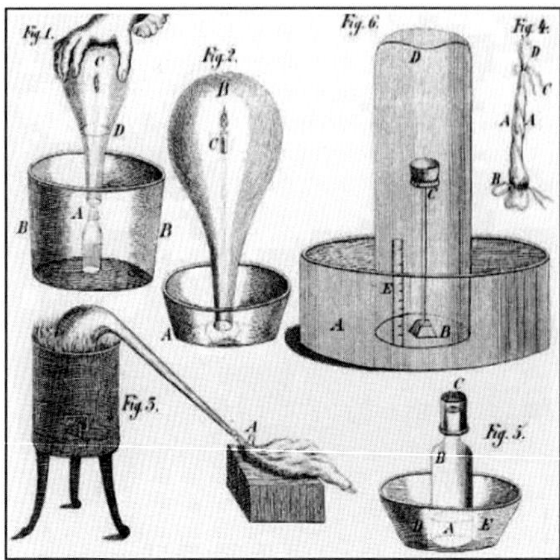

Figure 1.7 Experiments by C.W. Scheele to investigate the composition of air and to explain combustion and breathing processes. Source: Deutsches Museum, Munich, Germany.

1750–1850:	Industrial Revolution
1769:	James Watt (Figure 1.9) patents a steam engine
1790s:	Steam ship pioneer Samuel Morey invents a steam-powered paddle wheel
1791:	John Barber obtains a patent for a gas turbine
1792:	William Murdoch discovers the distillation of gas from coal and its use for lighting purposes
1794:	Concept for the first internal combustion engine by Robert Street
1800:	Phillippe Lebon patents an engine that uses compressed air and electricity for ignition
1801:	First coal powered steam engine
1806:	François Isaac de Rivaz invents a hydrogen-powered internal combustion engine

200 years ago . . .

1814:	George Stephenson builds a steam locomotive
1815–1819:	Sir Humphry Davy discovers catalytic combustion
1816:	Robert Stirling invents his hot air Stirling engine
1824:	Nicolas Léonard Sadi Carnot publishes that the maximum efficiency of a heat engine depends on the temperature difference between an engine and its environment
1834:	Joseph Morgan develops a machine that allowed the continuous production of molded candles
1837:	First American patent for an electric motor (US Patent 132)
1850:	Rudolf Clausius describes the first and second law of thermodynamics

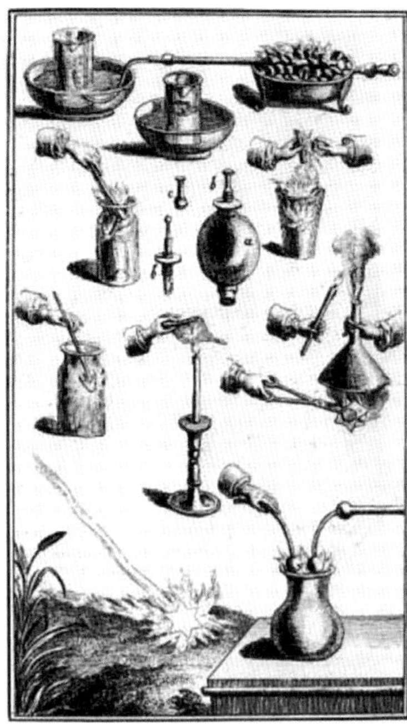

Figure 1.8 The gases which are formed in swamps are comparable to combustible gases from the lab. Source: Deutsches Museum, Munich, Germany.

~1855:	Robert Bunsen builds the Bunsen burner
1855:	Johan Edvard Lundstrom (Sweden) patents his safety match
1857:	Development of the kerosene lamp
1859:	John Tyndall discovers that some gases block infrared radiation. He suggests that changes in the concentration of the gases could bring climate change [4]

Figure 1.9 James Watt [10].

1860:	Étienne Lenoir and Nikolaus Otto build an internal combustion engine
1860:	Invention of fire extinguishers
1863:	J.D. Rockefeller opens an oil refining company in Cleveland
1863:	Julius Bernhard Friedrich Adolph Wilbrand invents trinitrotoluene (TNT) [9]
1866:	Swedish chemist Alfred Nobel invents dynamite by mixing kieselguhr with nitroglycerine [9]
1877:	Nikolaus Otto patents a four-stroke internal combustion engine (US Patent 194 047)
1880:	Thomas Alva Edison opens an electric power plant
1882:	James Atkinson invents the Atkinson cycle engine, which is used in some hybrid vehicles
1884:	Charles Parsons develops the steam turbine
1885:	Karl Benz builds a gasoline-powered car
1885:	Gottlieb Daimler patents the first supercharger
1892:	Rudolf Diesel patents the Diesel engine (US Patent 608 845)
1896:	Svante Arrhenius publishes the first calculation of global warming from human emissions of CO_2
1899:	Ferdinand Porsche creates the first hybrid vehicle
1903:	First flight by the Wright Brothers, Orville and Wilbur, at Kitty Hawk
1906:	Frederick Gardner Cottrell invents the electrostatic smoke precipitator

100 years ago . . .

1913:	René Lorin invents the ramjet
1915:	Leonard Dyer invents a six-stroke engine, now known as the Crower six-stroke engine named after its reinventor Bruce Crower
1920:	Robert H. Goddard develops the principle of a liquid-fueled space rocket
1923:	Fritz Pregl receives the Nobel Prize for combustion analysis
1929:	Felix Wankel patents the Wankel rotary engine (U.S. Patent 2 988 008)
1930s:	Global warming trend since late nineteenth century reported [4]
1936:	Maiden flight of the "Hindenburg" LZ 129 airship (volume 200 000 m^3, hydrogen-filled)
1954:	Foundation of the "Combustion Institute"
1957:	Russia launches "Sputnik I", the first artificial Earth satellite

50 years ago . . .

1969:	The USA land men on the moon, propelled by the Saturn V booster rocket developed by Wernher von Braun
1970s:	Electronically controlled ignition appears in automobile engines
1973:	Oil embargo and price rise bring first "energy crisis"
1975:	Catalytic converters are introduced on production automobiles in the US, co-invented by Carl Donald Keith
1979–1981:	Oil prices rise from \$13.00 to \$34.00/barrel
1981–2011:	The Space Shuttle, a partially reusable launch system and orbital spacecraft, is operated by the US National Aeronautics and Space Administration (NASA) for human spaceflight missions
1980s:	Electronic fuel injection appears on gasoline automobile engines
1986:	Word car population exceeds 500 million vehicles [11]

25 years ago . . .

1990:	First IPCC (Intergovernmental Panel on Climate Change) report says world has been warming and future warming seems likely
1990s:	Market introduction of hybrid vehicles that run on an internal combustion engine and an electric motor charged by regenerative braking
1990s:	CFD (computational fluid dynamics) is widely used as a tool for combustion simulation
1997:	Toyota introduces the Prius in Japan, the first mass-market electric hybrid car
1997:	The Kyoto Protocol to the United Nations Framework Convention on Climate Change (UNFCCC) sets binding obligations on industrialized countries to reduce emissions of greenhouse gases
1998:	50 year moratorium on mining and oil exploration in Antarctica approved

Current century . . .

2002:	Güssing's biomass gasification (demonstration plant with 8 MW$_{th}$ combined heat and power) in operation [12]
2003:	The US government announces its plan to build a near zero-emission coal-fired power plant for hydrogen and electricity production using carbon capture and storage (CCS)
2007:	European Union introduces new environmental regulations to reduce GHG (greenhouse gases) emissions by 20% by 2020

2008:	Wärtsilä builds the world's largest reciprocating engine, a two-stroke turbocharged diesel engine designed for large container ships with a power of 80 000 kW
2009:	Level of CO_2 in the atmosphere reaches 385 ppm [4]
2010:	1 000 000 000 cars on the road [11]
2013:	World's largest fluidized bed boiler (600 MW_{el}) is expected to be in operation in China
In future . . .	
2020–2030:	Peak oil expected [5]

1.3
Outlook

Combustion has been a science and engineering discipline for several hundred years, and it has driven the industrial revolution. Today, combustion plays an important role in our lives, from transportation to energy production, and it will continue to do so.

The focus of combustion research in the last decades has moved to pollution abatement, energy efficiency and alternative fuels (green combustion, zero emission combustion and near-zero emission combustion) [13]. Biomass – a renewable fuel – is expected to gain an increasing share over fossil fuels, as well as hydrogen as a clean energy carrier [14].

Today, energy is largely used inefficiently, because we deploy it as heat, where the limitations of Carnot apply, which are far from the thermodynamic limits. The Carnot efficiency is defined as

$$\eta = \frac{T_{\mathrm{h}} - T_{\mathrm{c}}}{T_{\mathrm{h}}}$$

where T_{h} is the absolute temperature of the hot body and T_{c} that of the cold body for a heat engine. So instead of talking about an "energy crisis" we could talk about a "heat crisis" of technical processes. Organisms, on the contrary, do not thermalize their energetic compounds, but rather utilize their chemical energy in a more efficient way: in a cascade of molecular-scale mechanisms that approach the reversible limit set by the difference in free energy. As these processes occur around room temperature, the irreversible losses ($T\Delta S$) are significantly smaller. A short-term technical realization of this concept is the *fuel cell* [15]. It is expected that fuel cells will gain importance as an efficient and clean "special" mode of combustion. There are various concepts and technologies, for example, proton exchange membrane (PEM) and molten carbonate (MC) fuel cells. Fuel cells can be combined with on-board fuel processing. *Gasification* can be used to obtain the feed for fuel cells from solid fuels.

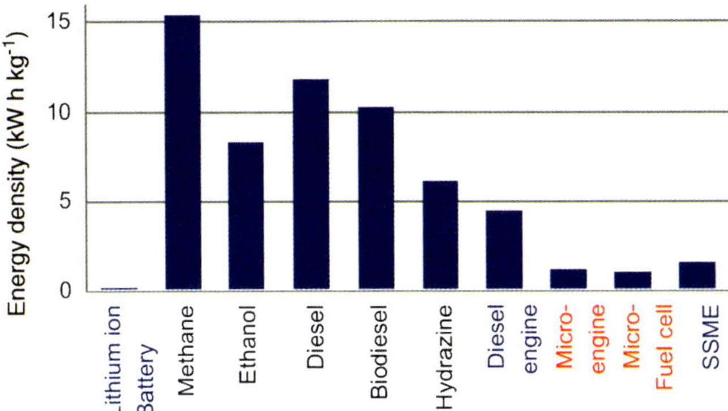

Figure 1.10 Comparison of specific energy densities of lithium ion batteries with hydrocarbon and oxygenated hydrocarbon fuels as well as different engines. SSME = space shuttle main engine. Reproduced with permission from [16].

When combustion processes are miniaturized, one talks about *"microscale combustion"*. This technology has attracted the interest of researchers recently. Microscale combustion differs fundamentally from "normal" combustion processes, since the vessel walls are closer than the quenching distance, so catalytic combustion processes are applied. Microscale combustion [16] might soon replace conventional batteries, as liquid fuels such as hydrocarbons have a high energy density, see Figure 1.10.

Figure 1.11 depicts a micro-gas turbine.

As mentioned, pollutant formation and reduction has become an important field of study in combustion in the recent decades as environmental awareness has increased. The study of health effects from combustion products [17] is also gaining importance.

Concerning fossil fuels, *oil sands* [18] will receive increased research attention. With respect to engines, *homogeneous charge compression ignition* (HCCI) [19] and various alternative ignition systems [20] are promising concepts, see Figure 1.12.

HCCI is difficult to control. The use of two fuels with different reactivities (such as gasoline and diesel) can help to improve this situation. Such a fuel is called *gasoline/diesel blend fuel* (GDBF) [22].

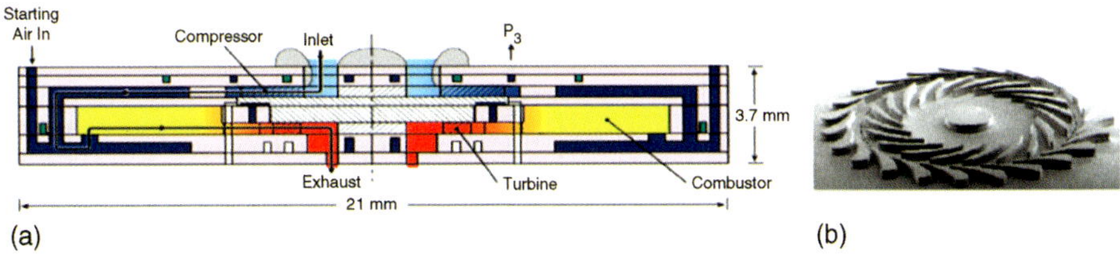

Figure 1.11 Scheme of a micro-gas turbine (a) and image of the turbine blade (b). Reproduced with permission from [16].

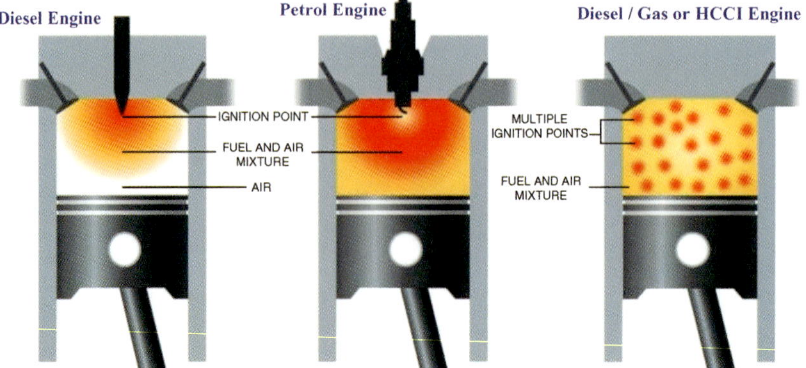

Figure 1.12 Concept of HCCI combustion with the aim of reducing soot and NO_x emissions. In HCCI and petrol (gasoline) engines, the fuel and air are mixed prior to combustion, which prevents soot emissions as found with diesel engines. The lean burn operation prevents the formation of NO_x [21].

Microdiagnostics and *combustion modeling* are other promising fields for future research. Another area of significant future research in the context of combustion is *energy storage*.

Today's burners and engines are mature, they will not solve the issues at hand. Major challenges in combustion research over the next decades, with a never-changing focus on safety and affordability, are *sustainability* (CO_2 reduction, energy efficiency, less pollutants) and *depletion of fossil fuels* (change to renewable resources).

There is additional potential in *knowledge exchange* and the *sharing of best practices* among combustion equipment operators.

It is also expected that *combustion synthesis* [23] will gain importance for the manufacturing of certain materials, see Figure 1.13 for an example.

Future research in combustion will help overcome several engineering limitations of today, as they are not fundamental limits imposed by nature.

We will see a transition from "conventional" fossil fuels to renewable fuels and also to "unconventional fossil fuels" (oil shale, tar sands, clathrates [25]), and eventually to all-renewable fuels, compare Figure 1.14 for the distribution of organic carbon on Earth.

The structure of a gas hydrate (clathrate) is shown in Figure 1.15.

Figure 1.16 depicts the location of hydrates for possible exploration on the world map.

Renewable fuels can be not only traditional fuels, such as wood, but also so-called energy crops, which can be burned directly or after conversion to a gaseous or liquid fuel by various technologies.

Figure 1.17 shows a projection of the oil and gas production in the US until 2035. One can see that unconventional oil and gas production will increase both in absolute numbers and relative contribution.

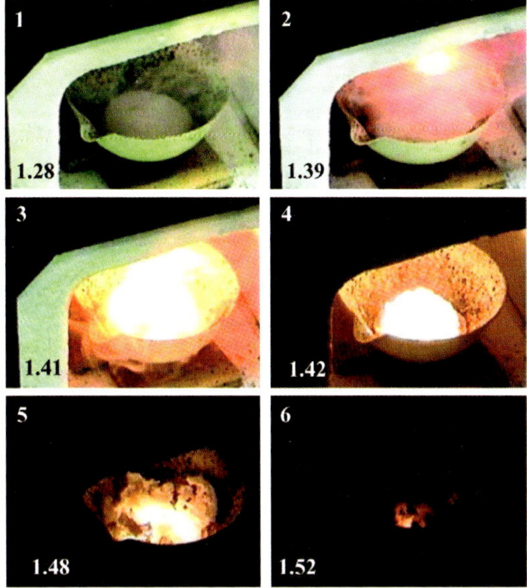

Figure 1.13 Combustion synthesis of Al_2O_3 in stoichiometric conditions at 600 °C. The number in the bottom right-hand corner refers to the time elapsed since the time the dish was placed in the oven (expressed in minutes and seconds) [24].

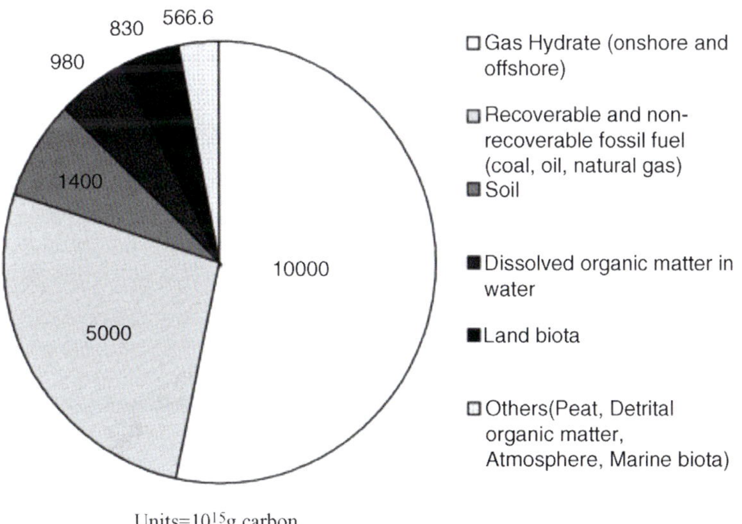

Figure 1.14 Distribution of organic carbon on earth (excluding dispersed organic carbon such as kerogen and bitumen) [25].

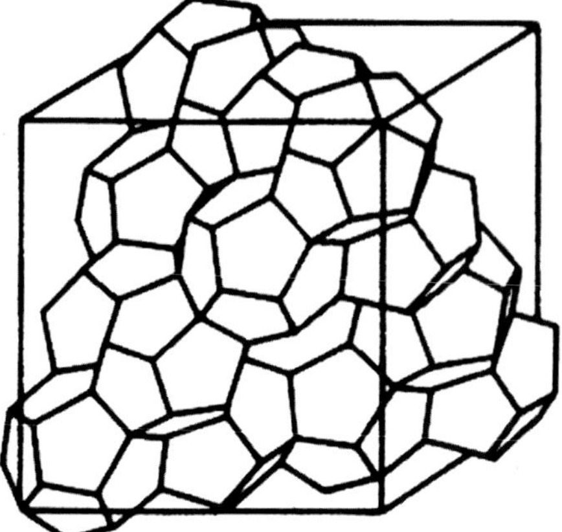

Figure 1.15 Unit cell of a gas hydrate [25].

Other emerging trends in combustion are:

- *Algae* for biomass combustion [27,28]
- *Gasification* for fuel cell technology
- Revival of the *external combustion engine* (Stirling), also in other areas such as solar power [29]
- Increasing significance of *methanol* as a transportation fuel [30].

In [31], a vision for process steam and process heat as "Combustion 2020" is formulated. An outlook on combustion research is provided in [32] and the IEA World Energy Outlook 2012 [5,26,33].

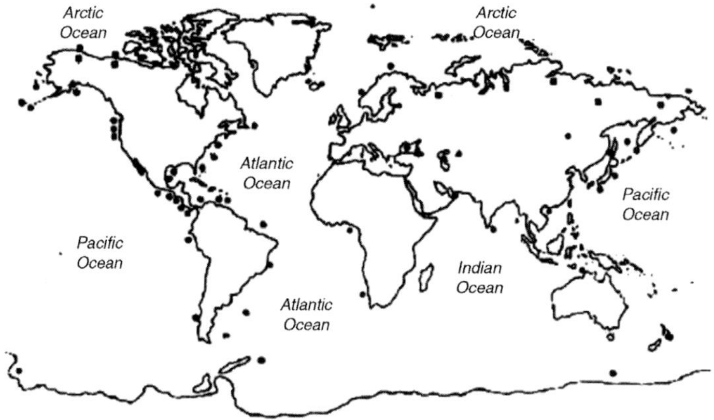

Figure 1.16 Map of *in situ* hydrate locations [25].

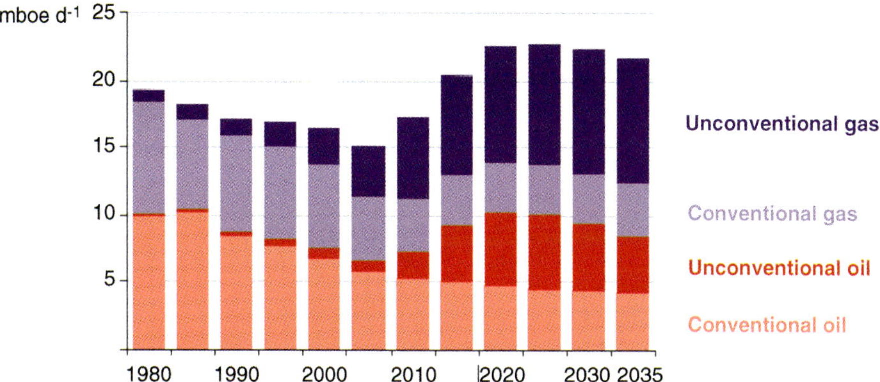

Figure 1.17 Trends in the US oil and gas production from "conventional" to "unconventional" according to the IEA Energy Outlook 2012 [26].

1.4
Web Resources

There are good resources on the internet to learn more about the history of combustion. A few of them are listed here for further exploration by the interested reader:

History Of Fire Milestone, One Million Years Old, Discovered In Homo Erectus' Wonderwerk Cave	http://www.huffingtonpost.com/ 2012/04/02/history-fire-million-homo-erectus_n_1397810.html
Alchemists, ancient and modern	http://www.economist.com/node/ 18226821
University of Bristol: Alchemy	http://www.chm.bris.ac.uk/ webprojects2002/crabb/history.html
IEA World Energy Outlook 2012	http://www.worldenergyoutlook.org/ media/weowebsite/2012/factsheets .pdf

References

1 Vattenfall (2013) http://www.vattenfall .co.uk/en/file/Natural_gas-ENG .pdf_16619005.pdf (accessed May 28, 2013).

2 Faraday, M. (2007) *The Chemical History of a Candle*, Book Jungle, ISBN: 978-1604241129.

3 Russum, D. (2013) http://www.geohelp .net/world.html (accessed May 28, 2013).

4 Spencer Weart & American Institute of Physics (2013) The Discovery of Global Warming, February 2013 http://www.aip

.org/history/climate/timeline.htm (accessed May 28, 2013).

5 World Energy Outlook 2012, IEA (2013) http://www.worldenergyoutlook.org/ publications/weo-2012, http://www .worldenergyoutlook.org/docs/weo2010/ weo2010_es_german.pdf (accessed May 28, 2013).

6 International Society of Explosives Engineers (2013) http://www.explosives .org/index.php/component/content/ article?id=69 (accessed May 28, 2013).

7 Cobb, C. and Goldwhite, H. (2001) *Creations of Fire: Chemistry's Lively History from Alchemy to the Atomic Age*, Basic Books, ISBN: 978-0738205946.

8 http://www.britannica.com/EBchecked/ topic/127367/combustion/285206/ History-of-the-study-of-combustion (accessed May 28, 2013).

9 Gregory, J.C. (1934) *Combustion from Heracleitos to Lavoisier*. Edward Arnold & Co, London.

10 Spear, B. and Watt, J. (2008) The steam engine and the commercialization of patents. *World Patent Information*, **30** (1), 53–58.

11 Sousanis, J. (2013) (15 August 2011) "World Vehicle Population Tops 1 Billion Units", *Wards Auto* http://wardsauto.com/ ar/world_vehicle_population_110815 (accessed May 28, 2013).

12 Hofbauer, H., Rauch, R., Bosch, K., and Koch, R., and Aichernig, C. (2002) Biomass CHP plant Güssing – A success story, in *Pyrolysis and Gasification of Biomass and Waste: Proceedings of an Expert Meeting, Strasbourg, 30 September–1 October* (ed. A.V. Bridgwater), CPL Press.

13 Chen, W.Y. Seiner, J. Suzuki, T. and Lackner, M. (eds.) (2012) *Handbook of Climate Change Mitigation*, Springer, ISBN: 978-1441979902.

14 Rifkin, J. (2003) *The Hydrogen Economy*, 1st edn, Tarcher, ISBN: 978-1585422548.

15 O'Hayre, R., Colella, W., Cha, S.-W., and Prinz, F.B. (2009) *Fuel Cell Fundamentals*, 2nd edn, John Wiley & Sons Inc., Hoboken, ISBN: 978-0470258439.

16 Ju, Y. and Maruta, K. (2011) Microscale combustion: Technology development and fundamental research. *Progress in Energy and Combustion Science*, **37** (6), 669–715.

17 Di Lorenzo, A. and D'Alessio, A. (1988) Trends in combustion technology in relation to health risk. *Annals of the New York Academy of Sciences*, **534**, 459–471.

18 Banerjee, D.K. (2012) *Oil Sands, Heavy Oil & Bitumen: From Recovery to Refinery*, Pennwell Corp., ISBN: 978-1593702601.

19 Zhao, F. (2003) *Homogeneous Charge Compression Ignition (HCCI) Engines: Key Research and Development Issues*, Society of Automotive Engineers, ISBN: 978-0768011234.

20 Lackner, M. (ed.) (2009) *Alternative Ignition Systems*, Process Eng Engineering GmbH, Vienna, ISBN: 978-3902655059.

21 Eco Gas (2013) http://www.eco-gas.com .au/wp-content/uploads/2009/06/ homogenous_mix1.jpg (accessed May 28, 2013).

22 Yu, C., Wang, J.-xin., Wang, Z., and Shuai, S.-jin. (2012) Comparative study on Gasoline Homogeneous Charge Induced Ignition (HCII) by diesel and Gasoline/ Diesel Blend Fuels (GDBF) combustion, Fuel, in press.

23 Lackner, M. (ed.) (2010) *Combustion Synthesis: Novel Routes to Novel Materials*, Bentham Science, eISBN 978-1-60805-155-7 http://www.bentham.org/ebooks/ 9781608051557/ (accessed May 28, 2013).

24 Civera, A., Pavese, M., Saracco, G., and Specchia, V. (2003) Combustion synthesis of perovskite-type catalysts for natural gas combustion. *Catalysis Today*, **83** (1–4), 199–211.

25 Lee, S.-Y. and Holder, G.D. (2001) Methane hydrates potential as a future energy source. *Fuel Processing Technology*, **71** (1–3), 181–186.

26 IEA (2013) http://www .worldenergyoutlook.org/pressmedia/ recentpresentations/ PresentationWEO2012launch.pdf (accessed May 28, 2013).

27 Haik, Y., Selim, M.Y.E., and Abdulrehman, T. (2011) Combustion of algae oil methyl ester in an indirect injection diesel engine. *Energy*, **36** (3), 1827–1835.

28 Agrawal, A. and Chakraborty, S. (2013) A kinetic study of pyrolysis and combustion of microalgae Chlorella vulgaris using thermo-gravimetric analysis. *Bioresource Technology*, **128**, 72–80.

29 Thombare, D.G. and Verma, S.K. (2008) Technological development in the Stirling cycle engines. *Renewable and Sustainable Energy Reviews*, **12** (1), 1–38.

30 Vancoillie, J., Demuynck, J., Sileghem, L., Van De Ginste, M., Verhelst, S., Brabant, L., and Van Hoorebeke, L. (2013) The potential of methanol as a fuel for flex-fuel and dedicated spark-ignition engines. *Applied Energy*, **102**, 140–149.

31 Office of Industrial Technologies (2013) *Combustion, Research and Development*, DOE (US Department of Energy), http://www.oit.doe.gov/combustion (accessed May 28, 2013).

32 Law, C.K. (2007) Combustion at a crossroads: Status and prospects. *Proceedings of the Combustion Institute*, **31** (1), 1–29.

33 http://www.worldenergyoutlook.org/media/weowebsite/2012/factsheets.pdf (2013).

2
Fuels

2.1
Introduction

Combustion is a series of exothermic chemical reactions between substances, including a combustible reactant, known as fuel, and an oxidizer, usually air or oxygen. The heat release is accompanied in most cases by light in the form of glowing or flame. Fuel in this context can be any material storing chemically bonded energy that can be extracted by the oxidation process after (self or external) ignition at a suitable temperature. The efficiency of the combustion process and the temperature to be maintained are affected by a number of parameters, including those of the environment, construction and operation; however, fuel characteristics are of major importance. In the following the most important properties of fuels are discussed for gaseous, liquid and solid fuels, respectively.

2.2
Gaseous Fuels

Gaseous fuels are simpler to ignite, handle and control than liquid or solid fuels. The molecular mixing of gaseous fuels and oxygen is very fast, especially when compared to the time of mass transport needed for the heterogeneous reactions of solid fuels. The combustion process of gaseous fuels is only limited by the velocity of mixing and the kinetics of the combustion reactions allowing for compact and intense burning. At atmospheric pressure a very short reaction time ($\sim$1 ms) can be achieved.

Burning stability means that the flame ignites easily and thereafter burns continuously and steadily. Burning stability depends on the burner geometry, the control of air and fuel flow that is maintained by the re-ignition points

1) above the fuel's minimum ignition temperature,
2) between the fuel's ignition limits, and
3) when the feeding-velocity equals the flame-velocity in the full operation range of the burner and under all circumstances.

Combustion: From Basics to Applications, First Edition. Maximilian Lackner, Árpád B. Palotás, and Franz Winter.
© 2013 Wiley-VCH Verlag GmbH & Co. KGaA. Published 2013 by Wiley-VCH Verlag GmbH & Co. KGaA.

Flame stabilization can be achieved through various means, including a continuously lit pilot flame(s), swirl, bluff body or other techniques that prevent flame extinction or lift off.

Gaseous fuels can be classified by their origin as

1) gases found in their final form in nature (e.g., natural gas, coal bed methane)
2) gases produced directly from various sources (e.g., hydrogen, biogas, landfill gas, synthesis gas)
3) by-product gases, where the primary aim of the process is not the generation of the gaseous fuel in question (e.g., coke oven gas, blast furnace gas, chemical processes)

The characterization as well as the combustion chemistry of the gaseous fuels can be generalized independently of the classification shown above, therefore we will not make a distinction in the following according to the origin of the fuel.

The most common gaseous fuel is natural gas, of which the main component is methane (CH_4). Valuable characteristics of natural gas are the relatively steady composition, high calorific value and combustion temperature, non-sooting flame, non-toxic nature and compressibility. Negative characteristics are: it is colorless, burns with a relatively low velocity, produces a non-luminescent flame, its exhaust is oxidative due to the water content, and it is highly explosive. The nominal heating value of natural gas is 35 MJ m^{-3}, although the exact value is a function of the actual composition. Certain sources are rich in inert components (CO_2, N_2), causing a reduced calorific value as low as 15 MJ m^{-3}. The composition of natural gas varies considerably between regions of origin, as shown in Tables 2.1 and 2.2.

Another important gaseous fuel in municipal use is PB (propane–butane) gas, obtained from raw natural gas by extracting the heavier components. PB gas can be stored and transported in liquid form. The chemical energy bound in PB gas is much higher than that of the natural gas: its calorific value is in the range 92–146 MJ m^{-3}. PB gas is used by home and business consumers lacking natural gas pipelines. PB gas is also utilized for enriching natural gas for high temperature technologies (e.g., glass production).

Table 2.1 Typical compositions of natural gas [1].

Name	Formula	Amount (vol.%)
Methane	CH_4	70–90
Ethane	C_2H_6	0–20
Propane	C_3H_8	
Butane	C_4H_{10}	
Carbon dioxide	CO_2	0–8
Oxygen	O_2	0–0.2
Nitrogen	N_2	0–5
Hydrogen sulfide	H_2S	0–5
Rare gases	Ar, He, Ne, Xe	trace

Table 2.2 Variations in composition of natural gas in Europe, vol. % [2–9].

Country	Field	CH_4	C_2H_6	C_3H_8	CO_2	N_2	H_2S
Denmark	Adda-1	71.63	6.29	5.78	—	0,3	—
	H-1x	88.75	4.25	1.41	0.55	0.9	—
Germany	Hannich-Nazza	15.10	0.50	—	0.9	83.2	—
	Fahner Hoehe 7	17.60	0.70	0.1	17.9	62.0	—
	Staakow	56.60	2.10	0.80	27.5	7.60	4.8
	Cappeln Z1	95.87	1.33	0.30	0.33	2.33	—
The Netherlands	K17-02B	61.92	6.9	3.15	1.06	20.60	2.8
	P01-01A	61.70	2.44	0.19	33.10	2.51	—
	Harlingen	95.79	3.50	—	—	3.61	—
Poland	Sulecin 21	1.80	0.24	0.18	97.6	0.04	—
	Rozansko-1	54.70	1.67	0.89	0.74	31.3	9.5
	Rokietnica-1	85.40	1.01	0.02	0.05	13.4	—
United Kingdom	Eskdale-2	98.80	0.90	0.30	—	—	—
	Gordon	82.00	—	—	—	16.0	—
Hungary	Kardoskút	77.01	2.00	0.94	15.77	3.00	—
	Beregdaróc (import)	97.91	0.81	0.28	0.05	0.81	—

The physical properties and combustion characteristics of PB gas components as well as other technological and technical gases are listed in Table 2.3.

2.2.1
Density

The density of the gas mixture (ρ_{mixt}) can be calculated from the density of each component (ρ_i) and their volume fraction φ_i (V_i/V):

$$\rho_{mixt} = \sum_{i=1}^{n} \varphi_i \rho_i \ \mathrm{kg\,m^{-3}} \tag{2.1}$$

With the exception of extreme conditions, gaseous fuels and combustion products can be assumed to behave according to the law of ideal gases. Therefore the density of a gas mixture as a function of temperature, T, and pressure, p, can be calculated as:

$$\rho_2 = \rho_1 \frac{T_1}{T_2} \cdot \frac{p_2}{p_1} \ \mathrm{kg\,m^{-3}} \tag{2.2}$$

2.2.2
Specific Heat Capacity

The specific heat capacity of the gas mixture ($c_{p\ mixt}$) can be calculated from the components' specific heat capacity values (c_{pi}) and the corresponding weight ratios of the components (mass fraction, w_i):

$$c_{p\ mixt} = \sum_{i=1}^{n} w_i \cdot c_{pi} \ \mathrm{kJ\,kg^{-1}\,K^{-1}} \tag{2.3}$$

Table 2.3 Density, molar mass and volume, gas constant of technical gases.

Name	Chemical Formula	Density kg m^{-3}	Molar Mass kg kmol^{-1}	Molar Volume m^3 kmol^{-1}	Gas Constant J kg^{-1} K^{-1}
Acetylene	C_2H_2	1.171	26.038	22.24	319.3
Argon	Ar	1.784	39.948	22.39	208.1
Benzene	C_6H_6	3.478	78.115	22.46	106.4
Isobutane	C_4H_{10}	2.668	58.12	21.78	143.1
n-Butane	C_4H_{10}	2.703	58.12	21.50	143.1
Ethane	C_2H_6	1.356	30.07	22.17	276.5
Ethylene	C_2H_4	1.261	28.054	22.25	296.4
n-Heptane	C_7H_{16}	4.459	100.21	22.47	83.0
n-Hexane	C_6H_{14}	3.840	86.178	22.44	96.5
Hydrogen	H_2	0.090	2.016	22.43	4124.7
Sulfur dioxide	SO_2	2.928	64.06	21.88	129.8
Hydrogen sulfide	H_2S	1.539	34.08	22.14	244.0
Air	-	1.293	28.96	22.40	287.1
Methane	CH_4	0.718	16.043	22.38	518.3
Nitrogen	N_2	1.206	28.013	22.40	296.8
Oxygen	O_2	1.429	31.999	22.39	295.8
n-Pentane	C_5H_{12}	3.457	72.151	20.87	115.2
Propane	C_3H_8	2.019	44.097	21.84	178.6
Propylene	C_3H_6	1.915	42.081	21.97	197.6
Carbon dioxide	CO_2	1.977	44.01	22.26	188.9
Carbon monoxide	CO	1.25	28.01	22.41	296.8
Steam	H_2O	0.804	18.015	23.46	461.5

In practice, specific heat capacity per unit volume of gas mixture is commonly used.

$$c_{p\,\mathrm{mixt}} = \sum_{i=1}^{n} \varphi_i \cdot c_{pi} \ \mathrm{kJ\,m^{-3}\,K^{-1}} \tag{2.4}$$

2.2.3
Molar Weight

The molar (molecular) weight of a gas mixture (M_{mixt}) can be calculated from the molar weights of the constituents (M_i):

$$M_{\mathrm{mixt}} = \sum_{i=1}^{n} \varphi_i \cdot M_i \ \mathrm{g\,mol^{-1}} \tag{2.5}$$

2.2.4
Gas Constant

For calculations, for example, related to burner design the gas constant of the gas mixtures (R_{mixt}) might be an important parameter. One can calculate the gas mixture's gas constant from the molar weight of the mixture (M_{mixt}) and the universal gas constant (R):

$$R_{\mathrm{mixt}} = \frac{R}{M_{\mathrm{mixt}}} \ \mathrm{J\,kg^{-1}K^{-1}} \tag{2.6}$$

The gas constant of gas mixtures can also be calculated from the weight ratios and the gas constant values (R_i) of the components:

$$R_{\mathrm{mixt}} = \sum_{i=1}^{n} w_i \times R_i \quad \mathrm{J\,kg^{-1}K^{-1}} \tag{2.7}$$

2.2.5
Thermal Conductivity

The thermal conductivity of gases is a feature depending on the material properties, temperature and pressure.

For gas mixtures, the thermal conductivity (λ_{mixt}) can be calculated based on the individual thermal conductivities (λ_i) according to the following equation:

$$\lambda_{\mathrm{mixt}} = \frac{\sum_{i=1}^{n} \lambda_i \cdot \varphi_i \cdot \sqrt[3]{M_i}}{\sum_{i=1}^{n} \varphi_i \cdot \sqrt[3]{M_i}} \ \ \mathrm{W\,m^{-1}K^{-1}} \tag{2.8}$$

2.2.6
Viscosity

During the flow of real gases shear stress (internal friction) cannot be neglected.

Shear stress arising between adjacent gas layers:

$$\tau = \eta \frac{\mathrm{d}w}{\mathrm{d}y} \ \ \mathrm{Pa} \tag{2.9}$$

In the correlation η is dynamic viscosity in Pa s; w is the flow velocity in $\mathrm{m\,s^{-1}}$; y is the distance coordinate perpendicular to the direction of flow in m. The dynamic viscosity of perfect gases in a pressure range from 10^3 to 10^6 Pa is independent of the pressure.

The viscosity of gas mixtures (η_{mixt}) cannot be calculated according to the simple mixing rules, especially in the case of high hydrogen content. The dynamic viscosity of gas mixtures can be determined according to the Herning–Zipperer equation:

$$\eta_{\mathrm{mixt}} = \frac{\sum_{i=1}^{n} \eta_i \cdot \varphi_i \cdot \sqrt[3]{M_i}}{\sum_{i=1}^{n} \varphi_i \cdot \sqrt[3]{M_i}} \ \ \mathrm{Pa\,s} \tag{2.10}$$

The accuracy of the calculation conducted with this equation decreases with increase in the mixture's hydrogen content (up to 10% hydrogen content the error is about 2%, above 25% the equation is not reliable).

Kinematic viscosity (v) can be expressed as a ratio of dynamic viscosity and density – both of which are strongly dependent on pressure and temperature.

$$v = \frac{\eta}{\rho} \ \mathrm{m^2\,s^{-1}} \tag{2.11}$$

The dynamic viscosity dependence on temperature can be taken into consideration with the Sutherland equation:

$$\eta = \eta_0 \frac{1 + \frac{C}{273}}{1 + \frac{C}{T}} \sqrt{\frac{T}{T_0}} \ \mathrm{Pa\,s} \tag{2.12}$$

where $p_0 = 101\ 325$ Pa and $T_0 = 273.15$ K ($= 0\,°C$), and η_0 is the corresponding dynamic viscosity, C is the Sutherland constant in K. Some of the common gases and their viscosity and conductivity data are listed in Table 2.4.

2.2.7
Heating Values

The higher heating value or HHV (also known as gross calorific value) is the heat released by burning a unit of dry fuel, when the starting temperatures of both the fuel and the air used for the combustion and the final temperature of the combustion product are 20 °C, the water content of the combustion product is in liquid form. The definition of lower heating value or LHV (also known as net calorific value) only differs in one aspect: the water content of the combustion product stays

Table 2.4 Heat conductivity and viscosity of gases (at 0 °C and 101.3 kPa).

Gas	Dynamic Viscosity, η (10^{-6} Pa s)	Kinematic Viscosity, v (10^{-6} m^2 s^{-1})	Heat Conductivity, λ (10^{-3} W m^{-1} K^{-1})	Sutherland Constant, C (K)
Acetylene	9.43	8.05	18.70	198.2
n-Butane	6.97	2.55	13.26	349
Ethane	8.77	6.45	18.96	287
Hydrogen	8.44	93.80	175.1	83
Sulfur dioxide	11.8	4.49	8.58	416
Hydrogen sulfide	11.82	7.68	13.14	331
Air	17.53	13.57	24.42	122
Methane	10.55	14.69	30.70	198
Nitrogen	16.34	13.55	24.31	107
Oxygen	19.62	13.73	24.66	138
Propane	7.65	3.82	15.24	324
Carbon dioxide	13.82	6.99	14.65	255
Carbon monoxide	16.93	13.55	23.26	102
Steam	8.66	10.77	23.73	673

in the gaseous form. Thus the difference between the HHV and the LHV is equal to the heat of evaporation. The higher the water content of the flue gas, the greater the difference between the HHV and the LHV. Accordingly, there is no difference between the HHV and the LHV if the combustion products do not contain water, for example, if the fuel is pure CO.

The higher and/or lower heating value of a dry gas mixture can be calculated using the mixing rules, that is, from the composition of the mixture and the corresponding heating values of the components:

$$HHV_{mixt} = \sum_{i=1}^{n} \varphi_i \cdot HHV_i \ kJ \ m^{-3} \tag{2.13}$$

$$LHV_{mixt} = \sum_{i=1}^{n} \varphi_i \cdot LHV_i \ kJ \ m^{-3} \tag{2.14}$$

Table 2.5 summarizes the lower and higher heating values of various combustible gases.

When both experimental and theoretical (calculated) heating values are available, generally the measured (experimental) data are preferred.

2.2.8
Ignition Temperature

During combustion the reaction is initiated by either an external ignition source or autoignition. They are both the result of energy transfer (see also Chapter 3, Section 3.4).

Autoignition occurs when the slow oxidation reactions across the whole volume accelerate suddenly and the slow oxidation process transforms into a sudden explosion, like burning. The autoignition temperature in a given system is defined as the lowest temperature at which the gas mixture in that given system exhibits self-acceleration of the reaction as described above. In order to prevent autoignition, the

Table 2.5 Heating values of pure gases.

Gas type	Higher heating value, HHV			Lower heating value, LHV		
	$MJ \ kg^{-1}$	$MJ \ kmol^{-1}$	$MJ \ m^{-3}$	$MJ \ kg^{-1}$	$MJ \ kmol^{-1}$	$MJ \ m^{-3}$
Hydrogen	142.380	287.024	12.796	120.470	242.856	10.827
Carbon monoxide	10.144	284.156	12.609	10.144	284.156	12.609
Methane	55.646	892.723	39.887	50.152	804.592	35.949
Ethane	52.034	1.564.652	70.558	47.652	1.432.887	64.616
Propane	50.510	2.227.355	101.980	46.513	2.051.097	93.910
n-Butane	49.090	2.853.126	132.691	45.347	2.635.555	122.572
n-Pentane	45.741	3.300.277	158.128	42.245	3.047.994	146.040
n-Hexane	48.829	4.207.957	187.502	45.240	3.898.647	173.720
n-Heptane	48.023	4.812.344	214.133	44.548	4.464.110	198.638
Ethylene	50.146	1.414.370	63.549	47.260	1.325.837	59.571
Benzene	42.474	3.317.833	147.708	40.772	3.184.863	141.788

container (the enclosing wall) of the fuel–oxidant mixture should be at a lower temperature than the autoignition temperature.

The ignition temperature is a feature of the fuel type and the igniting system. The gas–oxidizer mixture that has the lowest ignition temperature is called by various names, such as the minimum autoignition temperature, the minimum spontaneous ignition temperature and the self-ignition temperature. Usually the autoignition temperature reported in the literature is the minimum one.

The autoignition temperature depends on many factors, such as

- pressure,
- oxygen content,
- ignition delay,
- energy of the ignition source,
- flow conditions,
- volume of the combustion chamber.

Increasing the oxygen content and the presence of dust decreases the autoignition temperature [10]

2.2.9
Ignition Limits

Limits of ignition (at a given pressure) refer to concentration values, outside of which range ignition will not be possible. In other words, if the fuel concentration is above the upper limit or below the lower limit, the mixture cannot be ignited.

The ignition concentration limits of combustible gases and gas mixtures ($Z_{c\ mixt}$) are affected by the initial pressure and temperature, and their inert gas content and can be calculated as follows:

$$Z_{c\ mixt} = \frac{\sum_{i=1}^{n} \varphi_{c\ i}}{\sum_{i=1}^{n} \frac{\varphi_{c\ i}}{Z_i}} \tag{2.15}$$

where Z_i is the lower (or upper) ignition concentration limit of the components, and φ_{ci} is the volume ratio of combustible constituents to the combustible material of the mixture.

Table 2.6 lists ignition concentration limits of various gases in air at atmospheric pressure and ambient temperature (20 °C).

2.2.10
Laminar Flame Velocity

Experimental data confirm that, upon ignition of a combustible mixture, the reaction front travels with a definite velocity. The actual chemical reaction occurs in a thin layer separating the still unburned combustible gas mixture and the combustion product at any given moment. This layer is called the burning zone or flame front. The velocity of the combustion reaction can be characterized by the

Table 2.6 Ignition limits of combustible gases (in 20 °C air) and their laminar flame velocities.

Name	Formula	Ignition Limit (%)			Maximum laminar flame velocity (m s^{-1})
		lower	max.	upper	
Methane	CH_4	5.0	9.96	15.0	0.35
Ethane	C_2H_6	3.0	6.28	12.5	0.40
Propane	C_3H_8	2.2	4.54	9.5	0.39
Butane	C_4H_{10}	1.9	3.52	8.5	0.38
Pentane	C_5H_{12}	1.4	2.92	7.8	0.39
Hexane	C_6H_{14}	1.2	2.52	7.4	0.39
Heptane	C_7H_{16}	1.1	2.26	6.7	0.39
Acetylene	C_2H_2	2.5	9.50	80.0	1.50
Ethylene	C_2H_4	3.1	7.40	32.0	0.68
Carbon monoxide	CO	12.5	53.0	74.0	0.43
Hydrogen	H_2	4.0	42.0	75.0	2.80
Benzene	C_6H_6	1.3	—	44.0	—
Generator gas		16.0	—	64.0	—
Coke oven gas		5.0	—	33.0	—
City gas		5.0	—	38.0	—
Blast furnace gas		35.0	—	75.0	—
Natural gas		4.5	—	13.5	—

laminar flame velocity, that is, the velocity of the flame front, in the normal direction (perpendicular to the front itself).

The laminar flame velocity should not be mistaken for the velocity of burning. Laminar flame velocity is the physico-chemical property of the combustible gas for a given oxidizing agent, temperature and pressure and is independent of the actual equipment in which the gas is burned. On the other hand, the burning velocity is affected by the conditions of combustion (the construction of the actual equipment). Laminar flame velocity values are usually determined experimentally. The maximum laminar flame velocity of gases in air is shown in the last column of Table 2.6.

Among practical gaseous fuels, hydrogen has the highest laminar flame velocity while methane shows the lowest. Their maximum value corresponds to slightly fuel rich mixtures, that is, at a concentration below the stoichiometric ratio. Flame velocity is also dependent on the system's initial temperature: increasing temperature markedly increases flame velocity. The effective flame velocity has a great significance, especially for burner design (see Chapter 3, Section 3.1).

2.2.11
Wobbe Index

Although, most burners are designed for combusting a certain type of fuel, there are situations when a burner should operate efficiently with more than one fuel.

For both transport and the combustion itself, it is essential that the various fuels used in a specific burner be similar in nature. This interchangeability is best

Table 2.7 Wobbe indices of various practical fuels.

Fuel gas	Upper Wobbe index (MJ m^{-3})	Lower Wobbe index (MJ m^{-3})
Hydrogen	48.23	40.65
Carbon monoxide	12.80	12.80
Methane	53.28	47.91
Ethane	68.19	62.47
Ethylene	63.82	60.01
Acetylene	61.32	59.16
Propane	81.07	74.54
Propylene	77.04	71.88
Butylene-1	88.46	82.54
Natural gas	53.71	48.52
LPG	86.84	79.94

described by the Wobbe index (Wo) in J m^{-3} (or Wobbe number) that is calculated from the fuel's HHV in J m^{-3} and its specific gravity v, which is the ratio of the density of the fuel gas to the density of the dry air under the same conditions.

$$\text{Wo} = \frac{\text{HHV}}{\sqrt{v}} \tag{2.16}$$

The lower Wobbe index (Wo$_L$) can be calculated similarly from the LHV:

$$\text{Wo}_L = \frac{\text{LHV}}{\sqrt{v}} \tag{2.17}$$

The Wobbe index is a measure of the combustion energy output of a fuel in a given burner. If two fuels have similar Wobbe indices, their energy output will also be similar and so they can be substituted for each other without the need to adjust the pressure and/or the valve settings.

The Wobbe index is an important factor for fuel replacement situations. There are situations when even a short outage in the gas supply would not be tolerable (e.g., for a gas melting furnace) and in the case of natural gas outage a substitute fuel (propane–air mixture) must be used immediately. In order to operate seamlessly, it is essential that the main fuel and the substitute fuel have near identical Wobbe indices.

The Wobbe index is not additive; therefore the gas mixture's Wobbe index cannot be directly computed from the Wobbe indices of the components. For typical Wobbe indices refer to Table 2.7.

2.2.12
Methane Number

Internal combustion engines usually run on liquid fuels. The knocking tendency of liquid fuels is represented by the octane numbers. An analogous property of gaseous fuels is the methane number (MN).

Table 2.8 Methane number of various gaseous fuels.

Fuel Gas	Methane Number
Biogas	140
Landfill gas	140
Natural Gas	85
Synthesis gas	60
Propane	35
Water gas	30

Methane number is defined as the percentage of methane by volume blended with hydrogen that exactly matches the knocking behavior of the unknown gas mixture under specified operating conditions in a knock testing engine. For the range beyond 100 MN the reference fuel contains carbon dioxide instead of hydrogen. In this case the methane number of a mixture is $100 +$ the volume percentage of CO_2 in the reference methane–carbon dioxide mixture. Table 2.8 lists the MN of various practical gases. [11]

2.3
Liquid Fuels

Significant liquid fuels include various fuel oils for firing combustion equipment and engine fuels for on-site energy systems. Liquid fuels are usually not as easy to burn, treat and control as gaseous fuels. A liquid fuel can be mixed with oxygen only after atomization, therefore the burning process is limited by the atomization rate. In practice, the burning intensity of liquid fuels is smaller than that of gaseous fuels, the latter having higher calorific value (i.e., natural gas).

Since gasification (here meaning that a liquid phase evaporates and becomes a gas phase, not to be confused with the procedure of converting a solid fuel to gaseous components) is a vital part of the burning processes of most liquid fuels, the most important properties are those that affect gas formation. One of these properties is viscosity, important since it retards proper atomization or the development of droplets, a crucial step in the process of gas formation. Properties that affect storage and handling are also important because, unlike gaseous fuels that are usually transported through communal pipelines, liquid fuels have to be transported and stored by the consumer.

Stability factors (ignition temperature, ignition limits and flame velocity) are often unknown in the case of liquid fuels, on the other hand, flame stability issues are less common when liquid fuels are used.

The various chemical and physical properties of liquid fuels are given below [12].

2.3.1
Chemical and Physical Characteristics

Liquid fuels, with few exceptions, are mixtures of hydrocarbons derived by refining crude petroleum. In addition to hydrocarbons, crude petroleum usually contains small quantities of sulfur, oxygen, nitrogen, vanadium, other trace metals, and impurities such as water and sediment. Refining produces a variety of fuels and other products. Nearly all lighter hydrocarbons are refined into fuels (e.g., liquefied petroleum gases, gasoline, kerosene, jet fuels, diesel fuels, and light heating oils). Heavy hydrocarbons are refined into residual fuel oils and other products (e.g., lubricating oils, waxes, petroleum coke, and asphalt).

Crude petroleums from different oil fields vary in hydrocarbon molecular structure. Crude is paraffin-based (principally chain-structured paraffin hydrocarbons), naphthene- or asphaltic-based (containing relatively large quantities of saturated ring-structured naphthenes), aromatic-based (containing relatively large quantities of unsaturated, ring-structured aromatics), or mixed- or intermediate- based (between paraffin- and naphthene-based crudes). Except for heavy fuel oils, the crude type has little significant effect on the resultant products and combustion applications. Refer to Table 2.9 for typical values of liquid fuels.

2.3.2
Sulfur Content

The amount of sulfur in the fuel oil depends mainly on the source of the crude oil and, to a lesser extent, on the refining process. The normal sulfur content for the residual fuel oil (furnace oil) is of the order of 2–4%.

Table 2.9 Typical values for important liquid fuels.

		Gasoline	Diesel	Kerosene	Gas oil EL	Residual oil $S = 2.3\%$
Higher Heating value, HHV	$MJ\,kg^{-1}$	46.9	45.6	45.9	45.7	43.3
Lower Heating value, LHV	$MJ\,kg^{-1}$	44.1	42.9	43.7	42.8	40.9
Density at 15°C	$Kg\,dm^{-3}$	0.74	0.84	0.81	0.84	0.94
Viscosity at 20°C	$mm^2\,s^{-1}$	0.7	2.5	1.8	5.0	—
Composition						
Carbon, C	wt%	85.6	85.2	85.3	86.3	85.2
Hydrogen, H	wt%	14.3	14.7	14.1	13.4	11.1
Sulfur, S	wt%	0.05	0.05	0.2	0.3	2.3

Sulfur in fuels is undesirable because of the corrosiveness of sulfuric acid, formed during and after combustion, that condenses in cool parts of the chimney or stack, air pre-heater and economizer. Although low-temperature corrosion can be minimized by maintaining the stack at temperatures above the dew point of the flue gas, this limits the overall thermal efficiency of the combustion equipment. For certain industrial applications, the sulfur content of a fuel must be limited because of adverse effects on product quality. The applications include direct-fired metallurgy where work is performed in the combustion zone.

The emitted SO_2 in the flue gas can also react with the moisture in the atmosphere forming acidic rain.

2.3.3
Ash Content

The ash value is related to the inorganic material or salts in the fuel oil. The ash levels in distillate fuels are negligible. Residual fuels have higher ash levels. These salts may be compounds of sodium, vanadium, calcium, magnesium, silicon, iron, aluminum, nickel, and more.

Typically, the ash value is in the range 0.03–0.07%. Excessive ash in liquid fuels can cause fouling deposits in the combustion equipment. Ash has an erosive effect on the burner tips, causes damage to the refractories at high temperatures, and gives rise to high temperature corrosion and fouling of equipment.

2.3.4
Water Content

Water may be present in free or emulsified form in a fuel and can cause damage to the inside surfaces of the furnace during combustion, especially if it contains dissolved salts. It also can cause spluttering of the flame at the burner tip, possibly extinguishing the flame, reducing the flame temperature or lengthening the flame.

2.3.5
Carbon Residue

Carbon residue is obtained by a test in which the oil sample is destructively distilled in the absence of air. When commercial fuels are used in proper burners, this residue has almost no relationship to soot deposits, except indirectly when deposits are formed by vaporizing burners. Residual oil contains carbon residue of 1% or more.

2.3.6
Density and Specific Gravity

Density is defined as the ratio of the mass of the fuel to the volume of the fuel at a reference temperature of 15 °C.

Specific gravity is the heaviness of a substance compared to that of water at 4 °C, and it is expressed without units. In other words, the specific gravity is the density of fuel relative to the density of water at 4 °C. This might explain why this property is also called relative density. The specific gravity of water is defined as 1.

2.3.7
Viscosity

Because its molecules can slide around each other, a liquid has the ability to flow. The resistance to such flow is called the viscosity. Liquids, which flow very slowly, like glycerin or honey, have high viscosities. Those like ether or gasoline, which flow very readily, have low viscosities. Viscosity depends on the temperature and decreases with rising temperature; therefore in the case of liquids the numerical value of viscosity must always be quoted together with the corresponding temperature.

Viscosity is the most important characteristic in the storage and use of fuel oil. A high viscosity fuel may cause extreme pressures in the injection systems and will cause reduced atomization and vaporization of the fuel spray. Poor atomization may result in the formation of carbon deposits on the burner tips or on the walls. Therefore, pre-heating is necessary for proper atomization.

The viscosity of diesel fuel must be low enough to flow freely at its lowest operational temperature, yet high enough to provide lubrication to the moving parts of the finely machined injectors. Viscosity also determines the size of the fuel droplets which, in turn, govern the atomization and penetration qualities of the fuel injector spray.

Viscosity can be determined in laboratories with an instrument called a viscosimeter.

2.3.8
Pour Point

The pour point of a fuel determines the lowest temperature at which the fuel can be pumped through the fuel system. The pour point is 2 °C above the level at which oil becomes a solid or refuses to flow.

2.3.9
Cloud Point

Cloud point is the temperature at which wax crystals in the fuel (paraffin base) begin to settle out, resulting in the fuel filter becoming clogged. This condition exists when cold temperatures are encountered and is the reason why a thermostatically controlled fuel heater is required on vehicles operating in cold weather environments. Failure to use a fuel heater will prevent fuel from flowing through the filter and the engine will not run. Cloud point generally occurs at 4–7 °C above the pour point.

Table 2.10 Flash point and autoignition temperature of various liquid fuels.

Fuel	Flash Point ($^\circ$C)	Autoignition Temperature ($^\circ$C)
Gasoline	-45	>280
Diesel	>55	220
Kerosene (paraffin)	38	210
Ethanol (70%)	12	425
Vegetable oil (canola)	>282	N.A
Biodiesel	>120	>200

2.3.10
Flash Point

The flash point is an indication of how easily a chemical may burn. Materials with higher flash points are less flammable or hazardous than chemicals with lower flash points. The flash point of a fuel is the lowest temperature at which it can vaporize to form an ignitable mixture in air. Measuring a flash point requires an ignition source. At the flash point, the vapor may cease to burn when the source of ignition is removed.

The flash point is not to be confused with the autoignition temperature, which does not require an ignition source, or the fire point, the temperature at which the vapor continues to burn after being ignited. Neither the flash point nor the fire point is dependent on the temperature of the ignition source, which is much higher. See Table 2.10.

2.4
Solid Fuels

Solid fuels are more difficult to burn, handle and control than liquid or gaseous fuels. After the initial drying phase the volatiles are released from the fuel matrix and burn in the gas phase. The residual fuel, known as char, undergoes a heterogeneous combustion process, that is, the char is oxidized by the oxygen. The burning rates are usually low and a typically high combustion volume is necessary. However, in some cyclone firing equipment the intensity of gas and oil flames can also be achieved.

The usage of *waste or by-product fuels* and gasified solid fuels is becoming more widespread due to the rise in price of fuels. The factories that produce such material either use them on site as an energy source, or sell them as fuel. The difficulty of handling, the fluctuating amount of available waste, and the environmental pollution issues complicate the utilization of these fuels.

Most of the industrial burning processes require accurate temperature control and constant temperature. Solid fuels present a challenge, especially from the different qualities of wastes. These fuels are often burned in large combustion chambers, mainly in boilers and cement kilns. If it is favorable to use the solid fuel as a heat source that can be well controlled it is worth gasifying it. The forming synthesis gas (mainly CO and H_2) can be cleaned and then the combustion can be well controlled.

2.4.1
Origin of Solid Fuels

Millions of years ago trees and plants flourished in vast swamp areas. Their decay created peat beds, which were later covered with layers of sediment. Coalification of deposited materials occurred because of geochemical effects of temperature and pressure. Coalification is the conversion of the organic matter from wood to coal under pressure and temperature (see also Figure 2.1).

Depending on the duration of coalification, the properties of the coal change. The percentage of carbon increases and the oxygen content decreases as well as volatile components and moisture. This leads to the classification of coal into lignite (brown coal), bituminous coal and anthracite. Geologically, the peat is the youngest and anthracite the oldest solid fossil fuel.

Peat is an intermediate product of the conversion of wood into coal and chemically has composition and properties between biomass and low rank coal. It accumulates in swamps at the rate of about 3 m in 2500 years [13].

Lignite or brown coal is soft brown fuel which has inherent moisture content of around 60% and it belongs to the low rank coal. Low rank coal has lower energy content because of its low carbon content.

In the next stage as coalification proceeds in time, pressure and temperature increase the amount of carbon, creating high rank coal. High rank coal has higher energy content because of high carbon content.

Anthracite has the highest carbon content, heating value and least amount of volatile matter.

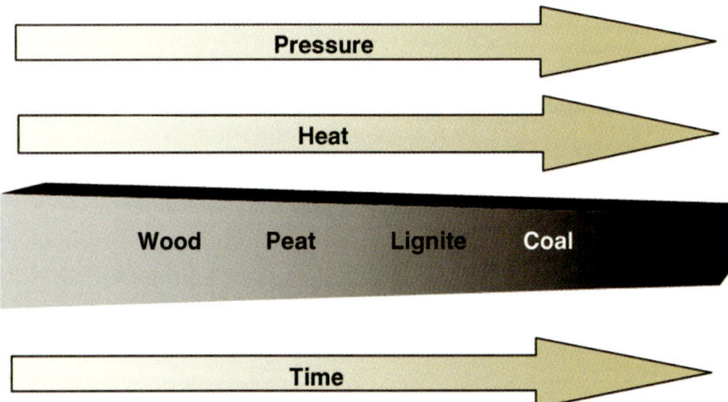

Figure 2.1 Schematic example of the coalification process.

2.4.2
Biomass

The term biomass covers a wide range of fuels, such as wood (see Figure 2.2), agricultural crops, forest and agricultural wastes and residues, and includes also the aquatic and marine biomass, such as algae and seaweed. Much solid waste, like paper, cardboard, waste wood, and so on, is biomass because it consists of lignocellulose materials.

Water availability and soil quality have the biggest influence on the yield and rate of growth of biomass.

Biomass is widely distributed around the world and is a renewable energy source which contains significant amounts of oxygen and, therefore, has lower energy value than coal. It is very reactive because it contains a large amount of volatiles.

Biomass material is mainly composed of cellulose, hemicelluloses and lignin. Cellulose is a linear polysaccharide of D-glucose but the hemicelluloses are branched polysaccharides and consist of several different monomer components, of which D-glucuronic acid is the main compound. Lignin has the highest carbon content because it contains a greater amount of aromatics. It is a crosslinked polymer of phenylpropane units.

The resource bases of biomass vary over a wide range. Its annual availability is approximately equal to that of coal but it depends significantly on the time when it is estimated. It depends on the rate of utilization and the rate of planting and growth. In contrast to coal, biomass is renewable and is located all around the world.

The main sources of biomass are from forest, agriculture, industries and other waste, and from parks and gardens.

Not all of the biomass can be used economically. The world biomass consumption is 1–2 Gtoe (gigatons of oil equivalent) [13] per year, including the non-commercial

Figure 2.2 Storage of wood logs in the forest.

component. The commercial biomass components are small but in some countries, such as Sweden, Germany and Brazil they are significant.

2.4.3
Waste or Opportunity Fuels

Municipal solid waste (MSW) is one of the biggest sources of solid waste, according to the European environment commission (Eurostat statistics) every person produces about 6 tonnes of solid waste each year.

Other sources of solid waste are from refuse industrial processes, such as manufacturing residue, rubber tires, reclaimed wood from demolition, and plastic waste. Solid waste also includes sewage sludge and animal dung.

Solid wastes are very inhomogeneous waste flows and have variable compositions.

Waste materials are very different from country to country as are also the facilities which are applied to use waste materials as fuel.

The total resource of combustible waste usable as a fuel is about 2.2 Gtoe [13]. From this amount 1.2 Gtoe is municipal waste and 1 Gtoe industrial waste [13].

2.4.4
Coal

In the nineteenth century the question of the availability of coal was raised because of the exponential use of coal and because it was used as a major source of energy for many centuries before oil and gas production.

The total resource of coal is about 18 000 Gtoe [13] but of course the proved reserve of coal is much lower because not all of the resources are recoverable. The reserves to production ratio R/P represents the lifetime of the coal resource. This ratio is based on the reserves remaining at the end of the year if the production continues at the previous year's rate. It is a varying ratio because the reserves and production differ from year to year. In the year 2011, based on BP data, Europe and Eurasia had the highest R/P ratio followed by North America, with R/P ratios of 220 [14].

In terms of coal distribution Europe and Eurasia, the USA, the Russian Federation, China and Australia have the most reserves.

Compared to the very fast increase in the production of fossil fuels (oil and gas), the growth in the global production of coal is not as fast and China today is the world's largest supplier.

The major sources of energy in tons oil equivalent are summarized in Table 2.11.

Compared with other fossil fuels, coal has a longer lifetime and in the not too far distant future it will play a bigger role for humans and will be one of the major sources of energy.

Coal, based on BP statistics, was again the fastest growing fossil fuel. The global production grew by 6.1%. The Asia Pacific region accounted for 85% of the global production growth, led by an 8.8% increase in China, the world's largest supplier [14].

Table 2.11 Utilization of energy sources.

Fuel	Usage (Gtoe y^{-1})
Coal	3.2
Oil	4.0
Natural gas	2.6
Nuclear	0.6
Renewable (hydroelectricity)	0.7
Biomass (estimated, mainly non-commercial)	1–2

2.4.5
Peat

There are large deposits of peat in Russia, Scandinavia, Germany, and Ireland. The resource is 461 300 Mt [13] peat which is approximately equivalent to the oil reserves in Saudi Arabia.

The use of peat has been declining but still has application for horticultural purposes.

2.4.6
Solid Fuels Characterization

The chemical and physical properties of solid fuels are characterized by analytical laboratory methods. Proximate and ultimate analyses are the standard test methods and are, together with the determination of the heating values (HHV and LHV), the most important characterization methods. The HHV is obtained by complete combustion of a fuel sample in a bomb calorimeter followed by cooling to 298 K. It can also be estimated by calculation, see Chapter 3, Section 3.1. Typical values for the heating values of a wide range of different fuels are given in Table 2.12.

Proximate or immediate analysis gives some information about the moisture, volatile matter, fixed carbon, and ash on a mass percent basis but it is not exact. The exact information about the solid fuel is determined by ultimate analysis.

A representative sample (homogenous) is required for the analysis but because the waste and biomass are inhomogenous materials this is a major difficulty and, therefore, special standards for biomass and waste have been established.

There are some standard test methods in most countries. The Deutsches Institut für die Normierung (DIN), the British Standard Institution (BSI), the American Society for Testing and Materials (ASTM) and the International Organization for Standardization (ISO) are some important standard organizations.

For the characterization of application as solid fuels a series of parameters are important and will be discussed in this chapter.

Table 2.12 Typical higher heating values (HHV) of various solid fuels [13,15].

	Higher Heating value, HHV (MJ kg^{-1})
Anthracite and bituminous coals (dry, ash free)	33–35
Lignite and sub-bituminous coals (dry, ash free)	24–32
Straw (dry), wheat, rice, sunflower	15–18
Grasses (dry), alfalfa, maize, switchgrass	17–18
Wood (dry) beech, fir, maple, larch, oak	17–20
Paper (dry, ash free)	12–16
Scrap tires (dry, ash free)	40
PE (polyethylene) (dry, ash free)	44–45
PET (polyethylene-terephthalate) (dry, ash free)	23
PP (polypropylene) (dry, ash free)	39
PVC (polyvinyl-chloride) (dry, ash free)	18
PS (polystyrene) (dry, ash free)	42
PA (polyamid) (dry, ash free)	33
Sewage sludge (dry, ash free)	18–25

2.4.7
Proximate Analysis

The proximate or immediate analysis give the first estimation and classification of a solid fuel and includes volatile matter, ash and water content.

Based on the DIN method, volatile matter is the part of the solid fuel which is released in gaseous form by slowly heating to 900 °C in the absence of oxygen. Solid fuels with larger amounts of volatile matter burn with a long flame and those with less volatile matter burn with a short flame. The amount of volatile matter has a significant effect on the dimension of the burning volume.

Ash is the part of solid fuel remaining after complete combustion of the sample in an oven at 815 °C until the weight is constant (for a biogenic sample a temperature less than 550 °C is used). The amount of ash affects the dimensions of the ash room in the boiler, ash transport, and the ash storage system. Ash contains elements such as silicon, aluminum, titanium, iron, calcium, magnesium, sodium, potassium, sulfur, and phosphorus which usually exist in their oxidized forms (e.g., SiO_2, Al_2O_3 and Fe_2O_3). There are some standard methods for determination of ash composition such as X-ray fluorescence (XRF), X-ray diffraction (XRD), atomic absorption spectroscopy (AAS), atomic emission spectroscopy (AES) combined with inductively-coupled plasma (ICP - AES). Another important characterization is the characterization of the ash melting behavior (see also Chapter 5). Significant concentrations of alkali metals, such as sodium and potassium, and of chlorides may lead to low melting ashes, for example, from biomass. Low melting ashes may lead to agglomerations, slagging and fouling of the combustor as well as corrosion. Finally, the content of trace metals may be of importance, leading to increased emissions of trace metals in the flue gas or in the fly ash or bottom ash of the combustion process (see Chapter 4, Section 4.1). Trace metals and impurities can be divided according to their mobility, highly volatile compounds are, for example, Hg and the halogens (Cl, F, Br); medium are, for example, Cd, Zn, and Pb; and low are Cr, Cu and Ni.

The water content in solid fuels can be divided into two kinds of moisture, known as gross moisture and hygroscopic moisture. Gross moisture is the part of the whole water content which may be evaporated by storage of the fuel in the air. The amount of gross moisture depends on the temperature and moisture in the environment. Hygroscopic moisture is the amount of water in the solid fuel, saved in capillaries and pores, and can be reduced by heating the sample to temperatures higher than 100 °C. The temperature 106 °C has been established by DIN. The high amount of water reduces the burning temperature and also raises the water dew point. Both of these effects are undesirable during the combustion; therefore, dried solid fuel is important for a commercial product.

2.4.8
Ultimate Analysis

In the ultimate analysis the contents of the elements are determined, carbon, hydrogen and oxygen being of especially high interest. In addition the sulfur, nitrogen, and in certain cases the chlorine contents of the fuels are measured (see Chapter 5). In Figure 2.3 the atomic ratios of oxygen to carbon (O:C) and hydrogen to carbon (H:C) are presented, defining different fuel classes (coalification diagram).

The total sulfur amount is usually between 0.3 and 2.5% and in some solid fuels up to 6% [16] is possible. Sulfur occurs in the organic and the inorganic part of the

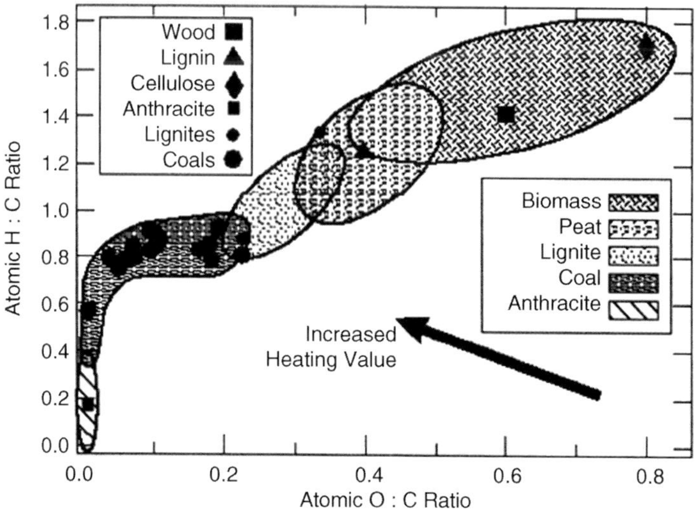

Figure 2.3 Coalification diagram showing compositional differences among biomasses [17].

Table 2.13 Ultimate and proximate analysis (n.a. not analyzed) of various fuels.

	Proximate Analysis				Ultimate Analysis (water ash free)					
	Raw Density (kg m^{-3})	Volatile Matter (wt%)	Moisture (wt%)	Ash Content (wt%)	Fixed Carbon (C-fix) (wt%)	Carbon (wt%)	Hydrogen (wt%)	Nitrogen (wt%)	Sulfur (wt%)	Oxygen (wt%)
coke	n.a.	10.6	2.2	0.8	86.4	87.42	2.93	1.55	5.67	2.43
bituminous coal I	1300	33	3	10.2	53.8	84.55	4.97	0.67	0.90	8.91
bituminous coal II	1520	30	6	2.4	61.6	86.8	4.27	1.48	0.50	6.95
sub bituminous coal I	1250	32.2	25.3	12.3	30.2	68.10	2.84	0.64	1.52	26.88
sub bituminous coal II	1100	24.5	52.7	4.89	17.91	65.46	4.51	0.47	0.24	29.33
peat	n.a.	59.9	12.36	2.59	25.15	58.69	4.70	1.60	n.a.	n.a.
beech wood	667	81.7	5.6	0.3	12.7	49.76	5.94	0.11	0.02	44.17
beech wood after devolatilization	n.a.	0	0	2.3	97.7	87.30	1.63	0.24	0.00	10.83
wood chips I	500	76	9.5	0.4	14.3	49.91	5.42	0.18	0.02	44.47
wood chips II	n.a.	82.54	1.65	0.46	15.35	49.33	6.01	0.22	0.00	44.43
wooden hardboard	880	77.1	6.2	0.4	16.6	51.03	5.88	0.16	0.02	42.91
wooden pressboard	730	73.3	8	0.9	18.7	49.69	5.68	2.89	0.05	41.68
straw pellets	n.a.	65.4	8.1	6.8	19.7	49.04	5.65	0.43	0.06	44.80
polyethylene	900	100	0	0	0	85.15	14.38	0.07	0.05	0.40
polypropylene	890	100	0	0	0	85.31	14.28	0.06	0.05	0.35
plastic/paper compound	n.a.	90.8	0	0.1	9.1	51.57	8.54	0.07	0.05	39.77
tire (waste)	1120	60.37	0	3.51	36.12	89.67	7.31	0.30	1.52	1.21
waste from malt industry	n.a.	n.a.	11.12	6	n.a.	47.98	6.64	4.99	0.38	40.01
sewage sludge	1140	60.6	7	26.7	32.4	56.11	7.50	6.30	0.80	29.28
sewage sludge after devolatilization	n.a.	0	0	45.2	54.8	38.44	1.30	3.65	0.57	56.04

solid fuel. In the inorganic part it occurs mainly as sulfide (pyrite) and sulfate (Na_2SO_4 and $CaSO_4$). The organic part originates from the plants. During combustion the sulfur appears either in the ash or reacts with oxygen and is emitted as SO_2 or in small amounts as SO_3. Both gases should be removed from the flue gas for environmental reasons (see Chapter 4).

The nitrogen content is typically 0.8–1.5% in bituminous coals and 0.2–0.8% in brown coals [16]. The oxygen reacts with the nitrogen of the fuel and may form nitrogen oxides such as NO_x (NO and NO_2) and N_2O (see Chapter 4).

See also Table 2.13 for typical values for proximate and ultimate analyses for various fuels.

2.4.9
Physical Properties

The physical properties of the solid fuels play an important role for the combustion, storage and transport characteristics of these fuels.

The particle size and distribution affects significantly the type of combustor applied. Fuel particles with a size of several cm will be used in grate firing systems or fluidized beds or rotary kilns, whereas for dust firing systems the typical particle size range is of the order of μm.

The bulk density in $kg\,m^{-3}$ is required for transport and storage of solid fuels and it is the ratio of the mass of the solid fuel stored to the volume. The DIN method uses a $0.2\,m^3$ volume for the determination. It depends also on particle size distribution and water content.

The energy density in $MJ\,m^{-3}$ is very important for storage, transport and the design of combustion boilers.

References

1 Typical Composition of Natural Gases, http://www.intellectualtakeout.org/library/chart-graph/typical-composition-natural-gas (access date: 2013-06-06).

2 Szunyog, I. (2009) Quality requirements of the application of biogases in natural gas public utility services in Hungary, PhD Thesis, University of Miskolc, Miskolc, Hungary.

3 Lokhorst, A. (ed.) (1997) *NW European Gas Atlas*, NITG-TNO, Haarlem.

4 May, F., Freund, W., Miller, E.P., and Dostal, K.P. (1968) Model experiments on the isotopic fractionation of natural gas components during migration. *Zeitschrift für Angewandte Geologie*, 14 (17), 376–381 (in German).

5 Weinlich, F.H. (1991) Genese und Verteilung der freien Gase im Stassfurtkarbonat der Lausitz –Teil 2: Die Verteilungsprinzipien der freien Gase. *Zeitschrift für Angewandte Geologie*, 37, 52–58.

6 Stahl, W., Faber, E., and Schoell, M. (1979) Isotopic gas-source rock correlations (abs.). *Geological Society of America Abstracts with Programs*, 11, 522.

7 Bandlowa, T. (1998) Erdgasführung im Karbon-Perm-Trias-Komplex der Mitteleuropaischen Senke. *Geologisches Jahrbuch*, 151, 3–65 (in German).

8 Lees, G.M., and Taitt, A.H. (1945) The geological results of the search for oilfield in Great Britain. *Quarterly Journal of the Geological Society*, 101, 255–278.

9 Bifani, R. (1986) Esmond gas complex, in *Habitat of Palaeozoic Gas in N.W. Europe* (eds J. Brooks, J.C. Goff, and B. van Hoorn), Geological Society Special Publication, London, pp. 209–221.

10 Kent, James A. (ed.) (2003) *Riegel's Handbook of Industrial Chemistry*, Kluwer Academic Publishers, New York.

11 Lackner, M. Winter, F., and Agarwal, A.K. (eds) (2010) *Handbook of Combustion, Vol. 3. Gaseous and Liquid Fuels*, Wiley-VCH Verlag GmbH, Weinheim.

12 Weston, K.C. (1992) Energy Conversion, Ch. 3: Fuels and Combustion, http://www.personal.utulsa.edu/~kenneth-weston/chapter3.pdf (access date: 2013-06-06).

13 Lackner, M. Winter, F., and Agarwal, A.K. (eds) (2010) *Handbook of Combustion, Volume 4: Solid Fuels*, Wiley-VCH Verlag, Weinheim, pp. 1–30.

14 BP Statistical Review of World Energy (2012), available online at http://www.bp.com/content/dam/bp/pdf/Statistical-Review-2012/statistical_review_of_world_energy_2012.pdf (access date: 2013-06-06).

15 BIOBIB, A database for Biofuels, Institute of Chemical Engineering, Vienna University of Technology (2013) http://www.vt.tuwien.ac.at/biobib (access date: 2013-06-06).

16 Effenberger, H. (2000) *Dampferzeugung*, Springer, pp. 20–40.

17 Jenkins, B.M., Baxter, L.L., Miles, T.R. Jr., and Miles, T.R. (1998) Combustion properties of biomass. *Fuel Processing Technology*, **54** (1–3), 17–46.

3
Combustion Principles

3.1
Basic Combustion Calculations

3.1.1
Determination of the Quantity of Normal and Oxygenated Air Necessary for Complete Combustion

3.1.1.1 Air Requirement of Gaseous Fuels

In order to determine the oxygen and air required for complete combustion, one must use the elementary combustion reactions of the combustible components of the fuel gas, as listed in Table 3.1 and shown in Figure 3.1.

For example, if the fuel contains 85% CH_4, 10% H_2 and 5% N_2, then only the stoichiometric reactions of methane and hydrogen should be taken into consideration:

$$CH_4 + 2O_2 \rightarrow CO_2 + 2H_2O \tag{3.1}$$

$$H_2 + 0.5O_2 \rightarrow H_2O \tag{3.2}$$

The combustion of 1 mole of CH_4 requires 2 mole of O_2, while 1 mole H_2 requires 0.5 mole of O_2.

It is well known that the molar volume of any ideal gas is the same: $22.41\,dm^3\,mol^{-1}$ or $22.41\,m^3\,kmol^{-1}$ at temperature 273.15 K and pressure 101.3 kPa.

Thus the complete combustion of $22.4\,m^3$ CH_4 requires $44.8\,m^3$ O_2, while the complete combustion of $22.4\,m^3$ H_2 requires $11.2\,m^3$ O_2.

In the example above, $1\,m^3$ gaseous fuel contains $0.85\,m^3$ CH_4 and $0.1\,m^3$ H_2, thus the total oxygen needed for the complete combustion of the mixture is

$$V_{O_2} = (0.85 \cdot 2) + (0.1 \cdot 0.5) = 1.75\,m^3$$

If the combustion is supplied by dry air, then its normal composition can be used and 21% oxygen and 79% nitrogen can be assumed. In this case the nitrogen content of the air is $79/21 = 3.762$ times the volume of the stoichiometric oxygen. Alternatively, one can say, that the air required for a complete combustion is $100/21 = 4.76$ times the oxygen required.

Combustion: From Basics to Applications, First Edition. Maximilian Lackner, Árpád B. Palotás, and Franz Winter.
© 2013 Wiley-VCH Verlag GmbH & Co. KGaA. Published 2013 by Wiley-VCH Verlag GmbH & Co. KGaA.

Table 3.1 Reactions and the products derived from the combustion of gaseous fuel components in the case of complete combustion at air factor $n = 1$, temperature 273.15 K and pressure 101.3 kPa.

Combustible Gas	Combustion Reaction	Oxygen $m^3\,m^{-3}$	Air Req. $m^3\,m^{-3}$	Flue gas Components $m^3\,m^{-3}$			Flue Gas Volume $m^3\,m^{-3}$	
		V_{O_2}	V_{air}	V_{CO_2}	V_{H_2O}	V_{N_2}	$V_{flue,\ wet}$	$V_{flue,\ dry}$
Carbon monoxide	$CO + 0.5\,O_2 = CO_2$	0.5	2.38	1	–	1.88	2.88	2.88
Hydrogen	$H_2 + 0.5\,O_2 = H_2O$	0.5	2.38	–	1	1.88	2.88	–
Methane	$CH_4 + 2O_2 = CO_2 + 2H_2O$	2.0	9.52	1	2	7.52	10.52	8.52
Ethane	$C_2H_6 + 3.5\,O_2 = 2CO_2 + 3H_2O$	3.5	16.65	2	3	13.15	18.15	15.15
Propane	$C_3H_8 + 5O_2 = 3CO_2 + 4H_2O$	5.0	23.80	3	4	18.80	25.80	21.80
Butane	$C_4H_{10} + 6.5\,O_2 = 4CO_2 + 5H_2O$	6.5	30.95	4	5	24.45	33.45	28.45
Pentane	$C_5H_{12} + 8O_2 = 5CO_2 + 6H_2O$	8.0	38.10	5	6	30.08	41.08	35.08
Hexane	$C_6H_{14} + 9.5\,O_2 = 6CO_2 + 7H_2O$	9.5	45.20	6	7	35.70	48.70	41.70
Heptane	$C_7H_{16} + 11\,O_2 = 7CO_2 + 8H_2O$	11.0	52.40	7	8	41.35	56.35	48.35
Octane	$C_8H_{18} + 12.5\,O_2 = 8CO_2 + 9H_2O$	12.5	59.50	8	9	47.00	64.00	55.00
Acetylene	$C_2H_2 + 2.5\,O_2 = 2CO_2 + H_2O$	2.5	11.90	2	1	9.40	12.40	11.40
Ethylene	$C_2H_4 + 3O_2 = 2CO_2 + 2H_2O$	3.0	14.30	2	2	11.28	15.28	13.28
Benzene	$C_6H_6 + 7.5\,O_2 = 6CO_2 + 3H_2O$	7.5	35.70	6	3	28.20	37.20	34.20

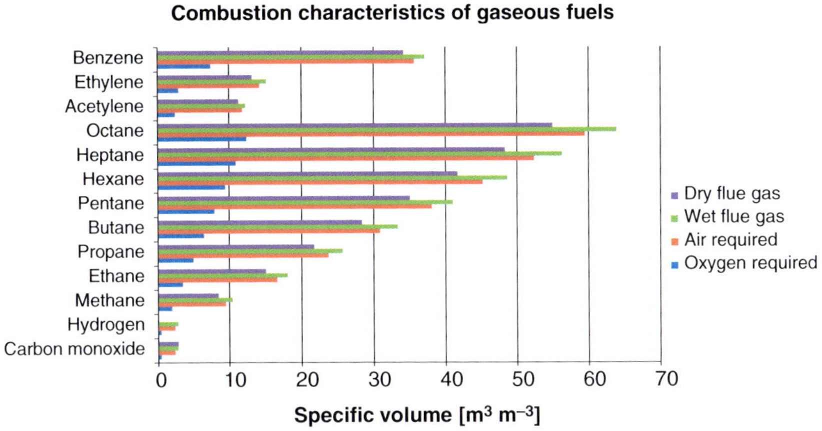

Combustion characteristics of gaseous fuels

Figure 3.1 Combustion characteristics of the most common gaseous fuels at stoichiometric ratios and complete combustion: oxygen and air required, wet and dry flue gas produced per unit volume of the fuel gas.

Consequently, in the example above the air requirement for $1\,m^3$ fuel is the following:

$$V_{air} = 1.75 + (1.7 \cdot 3.76) = 8.34\,m^3$$

or

$$V_{air} = 1.75 \cdot 4.76 = 8.34\,m^3$$

The general equation of the theoretical (stoichiometric) air requirement of the perfect combustion of a gaseous fuel mixture is – based on the stoichiometric reactions of all combustible components as shown in Table 3.1 – the following:

$$V_{air,stoich} = 0.0476\,(0.5 \cdot CO + 0.5 \cdot H_2 + 2 \cdot CH_4 + 3.5 \cdot C_2H_6 + 5 \cdot C_3H_8$$
$$+ \ldots + 2.5 \cdot C_2H_2 + \quad + 3 \cdot C_2H_4 + 7.5 \cdot C_6H_6 - O_2),\, m^3m^{-3}$$

$$(3.3)$$

In reality – due to imperfect mixing in practical cases –more air than the stoichiometric amount is needed, which is calculated as follows:

$$V_{air} = \lambda \cdot V_{air,stoich}\; m^3m^{-3} \tag{3.4}$$

In Eq. (3.4), λ is the air factor (also called excess air factor), the ratio of the actual air supply to the stoichiometric one. Its value is dependent on the mixing efficiency of the fuel and the combustion air. Typical values for gaseous fuels in industrial equipment are $\lambda = 1.05 \ldots 1.15$, however, even $\lambda = 1.5$ can be realistic for some communal and domestic equipment.

3.1.1.2 Air Requirement for the Combustion of Liquid and Solid Fuels

Since both solid and liquid fuels are usually measured by weight, their combustion calculations are the same. The calculation is analogous to that of the gaseous fuels

Table 3.2 Reactions and the products derived from the combustion of liquid and solid fuel components in the case of complete combustion at air factor $n = 1$, temperature 273.15 K and pressure 101.3 kPa.

Component	Combustion Reaction	Oxygen Req. $m^3\,kg^{-1}$	Air Req. $m^3\,kg^{-1}$	Flue Gas Components $m^3\,kg^{-1}$					Flue Gas Volume $m^3\,kg{-}1$	
		V_{O_2}	V_{air}	V_{CO}	V_{CO_2}	V_{SO_2}	V_{N_2}	V_{H_2O}	$V_{flue,wet}$	$V_{flue,dry}$
Carbon	$C + 0.5\,O_2 \rightarrow CO$	0.93	4.44	1.87	–	–	3.51	–	–	5.38
Carbon	$C + O_2 \rightarrow CO_2$	1.87	8.89	–	1.87	–	7.02	–	–	8.89
Sulfur	$S + O_2 \rightarrow SO_2$	0.70	3.33	–	–	0.7	2.63	–	–	3.33
Hydrogen	$H_2 + 0.5\,O_2 \rightarrow H_2O$	5.60	26.60	–	–	–	21.0	11.2	32.2	–

with the exception of the volume–mass conversion. The basic combustion reactions are listed in Table 3.2.

As an example, let us examine the combustion reaction of carbon

$$C + O_2 \rightarrow CO_2 \tag{3.5}$$

1 mole of carbon requires 1 mole of oxygen. Since 1 mole of carbon weighs 12 g and 1 mole of oxygen weighs 32 g and its volume (in normal state) is 22.41 dm^3, one can easily conclude that the complete combustion of 1 kg carbon requires $22.41/12 = 1.87\,\text{m}^3$ oxygen. The air required for the complete combustion of 1 kg of carbon is:

$$V_{air} = 1.87 \cdot 4.76 = 8.89\,\text{m}^3\text{kg}^{-1}$$

The general equation of the theoretical (stoichiometric) air requirement of the perfect combustion of a liquid or solid fuel is, based on the stoichiometric reactions of all combustible components as shown in Table 3.2:

$$V_{air,stoich} = 4.76 \cdot \frac{1}{100} \left[\frac{22.41}{12} C + \frac{22.41}{4} H + \frac{22.41}{32} (S{-}O) \right] \tag{3.6}$$

that is,

$$V_{air,stoich} = 0.0476 \cdot [1.87 \cdot C + 5.6 \cdot H + 0.7 \cdot (S{-}O)]\,\text{m}^3\text{kg}^{-1} \tag{3.7}$$

In practice, the actual air supplied is usually more than the stoichiometric value (see also gas combustion) and depends on the fuel, fuel preparation and combustion technology:

$$V_{air} = \lambda \cdot V_{air,stoich} \quad \text{m}^3\text{kg}^{-1} \tag{3.8}$$

Typical values for λ for some of the common situations are:

combustion of wood	$\lambda = 1.25 \ldots 1.35$
combustion of coal, coarse	$\lambda = 1.50 \ldots 1.80$
combustion of coal, fine	$\lambda = 1.20 \ldots 1.30$
combustion of oil	$\lambda = 1.15 \ldots 1.25$

3.1.1.3 Calculations for the Case of Oxygenated Air

With the introduction of cheaper technologies for the production of oxygen, the intensification of the combustion process, and thus the technology itself, is often achieved by the use of oxygenated air.

Obviously, in this case the nitrogen to oxygen ratio is changed. For example if the oxygen content of the air is boosted to 30%, then the nitrogen content decreases to 70%. In this case the stoichiometric air volume is $100/30 = 3.33$ times the oxygen volume required for complete combustion. Thus $1\,m^3$ of fuel gas and $1\,kg$ of carbon as a solid fuel, in the examples above require

$$1.75 \cdot 3.33 = 5.83\,m^3 m^{-3}$$
$$1.87 \cdot 3.33 = 6.23\,m^3 kg^{-1}$$

oxygen, respectively.

3.1.2
Calculation of the Volume and the Composition of the Flue Gas

3.1.2.1 Flue Gas of Gaseous Fuels

The volume and composition of the flue gas can be determined similarly to the calculation of the air requirement.

Let us examine again the combustion of the gas shown in the previous example. The mixture contains $85\%\,CH_4$, $10\%\,H_2$ and $5\%\,N_2$. The reaction equations are listed in Table 3.1.

It should be clear that the combustion of 1 mole of methane yields 1 mole of CO_2 and 2 moles of H_2O. The combustion of 1 mole of H_2 yields 1 mole of H_2O. Apart from CO_2 and H_2O, the flue gas will also contain nitrogen. Even though the fuel does not contain nitrogen, this component is present in the combustion air that enters the process. In the case of stoichiometric combustion of $1\,m^3$ fuel, the generated flue gas will contain the following components:

$$0.85\,m^3\,CO_2$$
$$2 \cdot 0.85 + 0.1 = 1.8\,m^3\,H_2O$$
$$1.75 \cdot 3.762 = 6.59\,m^3\,N_2$$

Thus the total volume of the flue gas is

$$0.85 + 1.8 + 0.05 + 6.59 = 9.29\,m^3 m^{-3}$$

The composition as a volumetric percentage is:

$$CO_2 = \frac{0.85}{9.29}\,100 = 9.15\%$$

$$H_2O = \frac{1.8}{9.29}\,100 = 19.40\%$$

$$N_2 = \frac{6.59 + 0.05}{9.29}\,100 = 71.45\%$$

If the combustion is not stoichiometric ($\lambda = 1$), then there will also be oxygen in the flue gas because the excess oxygen simply does not react with any fuel component and thus increases the flue gas volume. With the excess oxygen also comes excess nitrogen in the appropriate proportion.

A generalized equation for the calculation of the volume and the composition of the flue gas – using the equations of Table 3.1 – is:

Volume of the wet flue gas:

$$V_{flue,wet} = V_{CO2} + V_{H2O} + V_{O2} + V_{N2} \quad m^3/m^3 \tag{3.9}$$

where

$$V_{CO_2} = 0.01 \cdot (CO_2 + CO + CH_4 + mC_mH_n) \quad m^3 m^{-3} \tag{3.10}$$

$$V_{H_2O} = 0.01 \cdot \left(H_2 + CH_4 + \frac{n}{2} \cdot C_mH_n\right) \quad m^3 m^{-3} \tag{3.11}$$

$$V_{O_2} = 0.21 \cdot (\lambda - 1) \cdot V_{air,stoich} \quad m^3 m^{-3} \tag{3.12}$$

$$V_{N_2} = 0.79 \cdot V_{air} + 0.01 \cdot N_2 \quad m^3 m^{-3} \tag{3.13}$$

The dry flue gas does not contain the water vapor component

$$V_{flue,dry} = V_{N_2} + V_{O_2} + V_{CO_2} \, m^3 m^{-3} \tag{3.14}$$

A characteristic value for various fuels is the CO_2 content of its dry flue gas, generated under stoichiometric conditions. It is a maximum value, since fuel-lean combustion (excess air) dilutes the flue gas with O_2 and N_2, while fuel-rich (air deficient) combustion yields CO instead of some of the CO_2. This value can also be computed from the actual fuel composition

$$CO_{2max} = \frac{V_{CO_2}}{V_{CO_2} + V_{N_2,stoich}} \, 100\% \tag{3.15}$$

where:

$$V_{N_2,stoich} = 0.79 \cdot V_{air,stoich} + 0.01 N_2 \, m^3 m^{-3} \tag{3.16}$$

Typical values for CO_{2max}:

natural gas:	$11.8 - 12.0\%$
propane-butane:	$13.9 - 14.0\%$

3.1.2.2 Combustion Products of Liquid and Solid Fuels

The composition and volume of the flue gas is computed in an analogous manner. Thus, for wet flue gas

$$V_{flue,wet} = V_{CO_2} + V_{H_2O} + V_{SO_2} + V_{O_2} + V_{N_2} \, m^3 kg^{-1} \tag{3.17}$$

The amount of the individual components:

$$V_{CO_2} = 0.01 \cdot 1.87 \, C \, m^3 kg^{-1} \tag{3.18}$$

$$V_{H_2O} = 0.01 \cdot 11.2 H + 1.25 H_2O \, m^3 kg^{-1} \tag{3.19}$$

$$V_{SO_2} = 0.01 \cdot 0.7 \cdot S \ m^3 kg^{-1} \tag{3.20}$$

$$V_{O_2} = 0.21 \cdot (\lambda - 1) \cdot V_{air,stoich} \ m^3 kg^{-1} \tag{3.21}$$

$$V_{N_2} = 0.79 \cdot \lambda \cdot V_{air,stoich} \ m^3 kg^{-1} \tag{3.22}$$

The specific volume of the dry flue gas:

$$V_{flue,dry} = V_{CO_2} + V_{SO_2} + V_{O_2} + V_{N_2} \ m^3 kg^{-1} \tag{3.23}$$

The value of CO_{2max} can be determined using the same reasoning:

$$CO_{2max} = \frac{V_{CO_2}}{V_{CO} + V_{SO_2} + V_{N_2}} \tag{3.24}$$

3.1.2.3 The Effect of Oxygen Enrichment

The volume and the composition of the flue gas when oxygen enriched air is used is influenced by the actual O_2/N_2 ratio. All the rest should be trivial by now:

In the example above for the case of 30% oxygen content in the combustion air and 20% excess air ($\lambda = 1.2$) the composition of the flue gas is the following:

$0.85 \ m^3 \ CO_2$
$1.8 \ m^3 \ H_2O$
$1.75 \cdot 1.2 - 1.75 = 0.35 \ m^3 \ O_2$
$1,75 \cdot 1,2 \cdot 2,33 + 0.05 = 4.95 \ m^3 \ N_2$

The total volume of the combustion products is

$$V_{flue,wet} = 0.85 + 1.8 + 0.35 + 4.95 = 7.95 \ m^3 kg^{-1}$$

3.1.2.4 Effect of Temperature and Pressure (Ideal Gas Law)

This numeric value computed above is valid at normal state, that is, at the temperature $T_0 = 273.15$ K and pressure $p_0 = 101.325$ Pa. Obviously, the real temperature (and often the pressure as well) can be different than those of the normal state. Generally it is safe to assume that our gaseous components, including the flue gas, behave as an ideal gas.

The ideal gas law can be written as:

$$p \cdot V = n \cdot R \cdot T \tag{3.25}$$

where

p is the pressure (Pa),
V is the volume (m^3),
n is the amount of the gas (mol),
R is the ideal gas constant ($8.314 \ J \ mol^{-1} K^{-1}$) and
T is the absolute temperature (K)

Thus the actual (measurable) volume of the flue gas at T_{flue} temperature and p_{flue} pressure:

$$V_{flue} = \frac{p_0}{p_{flue}} \frac{T_{flue}}{T_0} V_0 \tag{3.26}$$

Getting back to the previous example, let us calculate the actual flue gas volume at the actual temperature of $700\,^\circ C$ and $20\,Pa$ overpressure:

$$V_{flue} = \frac{101.325\ Pa}{101.345\ Pa} \frac{973.15\ K}{273.15\ K} 7.95\ m^3\ kg^{-1} = 28.3\ m^3\ kg^{-1}$$

The composition by volumetric percentage is the following:

$$CO_2 = \frac{0.85}{7.95} 100 = 10.5\% \quad H_2O = \frac{1.8}{7.95} 100 = 23.2\%$$

$$O_2 = \frac{0.35}{7.95} 100 = 4.2\% \quad N_2 = \frac{4.95}{7.95} 100 = 62.1\%$$

The composition of an ideal gas mixtures does not change due to temperature or pressure variation, unless dissociation reactions are considered at the examined (e.g., flame) temperature.

Thus increasing the oxygen content in the air decreases the flue gas volume (and consequently the adiabatic flame temperature, as we will see soon).

3.1.2.5 Determination of the Actual Excess Air Factor

In order to examine the actual burning process, measurements are needed on the composition of the flue gas generated. From the actual measurements, one can calculate the real excess air factor. For this one should use the dry flue gas composition as most of the flue gas analyzers provide direct data on dry flue gas.

The following equation is based on the nitrogen balance of the fuel, combustion air and the flue gas, as well as on the excess O_2 content, appearing in the flue gas.

In the case of solid and liquid fuels the calculation is based on the nitrogen content provided by the stoichiometric and actual (real) combustion air volume.

$$\lambda = \frac{V_{air}}{V_{air,stoich}} = \frac{N_2}{N_2 - \frac{79}{21} O_2} = \frac{1}{1 - 3.762 \frac{O_2}{N_2}} \tag{3.27}$$

Most of the gaseous fuels already contain nitrogen, thus the fuel nitrogen must be taken into account in excess of the flue gas nitrogen:

$$\lambda = \frac{V_{air}}{V_{air,stoich}} = \frac{N_2 - \dfrac{N_{2,fuel}}{V_{flue,dry}}}{\left(N_2 - \dfrac{N_{2,fuel}}{V_{flue,dry}}\right) - 3.762 O_2} = \frac{1}{1 - 3.762 \dfrac{O_2}{N_2 - \dfrac{N_{2,fuel}}{V_{flue,dry}}}} \tag{3.28}$$

Incomplete combustion yields CO_2, CO, H_2, H_2O and N_2 components in the flue gas. The oxygen consumption decreases by $0.5(CO + H_2)$, thus the air volume changes by

$$\frac{4.76}{2}(CO + H_2) \tag{3.29}$$

while the nitrogen content changes by

$$\frac{3.76}{2}(CO + H_2) \tag{3.30}$$

The real excess air factor in this case can be calculated as:

$$\lambda = \frac{N_2}{N_2 - 3.76\left(O_2 - \dfrac{CO}{2} - \dfrac{H_2}{2}\right)} \tag{3.31}$$

3.1.3
Determination of the Combustion Temperature

If all of the heat generated by the combustion of a fuel is absorbed by the generated flue gas, then the flue gas reaches its theoretical combustion temperature. The calculation is based on a simple heat balance of an adiabatic combustion process, therefore this temperature is also called the adiabatic flame temperature.

$$T_{ad} = \frac{LHV}{V_{flue,wet} \cdot c_{p,flue}} \, {}^\circ C \tag{3.32}$$

Where LHV is the lower heating value of the fuel, $[kJ\,m^{-3}, kJ\,kg^{-1}]$, $V_{flue,wet}$ is the flue gas volume per unit amount of fuel, $[m^3\,m^{-3}, m^3\,kg^{-1}]$ and $c_{p,flue}$ is the average specific heat capacity of the flue gas at the combustion temperature, $[kJ\,m^{-3}\,{}^\circ C^{-1})]$

If the fuel and/or the combustion air are/is preheated, their heat content cannot be neglected. Thus:

$$T_{ad} = \frac{LHV + Q_{phys}}{V_{flue,wet} \cdot c_{p,flue}} \tag{3.33}$$

here the physical heat content of the fuel and the air is

$$Q_{phys} = \lambda \cdot V_{air,stoich} \cdot T_{air} \cdot c_{p,air} + T_{fuel} \cdot c_{p,fuel} \; kJm^{-3} \; (kJ\,kg^{-1}) \tag{3.34}$$

The specific heat content of the gas mixture can be calculated according to the simple mixing rule:

$$c_{p,fuel} = c_{p,CO_2} \cdot CO_2 + c_{p,H_2O} \cdot H_2O + c_{p,N_2} \cdot N_2 + \ldots \tag{3.35}$$

Heat capacity values for air and various components are listed in various technical tables as a function of temperature.

Besides the specific heat capacities, one can also use the enthalpy that is actually the product of the specific heat capacity and the temperature.

$$h = c_p \cdot T \, \text{kJm}^{-3} \tag{3.36}$$

Enthalpy values for various gases as a function of temperature are listed similarly in tables. The enthalpy of mixtures can be calculated by the mixing rule, as was the case for the specific heat capacity.

Since the specific heat capacity of gases is a function of the temperature and the temperature of combustion is to be computed, the computation usually requires an iterative approach.

Let us compute the theoretical adiabatic temperature of our gas mixture from the examples above.

The composition of the fuel is 85% CH_4, 10% H_2 and 5% N_2. Combustion is stoichiometric, thus the excess air factor is $\lambda = 1$.

Enthalpy of the combustion products:

$$h_0 = c_p \cdot T_{ad} = \frac{\text{LHV}}{V_{\text{flue,wet}}} = \frac{31605}{9.29} = 3410 \, \text{kJm}^{-3}$$

An engineering guess for the flame temperature: $T_{ad} = 2100\,°\text{C}$ (2373 K). At this temperature the enthalpies of the flue gas components are:

$$h_{CO_2} = 0.0915 \cdot 5186.81 = 474.5 \, \text{kJm}^{-3}$$
$$h_{H_2O} = 0.194 \cdot 4121.79 = 800.1 \, \text{kJm}^{-3}$$
$$\underline{h_{N_2} = 0.7145 \cdot 3131.96 = 2232.0 \, \text{kJm}^{-3}}$$
$$h_{2100} = \qquad\qquad\qquad 3506.6 \, \text{kJm}^{-3}$$

Since $h_0 < h_{2100}$, the real temperature is most probably smaller than 2100 °C. Let us iterate over the temperature and assume now $T_{ad} = 2000\,°\text{C}$. The enthalpies in this case are:

$$h_{CO_2} = 0.0915 \cdot 4910.51 = 449.7 \, \text{kJm}^{-3}$$
$$h_{H_2O} = 0.194 \cdot 3889.72 = 754.8 \, \text{kJm}^{-3}$$
$$\underline{h_{N_2} = 0.7145 \cdot 2970.25 = 2120.0 \, \text{kJm}^{-3}}$$
$$h_{2100} = \qquad\qquad\qquad 3324.5 \, \text{kJm}^{-3}$$

Since now $h_{2000} < h_0$, the real flame temperature must be between 2000 and 2100 °C. We can either iterate one more step or simply interpolate by the enthalpy values.

$$T_{ad} = 2000 + \frac{3410 - 3324.5}{3506.6 - 3324.5} 100 = 2047\,°\text{C}$$

For flame temperatures above 1800 °C, the effect of dissociation must also be accounted for. This phenomenon is based on the experience that both the CO_2 and the H_2O can disintegrate into oxygen as well as CO and H_2, respectively. Taking this effect into consideration our last and most exact equation for the flame temperature is the following

$$T_{ad} = \frac{LHV + Q_{phys} - Q_{diss}}{V_{flue,wet,dis} \cdot c_{p,flue}} \, °C \tag{3.37}$$

where

Q_{diss} is the dissociation heat from the disintegration of the flue gas CO_2 and H_2O.

Above 1500 °C a noticeable, and above 1800 °C a non-negligible dissociation occurs with the three atom components of the flue gas. These reactions are essentially the reverse of the combustion reactions and thus require heat that is equal to the heat generated in the forward reactions. The heat consumption results in decreasing flame temperature. The reactions are the following:

$$2CO_2 \rightarrow 2CO + O_2 \quad \text{with a heat consumption of} = 12645 \, kJm^{-3}CO_2$$

$$2H_2O \rightarrow 2H_2 + O_2 \quad \text{with a heat consumption of} = 10760 \, kJm^{-3}H_2O$$

The percentage of dissociation can be expressed with the following equations:

$$a = \frac{V_{CO_2 diss}}{V_{CO_2}} \cdot 100 \tag{3.38}$$

and

$$b = \frac{V_{H_2O diss}}{V_{H_2O}} \cdot 100 \tag{3.39}$$

that can be calculated from the reaction rate constants and the partial pressure values.

The rate of dissociation can be computed from the graphs shown in Figure 3.2 for CO_2 and Figure 3.3 for H_2O. Both graphs show the rates as a function of the temperature and the partial pressure.

The rate of dissociation can then be used to compute the heat effect of the reactions:

$$Q_{diss} = a \cdot V_{CO_2} \cdot 12\,645 + b \cdot V_{H_2O} \cdot 10\,760 \, kJ \, m^{-3} \tag{3.40}$$

where V_{CO_2} and V_{H_2O} are the specific volumes of CO_2 and H_2O, respectively in $m^3 m^{-3}$.

The theoretical (adiabatic) flame temperature cannot be achieved in real combustion applications. This is because for ignition and combustion time is

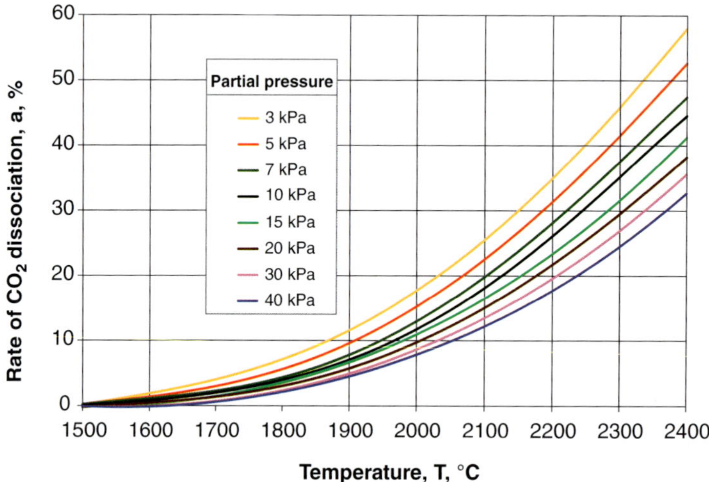

Figure 3.2 Rate of CO_2 dissociation as a function of temperature and partial pressure.

needed, and during this finite time the flame having a surface in space releases some of its heat to the environment. The real flame temperature is always lower than the theoretical one.

Through the intensification of mixing, the combustion process can be accelerated and thus the heat loss from the flame can be minimized. In this way the real flame

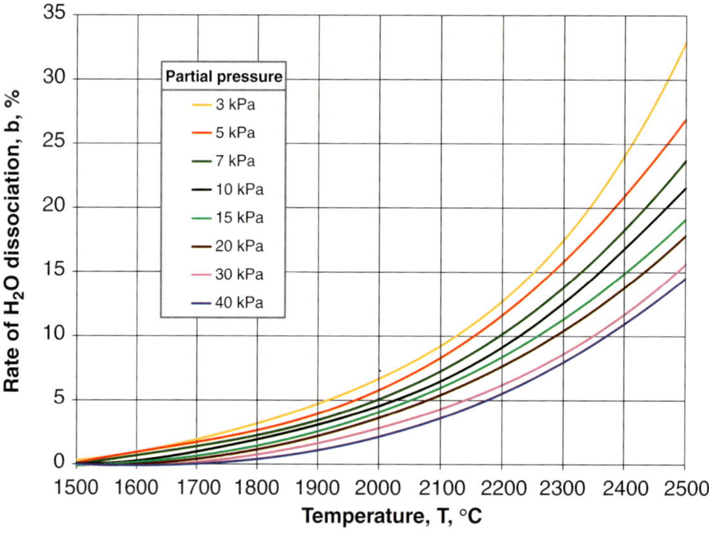

Figure 3.3 Rate of H_2O dissociation as a function of temperature and partial pressure.

temperature can approach the adiabatic flame temperature. The ratio of these temperature values is defined as the pyrometric efficiency:

$$\eta_{pyr} = \frac{T_{flame,measured}}{T_{ad,theoretical}} \tag{3.41}$$

The pyrometric efficiency, depending on the conditions of combustion, can vary between 0.65 and 0.85. For tunnel furnaces the value is approximately 0.80, for shaft kilns, 0.70, for rotary kilns, 0.70–75.

3.1.4
Heating Values

Heating values (calorific values) can be computed from the composition of the fuel. Only the combustible components should be taken into account. The calculation is done using the simple mixing rule for solid, liquid and gaseous fuels.

For solid and liquid fuels the composition is based on weight, thus the higher heating value (HHV) can be computed, from the carbon, hydrogen, sulfur and water content. The latter is important, because if the water content of the flue gas leaves the system in vapor form, the vaporization heat is lost, therefore the useful heat is less than the theoretical HHV. Using ultimate and proximate analysis results, the lower heating value (LHV) can be calculated as follows (composition values must be provided in weight fraction):

$$LHV = 33440 \cdot C + 115300 \cdot H + 9279 \cdot S - 2440 \cdot H_2O \;\; kJ\,kg^{-1} \tag{3.42}$$

If the HHV is given from laboratory measurement (bomb calorimetry), then the lower heating value can be calculated from the higher heating value, the fuel's hydrogen and moisture content, and the heat of water evaporation:

$$LHV = HHV - 2440 \cdot (9 \cdot H + H_2O), [kJ\,kg^{-1}] \tag{3.43}$$

where HHV is the higher heating value of the fuel $(kJ\,kg^{-1})$, H is the hydrogen content in weight fraction, H_2O is the water (moisture) content in weight fraction.

For gaseous fuels – since the reactants, the oxidizer and the reaction products are all gaseous – the calculation is even more straightforward. Since the fuels gas usually does not contain moisture directly, all water content must come from the hydrogen containing components. Using experimental data for the various combustible components the lower heating value can be computed using the following formula:

$$\begin{aligned} LHV = {}& 33940 \cdot CH_4 + 60799 \cdot C_2H_6 + 88438 \cdot C_3H_8 + 111042 \cdot C_4H_{10} \\ & + 137227 \cdot C_5H_{12} + 160949 \cdot C_6H_{14} + 57069 \cdot C_2H_4 + 53973 \cdot C_2H_2 \\ & + 10257 \cdot H_2 + 12048 \cdot CO \;\; kJm^{-3} \end{aligned}$$

$$\tag{3.44}$$

3.1.5
Laminar Flame Velocity

The laminar flame velocity is affected by the physical conditions of the combustion process, that is, burning characteristics of the fuel, air ratio, gas and air temperature, inert (CO_2, N_2) content, and the furnace temperature (via ignition time).

Let us calculate the laminar flame velocity of a natural gas fired flat flame burner. Known parameters:

Excess air factor: $\lambda = 1.05$
Natural gas composition: 90% CH_4, 5% CO_2, 5% N_2
Gas temperature: $T_{fuel} = 293\,K$
Air temperature: $T_{air} = 773\,K$
Heat capacities: $c_{p,O_2} = 1.52\,kJ\,m^{-3}\,K$ at 293 K
$\qquad\qquad c_{p,CO_2} = 2.26\,kJ\,m^{-3}\,K$ at 293 K
$\qquad\qquad c_{p,CH_4} = 1.88\,kJ\,m^{-3}\,K$ at 293 K
$\qquad\qquad c_{p,H_2O} = 1.77\,kJ\,m^{-3}\,K$ at 293 K
$\qquad\qquad c_{p,N_2} = 1.42\,kJ\,m^{-3}\,K$ at 293 K
$\qquad\qquad c_{p,air} = 1.293\,kJ\,m^{-3}\,K$ at 773 K

The calculation includes three steps:

1) Since the only combustible component in the fuel is methane, the first order estimate of the normal flame velocity of our fuel would be the value for CH_4 at the appropriate excess air. From Figure 3.4 at $\lambda = 1.05$ the CH_4 normal flame velocity is

$$w_0 = 0.38\,\mathrm{m\,s}^{-1}$$

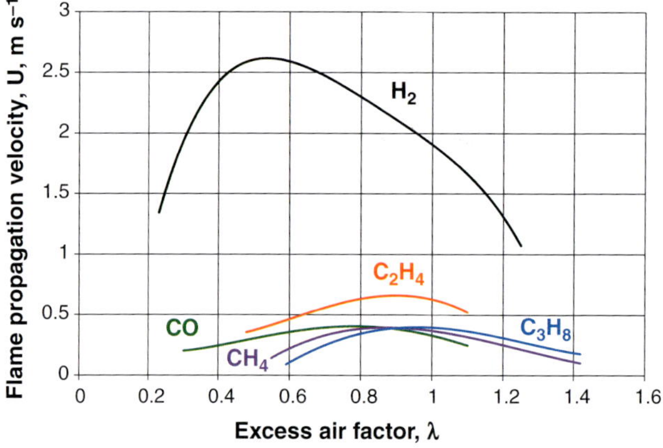

Figure 3.4 Laminar flame velocity as a function of the excess air.

2) Since our fuel contains non-combustible components, this inert gas content decreases the flame speed:

$$w = w_0 \cdot (1 - 0.8 \cdot N_2 - 1.6 \cdot CO_2 - 3 \cdot O_2)$$
$$= 0.38 \cdot (1 - (0.8 \cdot 0.05) - (1.6 \cdot 0.05)) = 0.33 \, \mathrm{m\,s^{-1}} \tag{3.45}$$

3) The temperature of the fuel–air mixture increases the flame speed.

Let us calculate the temperature of the mixture first, by using the appropriate heat capacity values from the tabular values of the components:

$$c_{p,\mathrm{fuel}} = CH_4 \cdot c_{p,CH_4} + CO_2 \cdot c_{p,CO_2} + N_2 \cdot c_{p,N_2}$$
$$= 0.9 \cdot 1.88 + 0.05 \cdot 2.26 + 0.05 \cdot 1.42 = 1.87 \, \mathrm{kJ\,m^{-3}K^{-1}} \tag{3.46}$$

The heat capacity of the fuel-air mixture can be calculated by taking the respective volumes into consideration:

$$c_{p,\mathrm{mix}} = (V_{\mathrm{fuel}} \cdot c_{p,\mathrm{fuel}} + V_{\mathrm{air}} \cdot c_{p,\mathrm{air}})/(V_{\mathrm{fuel}} + V_{\mathrm{air}})$$
$$= (1 \cdot 1.87 + 10 \cdot 1.293)/(1 + 10) = 1.35 \, \mathrm{kJ\,m^{-3}K^{-1}} \tag{3.47}$$

Thus the temperature of the mixture can be calculated:

$$T_{\mathrm{mix}} = (T_{\mathrm{fuel}} \cdot c_{p,\mathrm{fuel}} \cdot V_{\mathrm{fuel}} + T_{\mathrm{air}} \cdot c_{p,\mathrm{air}} \cdot V_{\mathrm{air}})/(c_{p,\mathrm{mix}} \cdot V_{\mathrm{mix}})$$
$$= ((293 \cdot 1.87 \cdot 1) + (773 \cdot 1.293 \cdot 10))/(1.35 \cdot 11) = 707 \, \mathrm{K} \tag{3.48}$$

The temperature effect can be considered as follows:

$$w_{\mathrm{real}} = w \cdot (T_{\mathrm{mix}}/T_0)^{1.8}$$
$$= 0.33 \cdot (707/293)^{1.8} = 1.89 \, \mathrm{m\,s^{-1}} \tag{3.49}$$

The actual flame propagation velocity at the given conditions for this furnace is $1.89 \, \mathrm{m\,s^{-1}}$, a fairly high value.

3.2
Heat-, Mass- and Momentum Transport and Balance

3.2.1
Transport

Transport is the basis for combustion and chemical reaction in general. Species have to be brought together, so that they can react with each other. However, there exist different transport mechanisms for the species and also transport mechanisms for heat as well as for the momentum. It is established that a concentration gradient leads to mass transport, called diffusion, and temperature gradients lead to heat transport, called heat conduction. Transport phenomena of mass, momentum and heat have notable analogies in their mathematical formulations; that means in the structure of their differential equations.

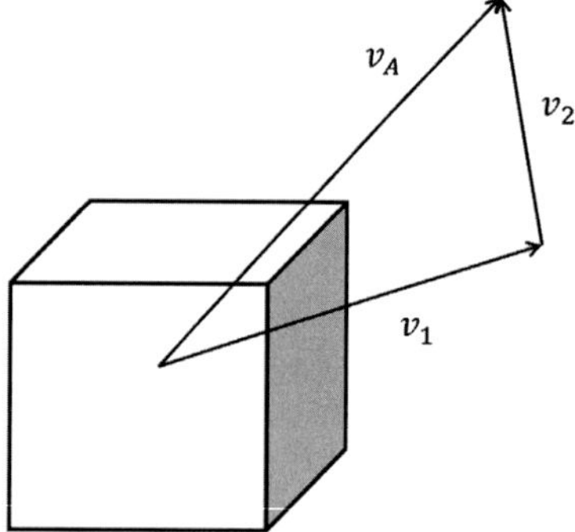

Figure 3.5 The transports mechanisms of convective and diffusive transport. Velocity vectors: $\vec{v}_1$ is the mass average velocity. $\vec{v}_2$ is the mass diffusion velocity. $\vec{v}_A$ is the velocity of species A.

3.2.2
Mass Transport

There are two main processes occurring when various different chemical species appear in a mixture. They cause the transport characteristics in this mixture. The two mechanisms which contribute to these transport processes are diffusion and convection.

Figure 3.5 shows the velocity vectors of a chemical species.

$$\vec{v}_A = \vec{v}_1 + \vec{v}_2 \tag{3.50}$$

3.2.2.1 Diffusive Mass Transport

Concentration differences lead to the diffusion of the species. The driving force is the concentration gradient and this is used in Fick's first law. It describes the molar or mass flow which is proportional to the concentration gradient. The Fick's first law of a binary system including species A and B is shown in Eq. (3.51) below. $\vec{N}_A$ is the molar flow of species A in $\mathrm{mol\,s^{-1}}$, D_{AB} is the diffusion coefficient of the binary system of A and B in $\mathrm{m^2\,s^{-1}}$, A is the area of consideration and $\frac{d\vec{c}_A}{dx}$ is the molar concentration gradient of species A in $\mathrm{mol\,m^{-4}}$.

$$\vec{N}_A = -D_{AB} \cdot A \cdot \frac{d\vec{c}_A}{dx} \ \mathrm{mol\,s^{-1}} \tag{3.51}$$

In order to explain how one can imagine a three-dimensional concentration distribution it is shown in Figure 3.6. With time the concentration peak will decrease due to its diffusion.

3.2.2.2 Convective Mass Transport

Due to a flow, for example, because a pump is used (forced convection) the species are transported with their velocity.

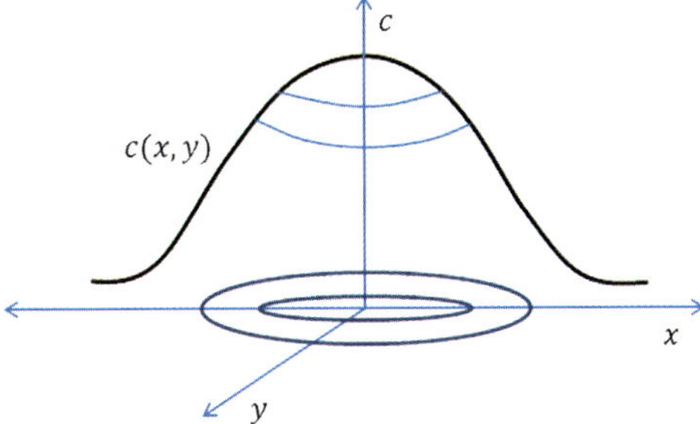

Figure 3.6 The concentration gradient at a certain time step in space.

Equation (3.52) describes the convective transport of species A, in which $\vec{m}_{A,\text{conv}}$ is the mass flow vector, ρ_A the density of the matter and $\vec{v}_1$ the velocity vector.

$$\vec{m}_{A,\text{conv}} = \rho_A \vec{v}_1 \ \text{kg}\,\text{s}^{-1}\,\text{m}^{-2} \tag{3.52}$$

3.2.3
Mass Transfer

The mass transfer is based on a concentration difference and is described in Eq. (3.53). $\dot{N} = \beta \cdot A \cdot (c_{A,1} - c_{A,2})$ is the molar flow of species A in mol s^{-1} and based on diffusive and convective transport mechanisms. Figure 3.7 shows a concentration profile of species A in combination with the velocity profile of the flow. A concentration boundary layer is formed.

$$\dot{N} = \beta \cdot A \cdot (c_{A,1} - c_{A,2}) \ \text{mol s}^{-1} \tag{3.53}$$

where β is the mass transfer coefficient in m s^{-1}, A is the area of exchange in m^2, $c_{A,1}$ and $c_{A,2}$ are the different molar concentrations of species A in mol m^{-3}.

A basic method to find the mass transfer coefficient β is the experiment. The use of similarity or model theory can reduce the number of experiments significantly.

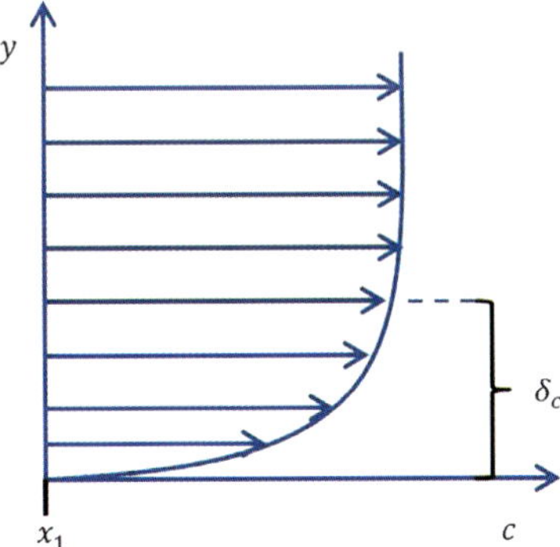

Figure 3.7 Concentration profile in a flow: δ_c shows the concentration boundary layer. Arrows indicate the velocity of the flow.

The solution of a physical problem will be independent of any unit and it can be represented by dimensionless variables.

To calculate the dimensionless parameters, the parameters are related to their corresponding characteristic parameters, for example, characteristic length (L_0) or characteristic concentration (c_0).

The Sherwood number will be calculated as below, where L_0 is the characteristic length and D_{AB} is the diffusion coefficient in $m^2\,s^{-1}$.

$$Sh = \frac{\beta \cdot L_0}{D_{AB}} \tag{3.54}$$

The Sherwood number (Sh) relates the total mass transfer to the diffusion and is a function of the Reynolds number (Re) and the Schmidt number (Sc). With the Sherwood number the mass transfer coefficient (β) can be calculated.

The Reynolds number is given by:

$$Re = \frac{V \cdot L_0}{\nu} \tag{3.55}$$

V is the characteristic velocity in $m\,s^{-1}$ and ν is the kinematic viscosity in $m^2\,s^{-1}$.

It can also be defined as the ratio of inertial forces to viscous forces.

The Schmidt number is given by:

$$Sc = \frac{\nu}{D_{AB}} \tag{3.56}$$

and relates the viscous diffusion rate to the mass diffusion rate.

3.2.4
Heat Transport

In order to characterize the mechanisms of heat transfer, three models are described. These are thermal conduction, convection and thermal radiation. By analogy with mass transfer, the transfer of heat takes place by diffusion and is called thermal conduction. Thermal radiation is not bounded to matter and can also appear in vacuum systems.

3.2.4.1   Heat Conduction

Equation (3.57) is known as Fourier's law and describes heat conduction. $\vec{q}_{conduction}$ is the heat flux in $[W\,m^{-2}]$, λ the thermal conductivity in $[W\,m^{-1}\,K^{-1}]$ and $\frac{\partial \vec{T}}{\partial x}$ is the temperature gradient in $[K\,m^{-1}]$.

$$\vec{q}_{conduction} = -\lambda \cdot \frac{\partial \vec{T}}{\partial x} \tag{3.57}$$

The thermal conductivity λ can be described as the ability of a material to conduct heat.

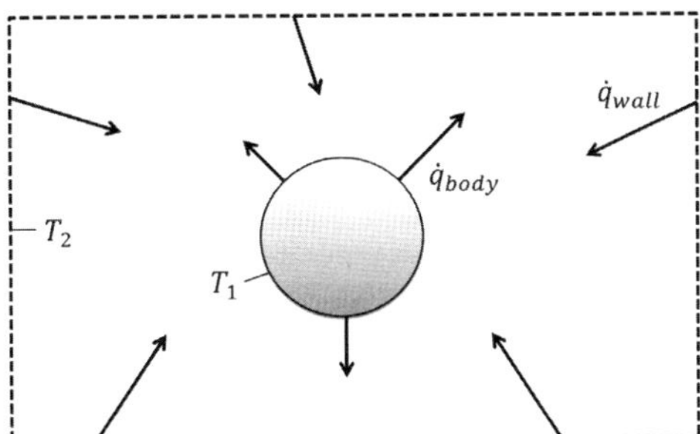

Figure 3.8 Radiation of a body at temperature T_1 in its surroundings with temperature T_2.

3.2.4.2 Thermal Radiation

In order to transport the energy by convection and conduction, a matter is required, but radiation does not need a material connection. When a body emits radiation, it is the internal energy that turns into electromagnetic waves. Gases and liquids absorb and emit the radiation inside their space, but solids absorb it within a thin layer of their surface (Figure 3.8).

Stefan–Boltzmann's law shows the maximum possible heat flux. This emitter is called a black body. $\dot{q}_{\text{radiation}}$ in $\mathrm{W\,m^{-2}}$ shows the black body radiation flux, σ is the Stefan–Boltzmann constant which is $5.67 \times 10^{-8}\ \mathrm{W\,m^{-2}\,K^{-4}}$], T the temperature in K.

$$\dot{q}_{\text{black body}} = \sigma.T^4 \tag{3.58}$$

The black body is an ideal absorber and ideal emitter.

The emissive power of real radiators can be described by using a correction factor. ε is the emissivity factor.

For a black body $\varepsilon = 1$.

$$\dot{q} = \varepsilon \cdot \sigma \cdot T^4 \tag{3.59}$$

When radiation hits a body, some of it will be reflected, some absorbed and some transmitted (refer also to Figure 3.9).

$$r + a + t = 1 \tag{3.60}$$

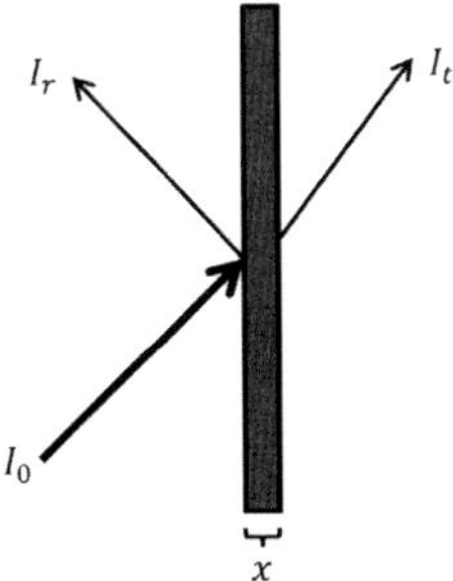

Figure 3.9 Radiation with an intensity of I_0 hits a body and is partly reflected (I_r), partly transmitted (I_t) and partly absorbed (not shown).

Where r is the reflectivity and is equal to $\dfrac{I_r}{I_0}$, a is the absorptivity and is equal to $= 1 - r - t$, and t is the transmissivity and equal to $t = \dfrac{I_t}{I_0}$.

The heat which is transmitted by radiation between two subjects happens not only from the subject at the higher temperature to the other one, but also from the cooler subject to the hotter. The net heat flow from the hotter subject to the body at lower temperature is of interest. The heat flow which is emitted by the radiator can be calculated with:

$$\dot{q}_{em} = \varepsilon \cdot \sigma \cdot T_1^4 \tag{3.61}$$

The heat absorbed by the radiator is described by Eq. (3.62), where a is the absorptivity of the radiator.

$$\dot{q}_{ab} = a \cdot \sigma \cdot T_2^4 \tag{3.62}$$

For a gray body it is assumed that $\varepsilon = a$.

The net heat flow is calculated using Eq. (3.63).

$$\dot{q}_{radiation} = \dot{q}_{em} - \dot{q}_{ab} = \varepsilon \cdot \sigma \cdot (T_1^4 - T_2^4) \tag{3.63}$$

In many applications heat transfer by convection and conduction must be considered in addition to the radiative heat transfer.

$$\dot{q} = \dot{q}_{cond.+conv.} + \dot{q}_{radiation} \tag{3.64}$$

$$\dot{q} = \alpha \cdot (T_1 - T_2) + \varepsilon \cdot \sigma \cdot (T_1^4 - T_2^4) \tag{3.65}$$

where α is the heat transfer coefficient for the convective and conductive heat transfer in $W\,m^{-2}\,K^{-1}$, refer also to Section 3.2.5.

Electromagnetic waves (Figure 3.10) spread out in a straight line at the speed of light (c_0) $3.00 \times 10^8\,m\,s^{-1}$ in vacuum. In a medium, while their frequency f remains unchanged, the velocity is lower than c_0.

The equation relating the wavelength λ in m and the frequency f in s^{-1} is:

$$c = \lambda \cdot f \tag{3.66}$$

The radiation reduces its energy when it passes through a gas or mixture of gases. This phenomenon happens because of the absorption and scattering of the

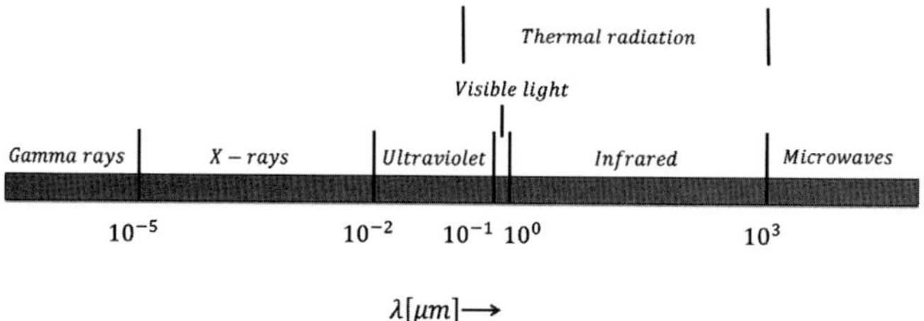

Figure 3.10 The electromagnetic wave spectrum showing the area of thermal radiation.

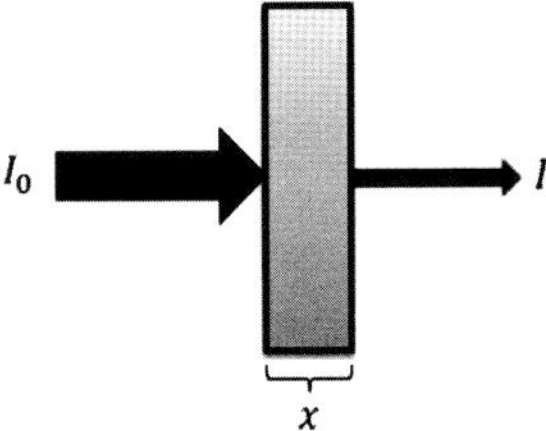

Figure 3.11 The absorption of the incident beam with intensity I_0. passing through a gas with a width of x in m, refer to Lambert–Beer's law.

waves through the gas molecules. If we assume that the radiation is only absorbed but not scattered, Lambert–Beer's law describes the reduction of the incident beam with an intensity of I in $W\,m^{-2}$ with increasing length x in m of the beam in the gas (Figure 3.11).

$$\frac{dI}{dx} = -k \cdot I \tag{3.67}$$

$$k = k' \cdot c \tag{3.68}$$

where k is the spectral absorption coefficient of the absorbing gas in m^{-1}, which also depends on the concentration of the absorbing gas c in $mol\,m^{-3}$. k' is the spectral absorption coefficient in $m^2\,mol^{-1}$.

Integration of Eq. (3.67) between $x = 0$ and x yields:

$$I = I_0 \cdot e^{-k \cdot x} \tag{3.69}$$

The amount of absorbance (A) can be calculated as follows:

$$A = \ln \frac{I_0}{I} = k' \cdot c \cdot x \tag{3.70}$$

The Lambert–Beer law is used as the basis for quantitative absorption spectroscopy.

In Figure 3.12 the radiation spectra of black bodies at different temperatures can be seen according to Planck's law. A radiation spectrum of a black body undergoes a maximum. The higher the temperature, the higher the total radiation according to the Stefan–Boltzmann law.

The maximum value of the wavelength of the radiation spectra show a maximum value which is described by Wien's displacement law:

$$\lambda_{max} \cdot T = 2897 \; \mu m \; K \tag{3.71}$$

The higher the temperature of the object, the more its radiation maximum λ_{max}, in μm, is shifted to lower wavelengths.

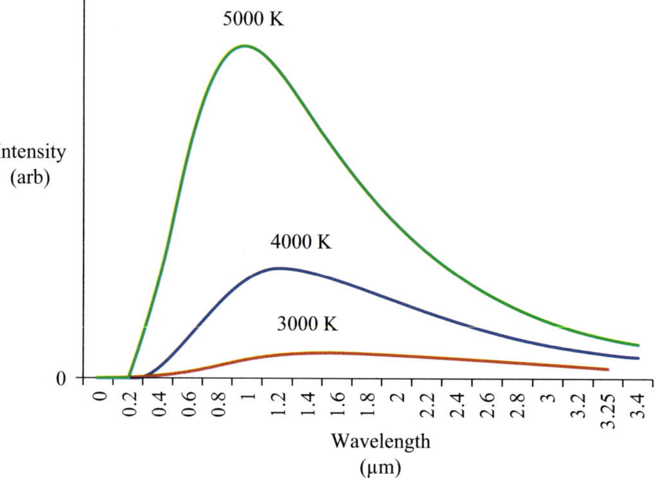

Figure 3.12 The radiation spectrum of a black body according to Planck's law for different temperatures.

3.2.5
Heat Transfer

While a fluid is flowing, the heat transfer $\dot{Q}$ (W m^{-2}) is by heat conduction due to the temperature gradient and by the movement of the fluid (convection). This is described by Eq. (3.72). (However, heat transfer by radiation may also be of importance, refer to Section 3.2.4.2.)

$$\dot{Q} = \alpha \cdot A \cdot (T_1 - T_2) \tag{3.72}$$

where α is the heat transfer coefficient in W m^{-2} K^{-1}, A is the area in m^2 and $T_1 - T_2$ is the temperature difference in K between the media.

Similar to the mass transfer case, where the flow affects the concentration boundary layer (see Figure 3.7), the temperature boundary layer is also affected and the heat transfer coefficient. Figure 3.13 shows the heat transfer between a sphere and its surrounding flow. The heat transfer coefficient is obtained from the dimensionless Nusselt number (Nu).

$$Nu = \frac{\alpha \cdot L_0}{\lambda} \tag{3.73}$$

where L_0 is the characteristic length in m and λ is the thermal heat conductivity in W m^{-1} K^{-1}.

The Nusselt number compares the convective heat transfer to the conductive heat transfer through the temperature boundary layer and is mainly a function of the

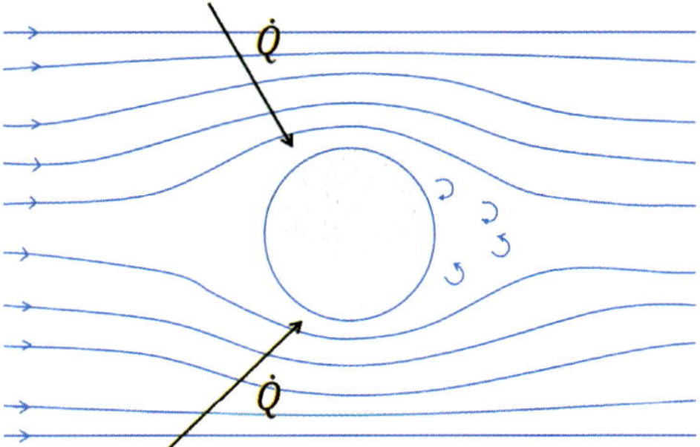

Figure 3.13 Sketch showing the heat transfer between a sphere and its surrounding flow.

Reynolds number (*Re*) and the Prandtl number (*Pr*) which is given by

$$Pr = \frac{\mu \cdot c_p}{\lambda} = \frac{\nu}{a} \tag{3.74}$$

$$a = \frac{\lambda}{c_p \cdot \varrho} \tag{3.75}$$

where μ is the dynamic viscosity in Pa s, a is the thermal diffusivity in $m^2\,s^{-1}$, c_p is the specific heat capacity in $J\,kg^{-1}\,K^{-1}$, λ is the thermal heat conductivity in $W\,m^{-1}\,K^{-1}$, and ϱ is the density of the fluid in $kg\,m^{-3}$.

As an example, the Nusselt number of a sphere or spherical particle in a flow of gas or liquid is obtained by the following equation:

$$Nu = 2 + 0.6\ Re^{\frac{1}{2}} Pr^{\frac{1}{3}} \tag{3.76}$$

3.2.6
Momentum Transport

The momentum (*P*) of a mass-carrying particle is a physical property and is formed by the product of mass (*m* in kg) and velocity (*v* in $m\,s^{-1}$) of the particle. In a closed system the momentum is subject to the conservation law.

$$P = m.v \tag{3.77}$$

Viscosity is a physical property of fluids caused by internal friction. In order to make viscosity comprehensible, an experiment is presented. A plate on a lake is moving constantly with velocity v and the flow profile shown in Figure 3.14 will be obtained after a certain time period, that is, after acceleration.

Due to adherence at the wall, the fluid at the plate has the same velocity as the plate itself.

In dealing with flow problems, the flow type used determines the flow characteristics and phenomena. The most common fluids used in chemistry and technique are water and air and represent Newtonian fluids. They fulfill the Newton's law of viscosity, which

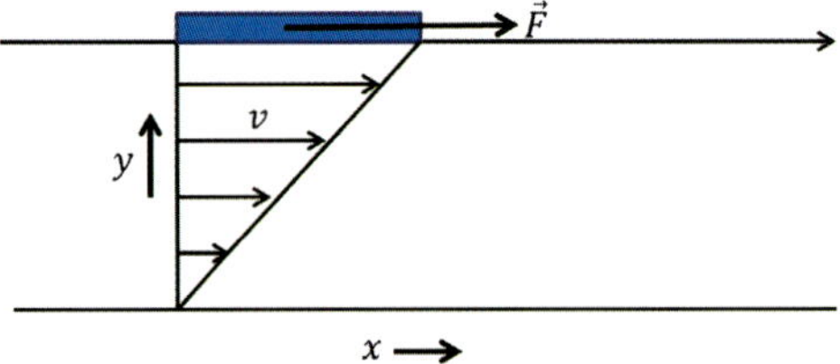

Figure 3.14 Velocity profile below a plate which is constantly drawn with force *F* over a lake.

is formulated in Eq. (3.78) where τ_{xy} is the shear force per unit area in Pa, v is the kinematic viscosity in $m^2\,s^{-1}$, ρ the density and $\dfrac{\partial \vec{v}}{\partial y}$ the velocity gradient in s^{-1}.

$$\vec{\tau}_{xy} = -v \cdot \rho \cdot \frac{\partial \vec{v}}{\partial y} \tag{3.78}$$

This law states that the shearing force is proportional to the shear rate, whereas the proportionality is a property of the fluid and is defined as dynamic viscosity. The correlation between dynamic and kinematic viscosity is given in Eq. (3.79) where μ is the dynamic viscosity.

$$v = \frac{\mu}{\rho} \tag{3.79}$$

There are other types of flow laws which are illustrated in Figure 3.15. Non-Newtonian fluids do show a more complicated correlation between shear stress and shear strain rate.

Typical examples of some non-Newtonian fluids are:

- Bingham plastics are rigid bodies when no or low stress is applied, but flow like a viscous fluid at high stress. For example, toothpaste.
- Dilatant (shear thickening) fluids' viscosity increases with the rate of shear strain. For example, sand + water.
- Pseudoplastic (shear thinning) fluids' viscosity decreases with the rate of shear strain. For example, fibrous solutions.

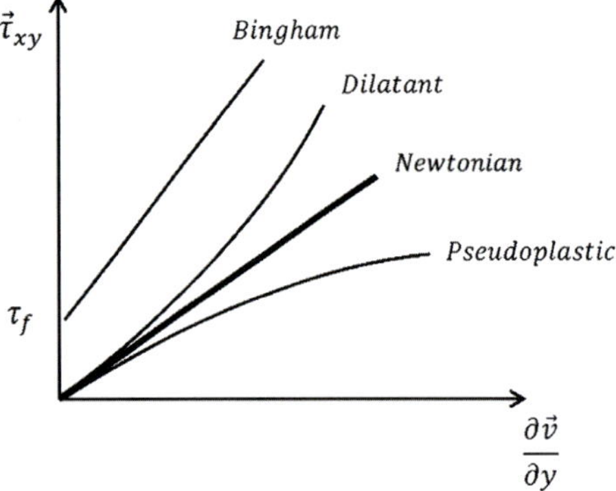

Figure 3.15 Flow characteristics of different media.

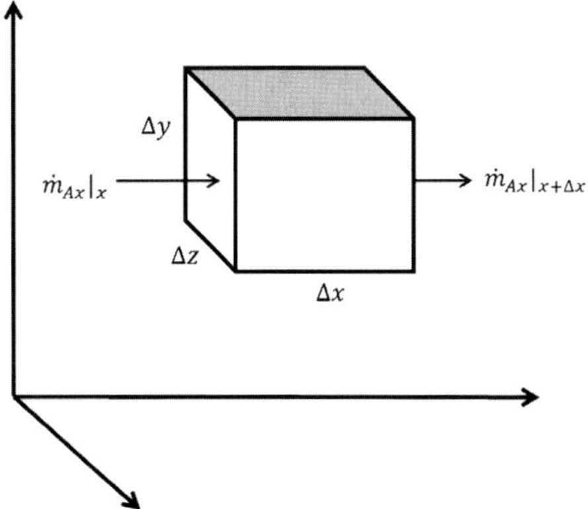

Figure 3.16 Balance space as a finite volume. $\dot{m}_{Ax}$ is the mass flow of species A based on area in kg s^{-1}m^{-2}.

3.2.7
Balance

In order to get balance equations for mass, energy or momentum, a balance volume, sketched in Figure 3.16, has to be defined.

The balance volume is enclosed by a real or imaginary surface, whereas macroscopic or microscopic mass, energy or momentum flows can pass through. In the case of temporal changing conditions, the balance is formed in general as follows:

$$\text{Temporal Change} = \text{Input} - \text{Output} \pm \text{Source/Sink} \tag{3.80}$$

There are four terms which are taken into account when balance equations are put together: The temporal change gives the total change within the finite volume in time. It can be an increase (e.g., in terms of mol s^{-1} for a species balance or in J s^{-1} for a heat balance) or a decrease, or it can be constant or zero. If it is zero then a steady state is given. The input gives the total input into the finite volume, for example, the input of species A or heat in time. The output gives the total output out of the finite volume. The source term or sink term represents, for example, the formation of species by chemical reaction in a species balance (mol s^{-1}) or the heat of formation of a chemical reaction in the heat balance (J s^{-1}).

Equation (3.80) leads to a mass balance when applied to a concentration field, to a heat balance when applied to a temperature field, and to a momentum balance when applied to a velocity field.

3.2.7.1 Mass Balance
We assume that we have a binary system, consisting of the species A and B. According to Eq. (3.80), the mass balance is calculated as:

$$\text{Temporal change}: \quad \frac{\partial \rho_A}{\partial t} \Delta x \Delta y \Delta z \, \text{kg s}^{-1} \tag{3.81}$$

$$\text{Input of A in } x \text{ direction}: \dot{m}_{Ax}|_x \Delta y \Delta z \tag{3.82}$$

$$\text{Output of A in } x \text{ direction}: \dot{m}_{Ax}|_{x+\Delta x}\Delta y \Delta z = \dot{m}_{Ax}|_x \Delta y \Delta z$$
$$+ \frac{\partial \dot{m}_{Ax}}{\partial x}\Delta x \Delta y \Delta z \tag{3.83}$$

$$\text{Rate of production of A by chemical reaction}: R'_A \Delta x \Delta y \Delta z \tag{3.84}$$

There are also output and input in other directions(y and z). After writing the entire mass balance and dividing by $\Delta x \Delta y \Delta z$ Eq. (3.85) is obtained.

$$\frac{\partial \rho_A}{\partial t} + \left(\frac{\partial \dot{m}_{Ax}}{\partial x} + \frac{\partial \dot{m}_{Ay}}{\partial y} + \frac{\partial \dot{m}_{Az}}{\partial z} \right) = R'_A \tag{3.85}$$

We can write the same input and output terms for component B. According to the law of conservation, it is valid to write:

$$R'_A + R'_B = 0 \tag{3.86}$$
$$\dot{m}_A + \dot{m}_B = \rho \cdot v \tag{3.87}$$

Equation (3.88) is called the equation of continuity and shows the total mass balance:

$$\frac{\partial \rho}{\partial t} + \nabla(\rho \cdot v) = 0 \tag{3.88}$$

Instead of writing balance equations in mass units, it is also possible to use molar units as below, where R_A is the molar rate of production of A per unit volume.

$$\frac{\partial C_A}{\partial t} + \left(\frac{\partial \dot{n}_{Ax}}{\partial x} + \frac{\partial \dot{n}_{Ay}}{\partial y} + \frac{\partial \dot{n}_{Az}}{\partial z} \right) = R_A \tag{3.89}$$

By substitution of Eq. (3.51) (Fick's first law) in $\dot{n}$, and under assumption of having no chemical reactions, the Fick's second law of diffusion is obtained:

$$\frac{\partial C_A}{\partial t} = D_{AB} \cdot \nabla^2 c_A \tag{3.90}$$

$$\text{with } \nabla^2 = \frac{\partial^2}{\partial x^2} + \frac{\partial^2}{\partial y^2} + \frac{\partial^2}{\partial z^2} \tag{3.91}$$

3.2.7.2 Heat Balance

The heat balance is, in principle, very similar to the mass balance. Using the general equation for the balance applied to a single burning particle as an example, leads to the following heat balance:

$$m \cdot c_p \cdot \frac{dT_2}{dt} = \alpha \cdot A(T_1 - T_2) + \varepsilon \cdot \sigma \cdot A(T_1^4 - T_2^4) + \Delta H_R(-r) \cdot V \quad \text{W or Js}^{-1} \tag{3.92}$$

where

m mass of the particle in kg

T_2 particle temperature (assuming an isothermal particle) in K

c_p specific enthalpy of the particle in $J\,kg^{-1}\,K^{-1}$

A surface area of the particle in m^2

The term $\alpha \cdot A(T_1 - T_2)$ considers thermal conduction and convection and summarizes the input–output terms regarding this transport mechanism. $\varepsilon \cdot \sigma \cdot A \cdot \left(T_1^4 - T_2^4\right)$, is the thermal radiation including again input–output terms, and $\Delta H_R(-r) \cdot V$ is the heat generation term (source or sink term depending on the reaction enthalpy ΔH_R) by chemical reaction (combustion).

3.2.7.3 Momentum Balance

A momentum balance can be done for a finite volume leading, for a Newtonian and incompressible fluid, to the following Navier–Stokes equation with constant density.

$$\underbrace{\frac{\delta v}{\delta t} + (v\nabla)v}_{\text{acceleration}} = \underbrace{-\frac{1}{\rho}\nabla P}_{\text{pressure forces}} + \underbrace{v\Delta v}_{\text{viscous forces}} + \underbrace{g}_{\text{gravitational forces}} \tag{3.93}$$

The Navier–Stokes equations are very important in flow simulations.

3.3
Elementary Reactions and Radicals

3.3.1
Elementary Reactions

There are many ways of classifying a chemical reaction. One option is the distinction between homogeneous and heterogeneous reactions. A reaction is homogeneous if all reactants are located in the same phase where the reaction will take place, as is the case in combustion processes.

Chemical reactions can be written as global reactions or on an elementary chemical reaction basis, the correlation between stoichiometry and rate expression is important here.

The reaction between oxygen and hydrogen (old: Knallgas reaction) is a good example to explain the differences between global reaction (non-elementary) and the elementary reactions.

$$2H_2(g) + O_2(g) \rightarrow 2H_2O(g) \tag{3.94}$$

Reaction (3.94) shows the simple net reaction (global reaction) which consists of a multitude of reactions, like chain reactions, among others. The mechanism is complex and has not yet been fully elucidated. Some of the steps involve radicals, which are atoms or molecules with unpaired electrons. They are highly chemically reactive. A few elemental reactions of the hydrogen–oxygen reaction forming radicals are shown in the following:

$$\text{Initiation} \quad H_2 + O_2 \rightarrow H_2O + O^\bullet \tag{3.95}$$

$$\text{Propagation} \quad OH^\bullet + H_2 \rightarrow H^\bullet + H_2O \tag{3.96}$$

$$\text{Branching} \quad O_2 + H^\bullet \rightarrow OH^\bullet + O^\bullet \tag{3.97}$$

$$O^{\bullet}+H_2 \rightarrow OH^{\bullet}+H^{\bullet} \tag{3.98}$$

$$\text{Termination} \quad H^{\bullet} + H^{\bullet} + M \rightarrow H_2 + M \tag{3.99}$$

$$H^{\bullet}+O_2+M \rightarrow HO_2^{\bullet}+M \tag{3.100}$$

A branching step is an elementary reaction which is characterized by the generation of more than one chain carrier, that is, a radical species. The third body (M) may be any stable species such as H_2, O_2, H_2O, N_2.

Compared to the rate laws of the elemental equations the rate of formation of radicals of the global reaction is complex, indicating a non-elementary reaction.

3.3.2
Reaction Rates

The rate of a reaction represents how fast a reaction takes place, for example, how fast product species are formed.

Consider the elemental chemical reversible reaction depicted as follows:

$$A + 2B \underset{k_{-1}}{\overset{k_1}{\rightleftharpoons}} C + D \tag{3.101}$$

The net rate of reaction (3.101) is given by equation (3.102), or in general notation equation (3.104) for the general reaction (3.103)

$$r = k_1 C_A C_B^2 - k_{-1} C_C C_D \tag{3.102}$$

$$\nu_A A + \nu_B B \underset{k_{-1}}{\overset{k_1}{\rightleftharpoons}} \nu_C C + \nu_D D \tag{3.103}$$

$$r = k_1 C_A^{\nu_A} C_B^{\nu_B} - k_{-1} C_C^{\nu_C} C_D^{\nu_D} \tag{3.104}$$

The reaction rate constants k_1 and k_{-1} refer to the forward reaction and reverse reaction, respectively. The rate law is a function of the reacting species and conditions like temperature, pressure, and concentrations of the species.

The correlation between reaction rate and concentrations is usually given in terms of power laws, which is exemplarily shown in Eq. (3.104). The exponents of the concentrations lead to the order of the reaction, whereas ν_A represents the order of the reaction with respect to reactant A, and ν_B the order with respect to reactant B. It is important to remember that the rate law expressions usually originate from experimental observations.

The rate of change of the amount of species i (R_i, in $\mathrm{mol\,s^{-1}\,m^{-3}}$) in j reactions is described in Eq. (3.105)

$$R_i = \sum_j \nu_{i,j} \cdot r_j \tag{3.105}$$

Where r_j is the specific reaction rate of a reaction j and is expressed in Eq. (3.106).

$$r_j = \underbrace{\frac{dN_i}{dt} \cdot \frac{1}{V}}_{R_i} \cdot \frac{1}{\nu_i} \tag{3.106}$$

3.3.3
Temperature Dependence

The temperature dependence of chemical reactions is usually described by the Arrhenius equation.

$$k_i = k_{0,i} \cdot e^{-\frac{E_i}{R \cdot T}} \tag{3.107}$$

Where $k_{0,i}$ represents the so-called frequency factor, R is the universal gas constant, and E_i is the activation energy of the reaction in $J\,mol^{-1}$.

In addition, and when necessary, an extended Arrhenius equation is used, usually in gas-phase chemistry. In this equation three parameters are applied to describe the temperature dependence:

$$k_i = \underbrace{k_{0,i} \cdot T^{\eta_i}}_{k_{i,1}} \cdot \underbrace{e^{-\frac{E_i}{R \cdot T}}}_{k_{i,2}} \tag{3.108}$$

The coefficient $k_{i,1}$ represents the frequency of collision and is a function of temperature only. In case of $\eta_i = 0$, the extended Arrhenius approach turns into the basic Arrhenius approach, the case of $\eta_i = 1/2$ is derived from the collision theory for bimolecular reactions, and $\eta_i = 1$ derives from activated complex or transition state theory.

The Arrhenius equation is used to determine the activation energy by measuring the rate of reaction at different temperatures. By linearizing the Arrhenius law a plot of $\ln(k_i)$ as function of $(1/T)$, called an Arrhenius plot (Figure 3.17), is formed as follows:

$$\ln(k_i) = \ln(k_{0,i}) - \frac{E}{R}\left(\frac{1}{T}\right) \tag{3.109}$$

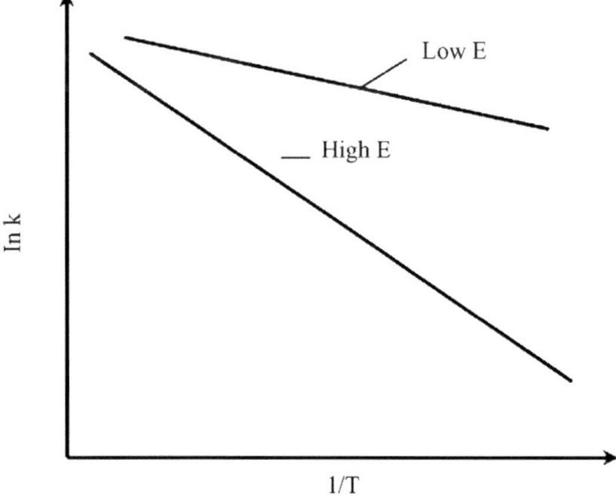

Figure 3.17 Temperature dependence of the reaction rate – Arrhenius plot.

In the Arrhenius plot the slope depicts the ratio $-E/R$, where a large slope characterizes a large activation energy and a small slope indicates a small activation energy.

3.3.4
Collision Theory

The Arrhenius' law is similar to the results of the collision theory, which says that reactants first have to physically meet in order to take part in a chemical reaction. Next, they have to collide with a certain energy, called the activation energy, which is high enough that the reaction can take place.

The average kinetic energy of a substance is proportional to its thermo-dynamic temperature, which in turn is proportional to the square of its speed. The consequences of this higher speed at higher temperatures are more frequent collisions, and the collisions are more energetic. The kinetic theory of gases leads to the collision rate of molecules in a gas. If bimolecular (molecules A and B) collisions are considered, the number of collisions of A with A per time (s) and volume (cm^3) (Z_{AA}) are calculated via equation (3.110). Collisions of molecule A with B per s per cm^3 Z_{AB} are calculated via equation (3.111). These equations calculate the rate equations of bimolecular reactions, provided every collision between a reactant molecule results in the conversion of reactants into product.

$$Z_{AA} = d_A^2 \underbrace{\frac{\mathcal{N}^2}{10^6}}_{n_A} C_A^2 \sqrt{\frac{4\pi kT}{M_A}} \tag{3.110}$$

$$Z_{AB} = \left(\frac{d_A + d_B}{2}\right)^2 n_A n_B \sqrt{8\pi kT\left(\frac{1}{M_A} + \frac{1}{M_B}\right)} \tag{3.111}$$

Z_{AA} number of collisions of A with A $(s^{-1}\,cm^{-3})$
Z_{AB} number of collisions of A with B $(s^{-1}\,cm^{-3})$
d_A diameter of an A molecule (cm)
d_B diameter of a B molecule (cm)
$\mathcal{N}$ Avogadros's number $= 6.023 \times 10^{23}\,mol^{-1}$
C_A concentration of A $(mol\,l^{-1})$
C_B concentration of B $(mol\,l^{-1})$
n_A $\mathcal{N} \cdot C_A/10^3$ number of molecules of A per cm^3
n_B $\mathcal{N} \cdot C_B/10^3$ number of molecules of B per cm^3
k Boltzmann constant $= 1.30 \times 10^{-16}\,erg\,K^{-1}$
T Temperature (K)
M_A molecular mass$/\mathcal{N} =$ mass of an A molecule (g)
M_B molecular mass$/\mathcal{N} =$ mass of a B molecule (g)

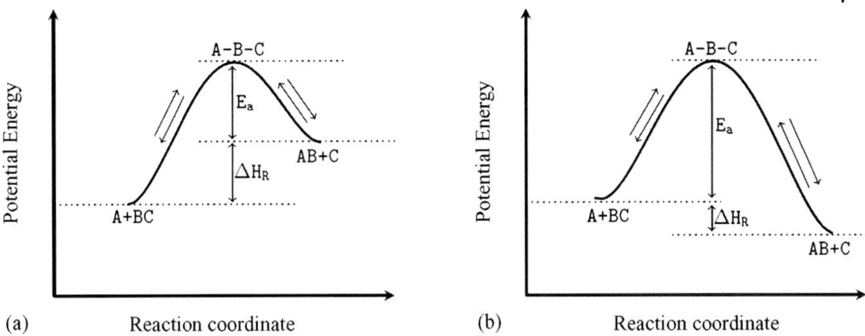

Figure 3.18 Comparison of an endothermic reaction (a) and an exothermic reaction (b).

As mentioned before, the molecules have to collide with a certain minimum energy, the activation energy. The activation energy can be visualized as a barrier to the energy transfer between reacting molecules. For the reaction (3.112) the energies involved in transformation of the reactants to products are depicted in Figure 3.18 when the reaction is endothermic (a) and exothermic (b).

$$A + BC \rightleftharpoons A\!-\!B\!-\!C \rightleftharpoons AB + C \tag{3.112}$$

Figure 3.18 shows the potential energy of the three molecule system A, B, C and the reaction progress. The chemical potential energy describes the energy of a species with regards to the arrangement of the atoms. The bonds between the atoms are a source of potential energy.

First the reactants A and BC are far apart. The system's energy at that point is just the bond energy of BC. By moving along the reaction coordinate, the reactants approach each other and the BC bond energy begins to break. At this point the energy of the system increases. At the maximum, intermolecular distances between A and B and between B and C are equal (A-B-C), which causes the high potential energy of this state. At the end of the reaction, the molecules A and B approach each other while decreasing the potential energy of the system by using it as bond energy, and C is far apart, refer also to [2].

3.3.5
Three-Body Reactions

Sometimes third bodies are required so that reactions can proceed. This body, called M, has to appear in the rate expression. Some third body species work better than others.

Some radical recombination reactions do not fulfill the Arrhenius approach. The energy, which is released by recombination of two radicals, is so large that it causes the product to decompose into its original radicals. Therefore, a third body is necessary to bind this energy so that the recombination can continue. Considering the chemical reaction (3.113), the rate law is given by Eq. (3.114).

$$H^\bullet + H^\bullet + M \underset{k_{-1}}{\overset{k_1}{\rightleftharpoons}} H_2 + M \tag{3.113}$$

$$r = k_1 C_{H^\bullet}^2 C_M - k_{-1} C_{H_2} C_M \tag{3.114}$$

Most three-body recombination reactions are pressure-sensitive. This means, that they occur at high pressure and rarely at low pressure.

3.3.6
Chemical Equilibrium

The chemical equilibrium is dynamic in nature where no net flows of species or energy appear.

The forward and reverse reaction rates of the elementary reaction are given in Eq. (3.115).

At equilibrium this equation can be combined to give the chemical equilibrium constant (K_c):

$$\frac{k_1}{k_{-1}} = K_C = \frac{C_C^{\nu_C} C_D^{\nu_D}}{C_A^{\nu_A} C_B^{\nu_B}} \tag{3.115}$$

3.3.7
Gibbs Enthalpy

In order to describe the chemical equilibrium from the thermodynamic aspect, the Gibbs energy (Δg^0) is introduced and applied to the chemical reaction as follows:

$$\Delta g^0 = \Delta h^0 - T \cdot \Delta s^0 \tag{3.116}$$

Δg^0 standard molar Gibbs enthalpy of reaction ($J\,mol^{-1}$)
Δh^0 standard molar enthalpy of reaction ($J\,mol^{-1}$)
T temperature (K)
Δs^0 standard molar entropy of reaction ($J\,mol^{-1}\,K^{-1}$)

At constant pressure and temperature a correlation between the Gibbs enthalpy (Δg^0) and the dimensionless equilibrium constant (K_c') can be derived from thermodynamic fundamental equations.

$$K_C' = e^{(-\Delta g^0/(R.T))} \tag{3.117}$$

R ideal gas constant $= 8.314\,J\,mol^{-1}\,K^{-1}$
K_C' dimensionless equilibrium constant

So if all the enthalpies of the species (h_i^0) are known as well as all the entropies (s_i^0) at the given conditions, the enthalpy difference and entropy difference can be calculated as follows:

$$\Delta h^0 = \sum v_i \cdot h_i^0 \tag{3.118}$$

$$\Delta s^0 = \sum v_i \cdot s_i^0 \tag{3.119}$$

then Δg^0 can be calculated and hence K_c'. Knowing K_c' and the stochiometric coefficients (v_i) K_c is usually obtained using the thermodynamic properties in pressure units as follows:

$$K_c = K_c' \cdot (P_{ref}/(R \cdot T))^{\sum v_i} \tag{3.120}$$

P_{ref} is the reference pressure (usually 1 atm).

3.3.8
Radicals

In order to understand what a radical is, it is important to know about the chemical bonds between the atoms. They consist of pairs of electrons, which are shared by two atoms. The chemical bond can be seen as the attraction between atoms, and the bond-dissociation energy is the measure of the strength of a chemical bond. The energetic state of a molecule is lower than the energy state of the molecule dissociated into its atoms. In order to split the two shared electrons of the bond by homolysis, sufficient energy amounts have to be supplied. The minimum energy amount involved has to be the bond-dissociation energy. In this way extremely reactive species, called radicals, are produced.

There are two ways to cleave a molecule (X–Y) into its atoms. By distributing the electrons unequally, the more electronegative part accepts electrons as the less electronegative part loses electrons. The cleavage results in two ions, X^+ and Y^-. This process is termed heterolytic cleavage or heterolysis (reaction (3.121)). The energy involved is called the heterolytic bond-dissociation energy.

$$X - Y \Rightarrow X^+ + Y^- \tag{3.121}$$

In comparison, the two electrons of one single bond can distinguish evenly. This process is termed homolytic cleavage or homolysis (reaction (3.122)) and generates high energy carrying species, called radicals ($X^\bullet, Y^\bullet$).

$$X - Y \Rightarrow X^\bullet + Y^\bullet \tag{3.122}$$

In general, radicals are an extremely reactive species, reacting rapidly with the majority of molecules. All radical initiation processes require sufficient amounts of energy which can be supplied by thermolysis, photolysis, or ionizing radiation (e.g., X-rays).

Some of the main reactions of radicals are listed in the following:

$$XY \Rightarrow X^\bullet + Y^\bullet \quad \text{Initiation} \tag{3.123}$$

$$X^\bullet + Y^\bullet \Rightarrow XY \quad \text{Combination} \tag{3.124}$$

$$X^\bullet + YZV^\bullet \Rightarrow XY + ZV \quad \text{Disproportionation} \tag{3.125}$$

$$XY^\bullet \Rightarrow X^\bullet + Y \quad \text{Fragmentation} \tag{3.126}$$

$$X^\bullet + YZ \Rightarrow XY + Z^\bullet \quad \text{Radical transfer} \tag{3.127}$$

$$X^\bullet + YZ \Rightarrow XYZ^\bullet \quad \text{Addition} \tag{3.128}$$

Radicals mostly exist as transient intermediates and never occur in high concentrations. That is the reason why radical transfer and addition is predominant causing chain reactions.

3.3.9
Development and Analysis of a Set of Reactions

In general, a chemical reaction consists of a sequence of many elementary reactions which defines the reaction mechanism of the net reaction. Especially in combustion, the number of elementary reactions can increase to several thousands.

As an example the complex reaction path of fuel-nitrogen to NO, N_2O and N_2 is depicted in Figure 3.19. It gives a general overview and shows the most important reaction routes. Another scheme is given in Figure 3.20 for methane oxidation. From such general reaction routes a simplified set of chemical reactions is developed, as given in Table 3.3 summarizing NO formation from fuel-nitrogen.

Complex reaction mechanisms can be analyzed with regard to the species flow (reaction flow analysis) showing the main reaction routes. Another method is sensitivity analysis showing the most sensitive reactions for the determination of the final product. Sensitivity coefficients are defined as follows:

$$S = \mathrm{d}P/\mathrm{d}K \tag{3.129}$$

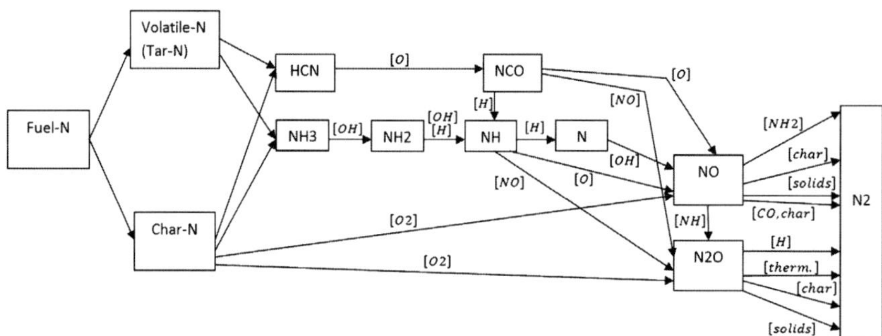

Figure 3.19 A complex set of reactions leading to NO_x from fuel nitrogen.

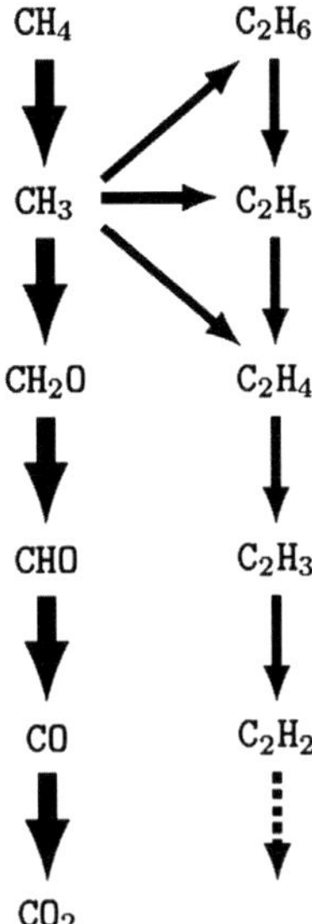

Figure 3.20 The main oxidation routes for methane combustion. The arrows indicate the most important routes. The width of the arrows corresponds to the species relative flow under the given conditions.

Table 3.3 A simplified fuel nitrogen conversion mechanism.

Used Reactions

No.				Catalyst
1	$NCO + O$	$\rightarrow$	$NO + CO$	
2	$NCO + OH$	$\leftrightarrow$	$NO + CO + H$	
3	$NH + O$	$\leftrightarrow$	$NO + H$	
4	$NCO + NO$	$\leftrightarrow$	$N_2O + CO$	
5	$NH + NO$	$\leftrightarrow$	$N_2O + H$	
6	$NH_2 + NO$	$\rightarrow$	$N_2 + H_2O$	
7	$N_2O + CO$	$\leftrightarrow$	$N_2 + CO_2$	
8	$N_2O + H$	$\leftrightarrow$	$N_2 + OH$	
9	$N_2O + OH$	$\leftrightarrow$	$N_2 + HO_2$	
10	$N_2O + O$	$\leftrightarrow$	$2\,NO$	
11	$N_2O + O$	$\leftrightarrow$	$N_2 + O_2$	
12	$N_2O + M$	$\leftrightarrow$	$N_2 + O + M$	
13	$HCN + O$	$\leftrightarrow$	$NCO + H$	
14	$NH_3 + O$	$\leftrightarrow$	$NH_2 + OH$	
15	$NH_2 + O$	$\leftrightarrow$	$NH + OH$	
16	$NO + CO$	$\rightarrow$	$1/2\,N_2 + CO_2$	bed mat.
17	$NO + 2/3\,NH_3$	$\rightarrow$	$5/6\,N_2 + H_2O$	bed mat.
18	$NH_3 + 3/4\,O_2$	$\rightarrow$	$1/2\,N_2 + 3/2\,H_2O$	bed mat.
19	NH_3	$\rightarrow$	$1/2\,N_2 + 3/2\,H_2$	bed mat.
20	$N_2O + CO$	$\rightarrow$	$N_2 + CO_2$	bed mat.

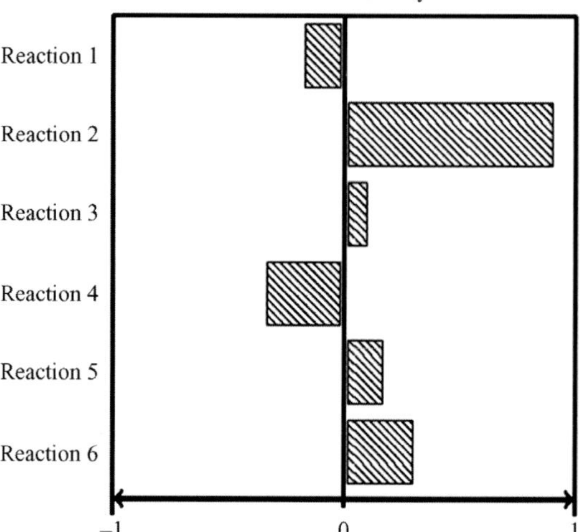

Figure 3.21 Sensitivity analysis of a set of reactions showing that reaction 2 is of high importance for the result because of its high sensitivity.

S sensitivity or sensitivity coefficient (units depend on the parameters used)
P any parameter, for example, concentration of a certain species C_i $(\mathrm{mol\,m^{-3}})$
K any parameter, for example, rate constant $(\mathrm{s^{-1}})$

Because of the difficulty with different units dimensionless sensitivity coefficients (or relative sensitivity coefficients, S_{rel}) are defined. They are based on standard parameters (here P_0, K_0) for example, initial concentrations.

$$S_{\mathrm{rel}} = (\mathrm{d}P/P_0) \cdot (K_0/\mathrm{d}K) \tag{3.130}$$

Sensitivity analysis is a powerful tool, increasing the understanding of the mechanism and its most sensitive parameters. Figure 3.21 shows a typical result of a sensitivity analysis. It reveals that the kinetic data for reaction 2 is very important for the final result. Therefore the kinetic data of reaction 2 should be well known.

3.3.10
Simplification of a Set of Reactions

Complex reaction sets usually lead to a stiff set of differential equations which are numerically difficult to solve. Two frequently used methods of simplification of reaction sets are the steady-state approximation and the partial equilibria. The approach of partial equilibria originates from the assumption of fast forward and reverse reactions. This simplification applies especially at high temperatures.

3.4
Ignition

3.4.1
Introduction

As mentioned earlier in this book, combustion is a ubiquitous technology. It accounts for approximately 85% of global primary energy production. *Ignition* is the process of starting combustion. It influences the quality of combustion, that is, speed, pollutant formation, probability of flame formation, and others [1]. In technical applications, such as in engines, reliable ignition that prevents misfiring is important, whereas the prevention of undesired ignition is critical for safety reasons in the process industries and other sectors [2,3]. This chapter presents the fundamentals of ignition (auto-ignition and induced ignition), plus some information on desired ignition processes including alternative technologies for ignition. Chapter 7 treats the prevention of undesired ignition.

3.4.2
Autoignition

Autoignition is the onset of a combustion process without support by an external ignition source. Since the autoignition temperature (AIT) depends on several parameters, such as fuel, stoichiometry, pressure, reaction volume and reaction vessel, precise conditions are necessary for its determination. AIT can be defined as follows: *"The autoignition temperature of flammable gases or liquids is the lowest wall-temperature measured in a glass flask by a method prescribed for its determination, at which the developing inhomogeneous gas/air or vapor/air mixture will just be stimulated to burn as a flame"* (see DIN 51794 and IEC 60079-4). A synonymous expression for the AIT is *"minimum ignition temperature"*.

The dependence of the autoignition temperature on the vessel size and vessel walls is related to heterogeneous chain termination reactions (quenching, see also mechanism of combustion, "radical chain reactions"). For gases, AIT is easiest to define. For liquids, AIT depends on the droplet size. The autoignition temperature of mists can be lower than the flash point of their respective liquid, caused, for example, by decomposition reactions.

At stoichiometric conditions and at high pressures, autoignition is favored. In oxygen, it may occur at temperatures up to 300 °C below that in air. Table 3.4 presents a few examples of autoignition temperatures in gases (all at stoichiometric mixing ratio and standard pressure), reprinted with permission from [4].

Auto-ignition is encountered in the following cases:

- Diesel engine (heating of the gas in the cylinders by compression prior to fuel injection)
- Fresh humid, decaying haystack (heat is confined and increases till ignition)

Table 3.4 Minimum ignition temperatures of common fuels. (Source: [4].

Fuel		Minimum Ignition Temperature ($^\circ$C)	
		in air	in O$_2$
Butane	C$_4$H$_{10}$	405	283
Carbon monoxide	CO	609	588
Ethane	C$_2$H$_6$	472	
Hydrogen	H$_2$	572	560
Methane	CH$_4$	632	556
Propane	C$_3$H$_8$	493	468

Autoignition can occur with gaseous, liquid and solid fuels, and it can be intentional or unintentional.

3.4.3
Induced Ignition

In induced ignition, an ignition source is used to start combustion. Ignition sources can be hot surfaces, sparks, flames and others. The ignition source can be activated on purpose, such as in an Otto engine, or it can be caused unintentionally, for example, an electrostatic discharge. In contrast to autoignition, induced ignition of gases does not require a minimum temperature.

For liquids, there are two temperatures of interest: *flash point* and *fire point*.

By definition, the *flash point* is the lowest temperature at which the vapor formed above a pool of the liquid can be ignited in air at a pressure of 1 atmosphere under specified test conditions. The "closed cup" and the "open cup" method are used. Small impurities of low-boiling flammable liquids or gases will lower the flash point, and the flash point of a mixture may be lower than that of the individual components. The flash point measured in a closed cup is used for the classification of flammable liquids in important regulations such as GefStoffVO, VbF, ADR, RID, IMDG Code, UN Recommendations and IATA code. At the flash point, the flame will stop as soon as the ignition source is removed. By contrast, the *fire point* is the lowest temperature at which the sample will support combustion for at least 5 s. Above the fire point, the temperature is high enough to keep the flame alive with freshly formed vapors. Hence, the fire point is always a few degrees above the flash point. The fire point depends on the vapor pressure of the respective substances. Figure 3.22 shows a lab experiment to check the flash point of a liquid.

In Table 3.5 data for gasoline and diesel are provided to illustrate the difference between flash point (induced ignition) and autoignition temperature. For details, the following norms can be consulted: ASTM D93, IP34, ISO 2719, DIN 51758, JIS K2265 and EN ISO 3679.

Although the flash points of diesel and gasoline differ significantly, the autoignition temperatures are close to each other.

Figure 3.22 This liquid is too cold, the flash point is not reached.

Table 3.5 Comparison of flash point and autoignition temperature.

	Flash Point (°C)	Autoignition Temperature (°C)
Gasoline	−20	240
Diesel	55	220

The spread between flash point and fire point depends on the volatility (vapor pressure) of the fuel. As shown in Table 3.6 for two examples, a higher flash point will usually give a higher delta to the fire point, too. The fire point is always above the flash point.

Liquid fuels are often burned as *sprays*. In this mode, one can distinguish between three autoignition processes: droplet ignition, group ignition and spray ignition [5], see Figure 3.23.

For solid fuels, typical autoignition temperatures are provided in Table 3.7.

Important to note is that the autoignition temperature of decomposition products from solids often lies significantly below that of the solids themselves. A good example is a pile of wet grass. Due to the confinement, the heat production results in a runaway reaction where self-heating finally ignites the decomposition products of the hay.

Dusts are characterized by a high surface/volume ratio. Therefore, fine dusts, typically $< 100\,\mu$m, when suspended in air, can be ignited easily, as opposed to large chunks of the same solids, see also Chapter 7. Hybrid mixtures, that is, compositions that contain two phases, tend to ignite more easily than the pure constituents.

Table 3.6 Flash point and fire point for two liquid hydrocarbons.

Substance	Flash Point (°C)	Fire Point (°C)	Difference (°C)
Benzene	−11	−9	2
Lubrication oil	148	190	42

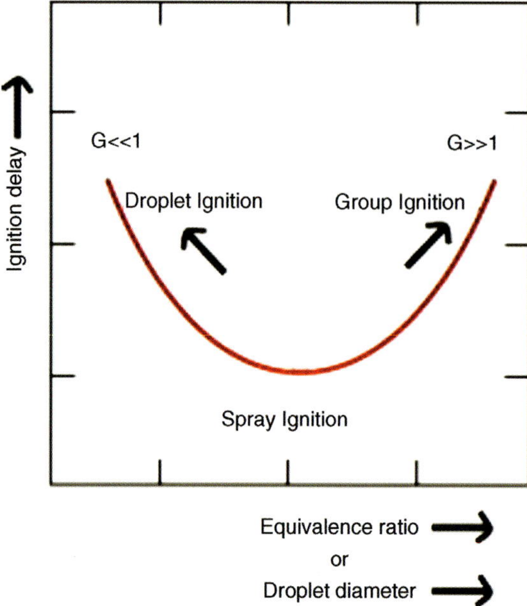

Figure 3.23 Qualitative description of the three autoignition modes in sprays. Reproduced from [5].

Table 3.7 Autoignition temperature of solids.

Material	Autoignition Temperature (°C)
White phosphorus	60
Newspaper	240
Cotton wool	260
Hardwood	295
PVC	530

3.4.4
Theoretical Models for Ignition

There are two commonly used models for ignition:

- Theory of *Semenov* (homogeneous ignition, thermal explosion)
  It is useable as long as there is no temperature gradient. An example for a thermal explosion is the reaction: $Cl_2 + H_2$
- Theory of *Frank-Kamenetskii* (inhomogeneous ignition)
  This model allows a temperature gradient to be taken into account, which occurs when there is a high resistance to heat transfer in the reacting system, or when the system has reactants with a low thermal conductivity and/or highly conducting walls.

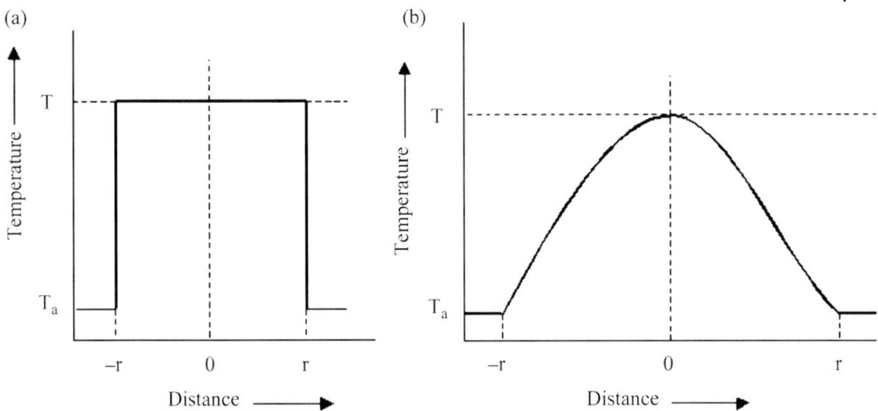

Figure 3.24 Temperature profiles in the models of Semenov (a) and Frank–Kamenetskii (b).

The temperature profiles for the two models are shown schematically in Figure 3.24.

Details can be found in [6].

3.4.5
Explosives

Explosives [7] are characterized by a very fast oxidation reaction. They detonate rather than burn by consuming oxygen within the compound rather than from the ambient air, and the expansion of the hot evolving gases gives off energy. Explosives can be grouped into two basic categories: low explosives and high explosives. The former tend to deflagrate, the latter to detonate. Low explosives are used in fire arms and space rockets. Black powder (gun powder) was a very common low explosive that is still used in fireworks. However, it is quite unsafe and has, therefore been widely replaced. The first high explosive used for commercial blasting operations was nitroglycerine, also called "blasting oil." Nitroglycerine is an unstable chemical and therefore dangerous to use. This was improved when Alfred Nobel invented dynamite by mixing nitroglycerine with kieselguhr. Today, explosives are used in mining, building demolition, pyrotechnics, and construction. They also inflate airbags and activate seatbelt pretensioners.

Homogeneous, that is, one-component explosives, have unstable chemical structures that contain $O-O$ bonds (peroxides) or $-N_3$ (azides), for instance. The structure of trinitrotoluene (TNT) is shown in Figure 3.25.

An important value of explosive materials such as liquid peroxides is the SADT (self-accelerating auto decomposition temperature), a term that is self-explanatory and important for storage and transportation. At 10 K below the SADT, one speaks about the "emergency temperature".

So-called *hypergolic mixtures* react spontaneously upon contact of the two reaction partners. They will produce reliable ignition and hence are used, for example, in

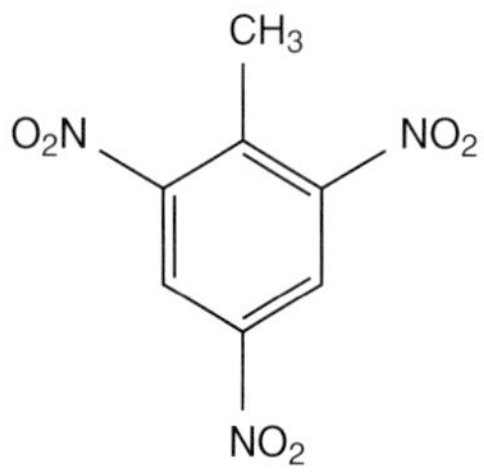

Figure 3.25 Formula of the explosive TNT (2,4,6-trinitrotoluene). Upon detonation, TNT decomposes according to $2\,C_7H_5N_3O_6 \rightarrow 3\,N_2 + 5\,H_2 + 12\,CO + 2\,C$. The reaction is exothermic but has a high activation energy.

rockets. An example of such a mixture is derivatives of hydrazine with N_2O_4, for example, UDMH (2-dimethylhydrazine).

3.4.6
Flammability Limits

The range in which a mixture of fuel and oxidizer will ignite is called the flammability range, flanked by lower and upper flammability limits (synonymous: explosive limits, ignition limits). Typically, they are given as vol%, see Table 3.8.

One can see from this table that hydrogen has a very wide flammability range. Its danger upon accidental release, however, is significantly reduced because it will diffuse very fast, as opposed to, for example, propane and butane, which are heavier than air (This is why LPG-powered cars are forbidden in some underground parking areas; LPG, liquefied petroleum gas, is a mixture of propane and butane, which can be liquefied at room temperature under pressure). In oxygen, all gases have wider flammability ranges than in air or in dilution with another inert gas. Outside the flammability limits, a gaseous fuel/oxidizer mixture is too "fat" or too "lean" for ignition.

Table 3.8 Flammability limits of common fuels at standard temperature and pressure. (Source: [4].)

Fuel		Flammability Limits (% of fuel gas by volume)			
		in Air		in O_2	
		Lower	Upper	Lower	Upper
Butane	C_4H_{10}	1.86	8.41	1.8	49
Carbon monoxide	CO	12.5	74.2	19	94
Ethane	C_2H_6	3.0	12.5	3	66
Hydrogen	H_2	4.0	74.2	4	94
Methane	CH_4	5.0	15.0	5.1	61
Propane	C_3H_8	2.1	10.1	2.3	55
Propylene	C_3H_6	2–4	10.3	2.1	53

Note: Flammability limits can be given by volume or by mass (conversion possible via density). For dusts, only the lower flammability limit is of practical interest.

The flammability limits depend on the pressure of the mixture. For mixtures of several gases, the flammability limits can be estimated with the *Rule of Le Chatelier*:

$$IL_{\text{mixture}} = \frac{1}{x_1 \cdot \left(\dfrac{1}{IL_1}\right) + x_2 \cdot \left(\dfrac{1}{IL_2}\right) + x_3 \cdot \left(\dfrac{1}{IL_3}\right) + \ldots + x_n \cdot \left(\dfrac{1}{IL_n}\right)} \tag{3.131}$$

where IL_i stands for the "ignition limit", x_i for the mole fraction and $i = 1, 2, 3, \ldots$ n for the number of gas species. The rule of Le Chatelier provides a fairly good estimation (exception: acetylene/air).

Here is an example:

The upper ignition limit of C_3H_8 is 10.1%, and for H_2 it is 74.2%. Therefore, the upper explosive limit of a mixture that contains 40% propane and 60% hydrogen can be estimated to be 21%:

$$UL_{\text{mixture}} = \frac{1}{0.4 \cdot \left(\dfrac{1}{10.1}\right) + 0.6 \cdot \left(\dfrac{1}{74.2}\right)} = 21.0\%$$

3.4.7
Minimum Ignition Energy

The minimum ignition energy (MIE) of flammable gases, vapors or dust clouds is defined as the minimum value of the electric energy, capacitively stored in a discharge circuit which – upon discharge across a spark gap – ignites the standing mixture in the most ignitable composition (see ASTM E582-76). MIE is given in mJ. Typical values of MIE are 10^{-3} to 10^{-1} mJ for explosives, hydrogen and hydrocarbons in oxygen. (Figure 3.26) Static electricity is on this level and can hence cause an ignition of several gas/air mixtures. Typical dust/air mixtures can be ignited between 1 mJ and 1–10 J, that is, only with much higher energies. The finer a dust is, the easier it will be to ignite.

Compare with flammability limits: As they are approached, MIE increases sharply. The lowest value of MIE is around the stoichiometric mixture. The heavier a fuel becomes, the more MIE will move into the fuel-rich area as larger molecules are diffusing to the ignition spot more slowly.

3.4.8
Quenching and Maximum Experimental Safe Gap (MESG)

The term *"quenching"* denotes the sudden inactivation of radicals. When radicals, which are necessary for sustaining a combustion reaction, reach a (cold) surface like a vessel wall, they react and are no longer available. This is then a heterogeneous chain termination process, as opposed to a "homogeneous chain termination" reaction which can occur in the gas phase as a three-body collision. Quenching leads

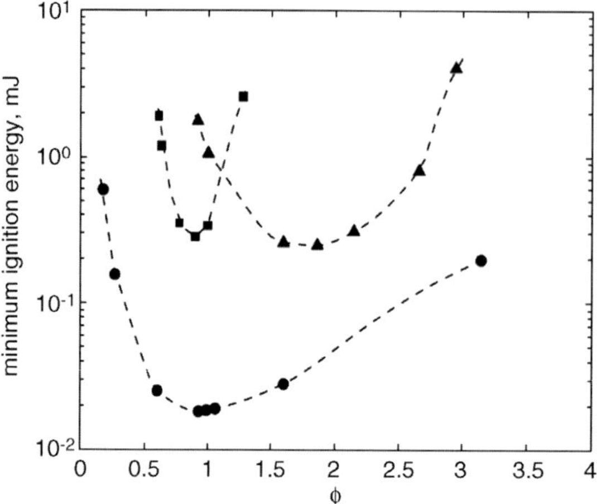

Figure 3.26 Minimum ignition energies of (●) hydrogen/air, (■) methane/air and (▲) heptane/air mixtures in relation to ϕ (the equivalence ratio) at atmospheric pressure. Reproduced with permission from [8].

to soot and CO formation in internal combustion engines, but it can also be used for safety purposes, as is illustrated by Davy's lamp (see Figure 3.27). In this device, a fine grid of metal prevents a flame inside from igniting potentially explosive gases on the outside.

Davy's lamp was used in coal mines where methane that had desorbed from the coal seams led to several catastrophic explosions in the past. It brought a tremendous

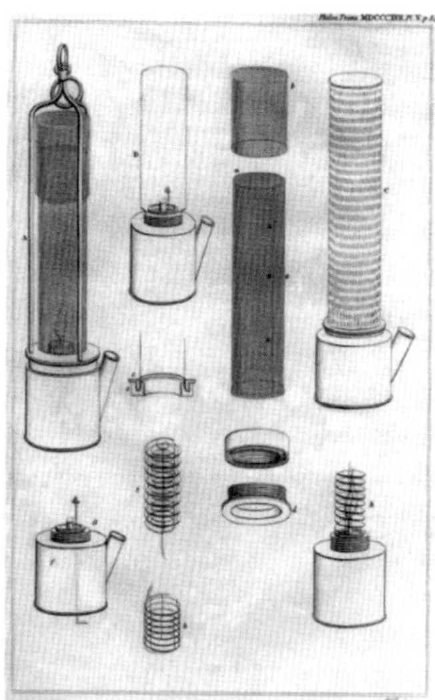

Figure 3.27 Davy's miners' lamp. Source: [9].

gain for miners' safety. An important term in this context is the so-called *maximum experimental safe gap* (MESG). A fine wire mesh in a tube filled with a combustible fuel/air mixture will not let a flame pass through below a certain minimum wire-to-wire distance (the MESG), the reason being flame quenching: the metal wires extract heat and quench the radicals at their surface.

MSEG depends on the gas. Note that turbulent explosions can be transmitted through openings smaller than the laminar flame quenching distance! The norm on MESG is IEC 60079. MESG is an important safety parameter in process engineering. It is used to design flame arrestors.

3.4.9
*p*T Explosion Diagram

The so-called *pT* explosion diagram (*p* stands for pressure and *T* for temperature) is very instructive to understand ignition. It is depicted in Figure 3.28 for the O_2/H_2 system. One can see an S-shaped curve over the temperature versus pressure chart.

To understand the *pT* explosion diagram, let us pick the following point in Figure 3.28: 800 K, 1.3 mbar (523 °C, 1 mm Hg) – there is no explosion. Instead, O_2 and H_2 only react slowly with each other, resulting in a gradual mixture consumption, temperature increase and water formation.

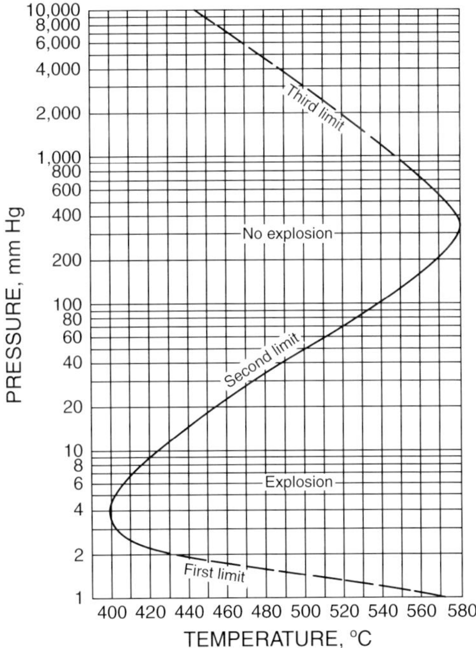

Figure 3.28 *pT* explosion diagram for H_2 in oxygen. Test conditions: stoichiometric mixture, spherical vessel with volume 7.4 cm³ coated with KCl. Reproduced with permission from [6].

Under these conditions, radicals diffuse to the vessel walls where they are quenched. According to Fick's law, the diffusion coefficient is inversely proportional to pressure, so the low pressure of 5 mbar lets the radicals dissipate away.

$$\text{Fick's law}: J = D^* \delta c / \delta t \tag{3.132}$$

with J being the mass flow in $\mathrm{mol\,m^{-2}\,s^{-1}}$, D the diffusion coefficient, c the concentration and t the time. As the diffusion coefficient is inversely proportional to pressure ($D \sim 1/p$), diffusion will be more important at low pressures as in the above-described situation at 5 mbar.

Let us now proceed in the diagram to higher pressures, for example, 800 K and 50 mbar (37.5 mm IIg): one observes that above approximately 2 mbar, the mixture will explode. Explanation: After passing the so-called "first explosion limit", diffusion cannot keep pace with the formation of radicals in the gas phase, and spontaneous ignition occurs. Since there is a competition between chain branching (radical formation) and radical quenching, the properties of the vessel wall show a strong influence in this first explosion limit.

Let us further increase the pressure to >150 mbar (112.5 mm Hg), 800 K: again there is no ignition as the "second ignition limit" is passed. In this region, the competition between chain branching reactions and chain termination reactions is dominant.

At low pressures, the following reaction is favored : $\mathrm{H^{\bullet} + O_2 \rightarrow OH^{\bullet} + O^{\bullet}}$
At higher pressures, this reaction gains importance : $\mathrm{H^{\bullet} + O_2 + M \rightarrow HO_2^{\bullet} + M}$

M is a third gas molecule (quenching partner).

$\mathrm{HO_2^{\bullet}}$ exhibits a much lower reactivity than $\mathrm{OH^{\bullet}}$, hence this reaction corresponds essentially to chain termination. Three-body-collisions are favored at high pressures. The third body is necessary to carry away the recombination energy in homogeneous reactions, whereas the reaction wall plays this role in heterogeneous reactions.

Now as the pressure is further increased on our journey through Figure 3.28, one reaches the "third explosion limit" and the realm of "thermal ignition". In this region, there is a competition between heat generation ($\sim x^3$) and heat dissipation ($\sim x^2$).

Note: For hydrocarbons, the pT explosion diagram is more complex. There are, around the third ignition limit, zones of "cool flames" and multi-stage ignition.

Explanation of *cool flames*:

A cool flame is one that burns at about a maximum temperature of 120–600 °C, which is significantly below ordinary combustion temperatures. A typical temperature increase upon ignition of a cool flame is a few 10 °C only. Cool flames are responsible for engine knock with low octane-grade fuels in internal combustion engines. A cool flame is a faint, luminous combustion that proceeds slowly and with little heat emission under incomplete combustion. The phenomenon was first observed by Sir Humphry Davy in 1817, who discovered that he did not burn his fingers and could not ignite a match in a cool flame. A cool flame reaction never proceeds to complete combustion. Instead, fuel and oxidizer break down and recombine to a variety of stable chemical compounds, including alcohols, acids, peroxides, aldehydes, and carbon monoxide. Cool flames have an oscillating character and can sustain for a long period of time.

In engines, cool flames are associated with two-stage-ignition and characterized by an NTC (negative temperature coefficient). Whereas in normal reactions, the reaction rate increases with temperature, the opposite is observed under some engine conditions. *Degenerate branching* is a chain termination process as the temperature is increased: the precursor substances are thermally unstable, therefore, they are inactivated and the chain branching reaction is deprived of its foundation. More details on cool flames can be found in [10], images in [11].

3.4.10
Ignition Delay Time

When an explosive gas/air mixture is ignited, the pressure rise is not noticed immediately. Instead, a certain time, the so-called ignition delay time, has to elapse. In this time, the radical pool is being built up. In the beginning, the flame kernel is still very small. As it grows, it consumes more and more combustible material and finally produces a noticeable signal. The definition of onset or start of combustion is arbitrary. Typical definitions are 5% of maximum pressure reached or 2% of fuel consumed, and so on.

The ignition delay time can be considered an induction time. Figure 3.29 provides an example.

The ignition delay time depends on the temperature as follows: $\tau = Ae^{\frac{B}{T}}$, see Figure 3.30. A is the pre-exponential factor, B another constant, and T the temperature.

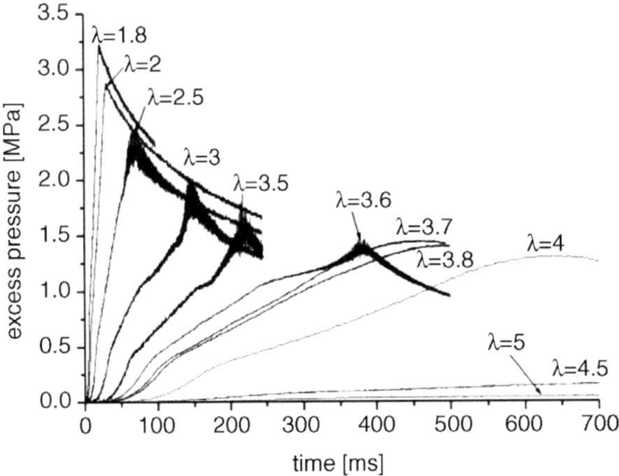

Figure 3.29 Ignition delay time of various hydrogen/air mixtures in a closed reaction vessel. One can see that the leaner the mixture, the longer the ignition delay time.

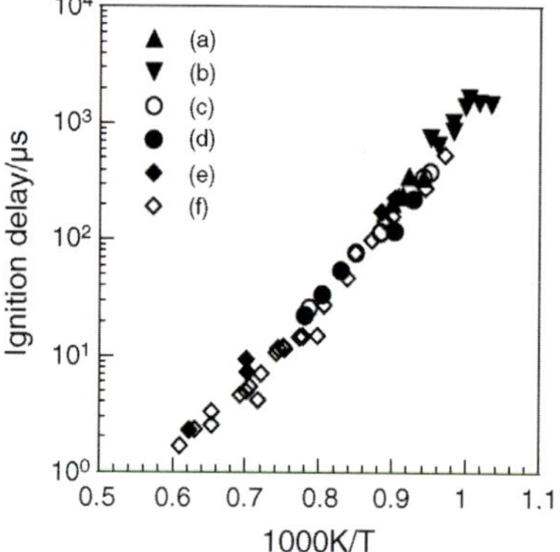

Figure 3.30 Ignition delay of kerosene–air mixtures; a: Jet A at 30 atm, b: JP-8 at 30 atm, c: Jet A at 20 atm, d: jet A at 10 atm, e: kerosene at 10 atm, f: Jet A at 10 atm. Reproduced with permission from [12].

3.4.11
Ignitability

Gasoline engines use induced ignition to bring about combustion; *spark plugs* (Figure 3.31) provide a "trigger" at the right point in time (time is measured in "degree crank angle" by automotive engineers). Essentially, a spark plug consists of two closely spaced electrodes between which a high voltage is applied so that a plasma is generated.

Each engine has a so-called "characteristic map", which defines the best ignition time based on load and engine speed. An example is given in Figure 3.32.

Conversely, Otto engines rely on auto-ignition of diesel injected into the engines' combustion chambers. Gasoline must not ignite too spontaneously, as this would

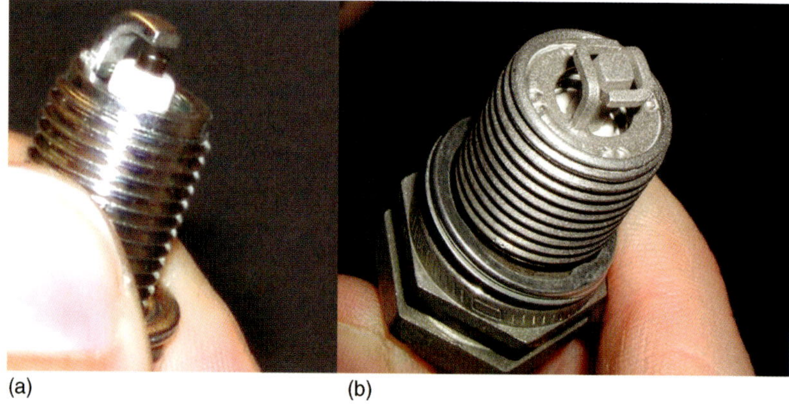

Figure 3.31 Spark plug for induced ignition in an automotive Otto engine (a) and a large, stationary gas engine (b).

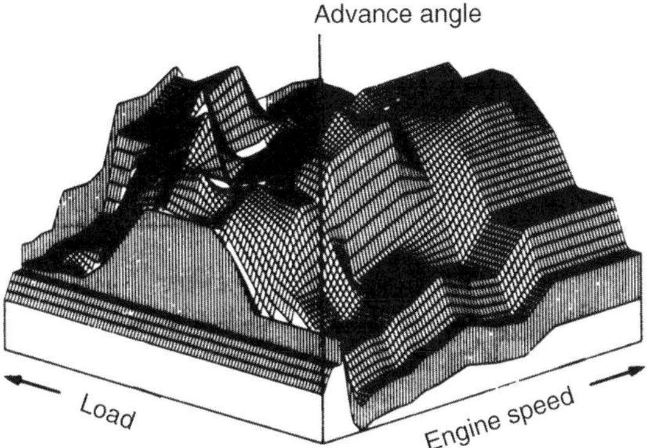

Figure 3.32 Characteristic map of an engine, showing the optimum ignition time. Source [13].

produce so-called "engine knock" which is accompanied by dangerous pressure spikes (see Figure 3.33). On the other hand, diesel has to be rather easy to ignite at the desired temperature and pressure. A measure of the "ignitability" of fuels is given by the octane number in the case of gasoline, the cetane number in the case of diesel and the methane number in the case of gaseous fuels.

Diesel engines, which are based on autoignition, use a glowplug (Figure 3.34). This is a small heating device which helps overcome excessive heat losses at engine start. Glow plugs reach temperatures of around 850 °C. Preheating takes 10 s or more for older engines and a few seconds only in newer ones.

3.4.12
Octane Number

The octane number (ON, octane rating) is a measure of the quality of automotive and aviation fuels. The lower the octane number, the more easily the fuel will autoignite.

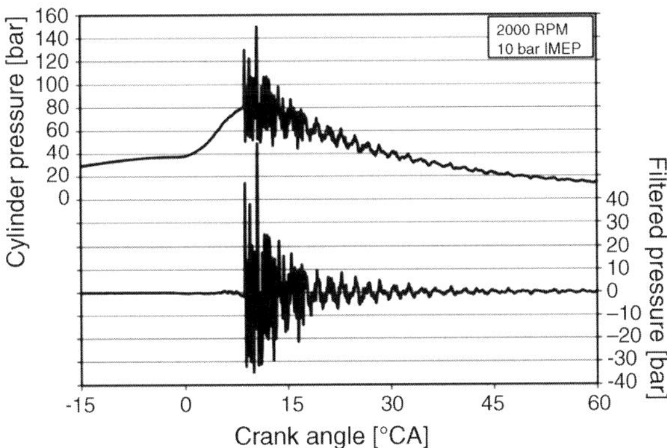

Figure 3.33 Typical cylinder pressure trace for heavy knocking in an engine. RPM = rotations per minute, IMEP = indicated mean effective pressure. Reproduced with permission from [14].

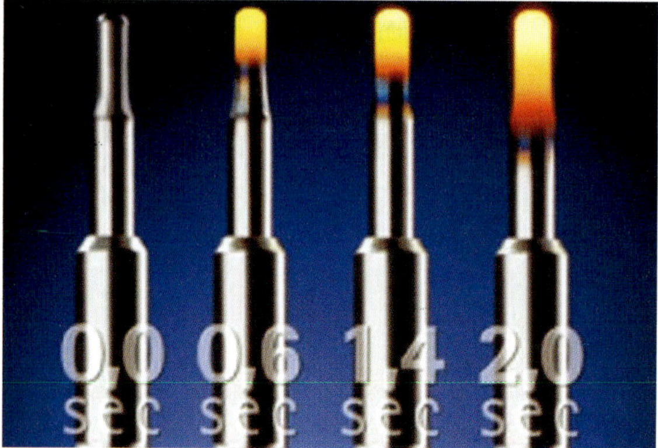

Figure 3.34 Glow plug for diesel engines. Reproduced from [15].

ON can be measured in a test engine and is defined by comparing a given fuel with the mixture of 2,2,4-trimethylpentane (iso-octane) and n-heptane with the same knocking properties as the fuel under investigation. Pure iso-octane is set at 100 (low knocking propensity) and heptane at 0 (high knocking properties). The selection of n-heptane as the zero point of the scale was made because of its availability in high purity at the time of introduction. "*Knocking*" is an undesired, fast combustion process characterized by high pressure spikes to due sudden autoignition of fuel/air in Otto engine cylinders. It can destroy an engine.

Avgas (aviation gasoline) is comparable to automotive gasoline (mogas, motor gasoline) except that it is used as aviation fuel for piston-engine aircraft. Avgas still contains tetraethyl lead (TEL) as anti-knock agent. For mogas, MTBE (methyl *tert*butyl ether) is used to enhance combustion stability.

Depending on the test method, one can distinguish between RON and MON (RON = research octane number, MON = motor octane number). Otto engines generally tolerate fuel with a higher-than-specified octane number.

3.4.13
Cetane Number

The cetane number (CN, cetane rating) is a measure of a diesel fuel combustion performance, more specifically the fuel's ignition delay. Higher cetane fuels have shorter ignition delay times. Generally, diesel engines use fuels with CN between 40 and 55. Biodiesel has a typical CN of 46 to 52.

Cetane is a mix of unbranched, aliphatic alkane molecules. It ignites very easily under compression, so it was assigned a cetane number of 100, whereas alpha-methylnaphthalene was given a cetane number of 0. Real diesel fuels are indexed to cetane as to how well they ignite under compression. The cetane number therefore measures how quickly the fuel starts to burn (autoignites) under diesel engine conditions. Other quality parameters of diesel fuel are its density, lubricity, cold-flow properties and sulfur content.

Comment: For gaseous fuels, an important quality parameter is the methane number.

3.4.14
Ignition in Various Combustion Devices

Power plants are started with pilot burners. Gasoline engines use electric spark plugs (see above). Although spark plugs have been in use for more than a century, they have inherent disadvantages so that researchers have been looking for alternative modes of induced ignition.

An alternative ignition system is laser-induced ignition (*laser ignition*) [16,17] (see Figure 3.35).

An obvious advantage of optical ignition is the free choice of the ignition location. Conventional spark plugs can only ignite the fuel/air mixture at the vessel wall, because otherwise the protruding metal would disturb the fluid flow. Due to system costs and difficulties with a suitable window material, no commercial laser ignition system has made it into automotive cars yet. However, several missiles are fired using laser ignition.

Other alternative ignition systems for automotive engines are (see Figure 3.36)

- High frequency ignition [19]
- Microwave ignition [18,20]
- Corona ignition [18]
- Plasma-assisted ignition [21,22,18]

Two further ignition concepts for automotive engines are "*skip cycle ignition*" [23] and HCCI (*homogeneous charge compression ignition*) [24]. In R&D, *shock tube ignition* [25] is also utilized. See [26] for a summary on alternative ignition systems.

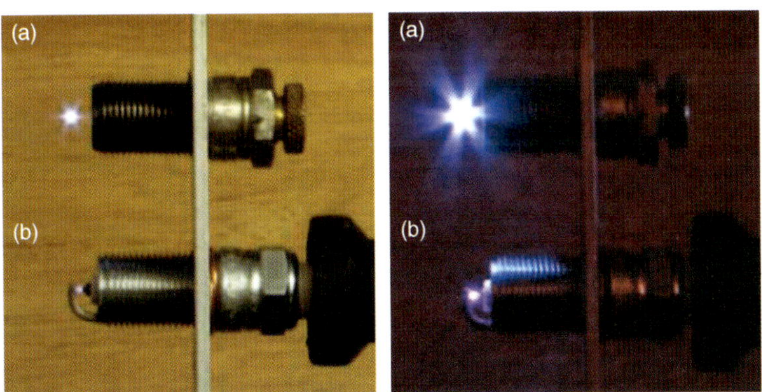

Figure 3.35 Comparison of (a) an optical spark plug (laser ignition) and (b) conventional, electrical spark plug. Reproduced with permission from [17].

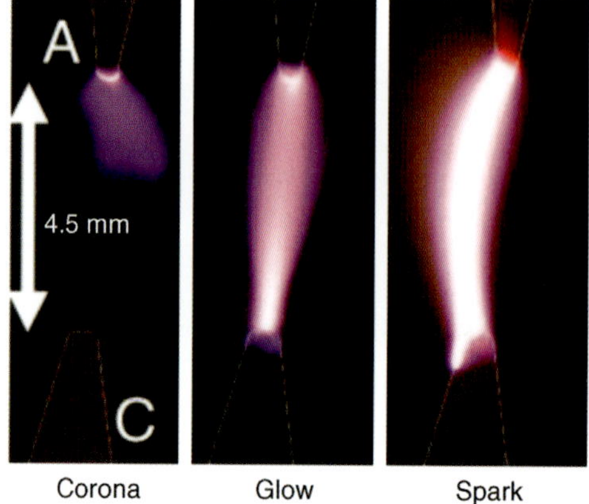

Figure 3.36 Appearance of corona, glow and spark discharges. Reproduced from [18].

3.4.15
Undesired Ignition

For safety reasons, undesired ignition and, worse, undesired explosion, of combustible materials has to be avoided. Basically, not only fuels, but all combustible materials are at risk for undesired ignition. The larger the stored volume, the higher the risk for damage and loss in the case of a fire will be. Figure 3.37 shows an example of an undesired ignition with huge consequences: the Buncefield fire in a UK oil storage terminal in December 2005.

Safety starts with proper process design [3]. The flammability is a critical property for building materials, as specified by DIN 4102 and EN 13501. A classic textbook on ignition in this context is [28]. See also [29,30] and Chapter 7 in this book.

Figure 3.37 Buncefield site prior to incident and showing the major fires that followed the explosion. Reproduced with permission from [27].

References

1 Knothe, G., Matheaus, A.C., and Ryan III, T.W. (2003) Cetane numbers of branched and straight-chain fatty esters determined in an ignition quality tester. *Fuel*, **82** (8), 971–975.

2 Babrauskas, V. (2003) *Ignition Handbook: Principles and Applications to Fire Safety Engineering, Fire Investigation, Risk Management and Forensic Science*, Fire Science Publications, ISBN: 978-0972811132.

3 Kjellén, U. (2007) Safety in the design of offshore platforms: Integrated safety versus safety as an add-on characteristic. *Safety Science*, **45** (1–2), 107–127.

4 Reed, J.R. (1986) *North American Combustion Handbook*, 3rd edn, vol. 1, North American Manufacturing Corporation, Cleveland, OH, USA.

5 Bouali, Z., Pera, C., and Reveillon, J. (2012) Numerical analysis of the influence of two-phase flow mass and heat transfer on n-heptane autoignition. *Combustion and Flame*, **159** (6), 2056–2068.

6 Warnatz, J., Maas, U., and Dibble, R.W. (2006) *Combustion: Physical and Chemical Fundamentals, Modeling and Simulation, Experiments, Pollutant Formation*, 4th edn, Springer, ISBN: 978-3540259923.

7 Meyer, R. (1987) *Explosives*, 3rd edn, Wiley-VCH Verlag GmbH, ISBN: 978-3527265992.

8 White, C.M., Steeper, R.R., and Lutz, A.E. (2006) The hydrogen-fueled internal combustion engine: a technical review. *International Journal of Hydrogen Energy*, **31**, 1292–1305.

9 Davy, H. (1817) Some new experiments and observations on the combustion of gaseous mixtures, with an account of a method of preserving a continued light in mixtures of inflammable gases and air without flame. *Philosophical Transactions of the Royal Society of London*, **107**, 77–85, http://www.jstor.org/stable/107574 (accessed June 1, 2013).

10 Bonner, B.H. and Tipper, C.F.H. (1965) The cool flame combustion of hydrocarbons I—Cyclohexane. *Combustion and Flame*, **9** (3), 317–327.

11 Pearlman, H. (2007) Multiple cool flames in static, unstirred reactors under reduced-gravity and terrestrial conditions. *Combustion and Flame*, **148** (4), 280–284.

12 Dagaut, P. and Cathonnet, M. (2006) The ignition, oxidation, and combustion of kerosene: A review of experimental and kinetic modeling. *Progress in Energy and Combustion Science*, **32** (1), 48–92.

13 Sachenbacher, M. and Williams, B.C. (2005) *Automated Model Generation Using Qualitative Abstraction*, CSAIL, MIT Computer Science and Artificial Intelligence, Laboratory, http:// publications.csail.mit.edu/abstracts/ abstracts05/mers2/mers2.html#fig1 (accessed June 1, 2013).

14 Verhelst, S. and Wallner, T. (2009) Hydrogen-fueled internal combustion engines. *Progress in Energy and Combustion Science*, **35** (6), 490–527.

15 http://www.turbocompressori.net/glow_plugs .htm (2012) (accessed 1 June 2013).

16 Puhl, M. (2011) *Corona and Laser Ignition in Internal Combustion Engines: A Comparison to Conventional Spark Plug Ignition*, VDM Verlag Dr. Müller, ISBN: 978-3639323115.

17 Morsy, M.H. (2012) Review and recent developments of laser ignition for internal combustion engines applications. *Renewable and Sustainable Energy Reviews*, **16** (7), 4849–4875.

18 Starikovskiy, A. and Aleksandrov, N. (2013) Plasma-assisted ignition and combustion. *Progress in Energy and Combustion Science*, **39** (1), 61–110.

19 Bae, C., Lee, J., and Ha, J. (1998) High-Frequency Ignition Characteristics in a 4-Valve SI Engine with Tumble-Swirl Flows, SAE Technical Paper 981433. doi: 10.4271/981433.

20 Meir, Y. and Jerby, E. (2012) Thermite powder ignition by localized microwaves. *Combustion and Flame*, **159** (7), 2474–2479.

21 Starikovskiy, A. and Aleksandrov, N. (2013) Plasma-assisted ignition and combustion. *Progress in Energy and Combustion Science*, **39** (1), 61–110.

22 Starikovskii, A.Yu. (2005) Plasma supported combustion. *Proceedings of the Combustion Institute*, **30** (2), 2405–2417.

23 Kutlar, O.A., Arslan, H., and Calik, A.T. (2007) Skip cycle system for spark ignition engines: An experimental investigation of a new type working strategy. *Energy Conversion and Management*, **48** (2), 370–379.

24 Zhao, F. (2003) *Homogeneous Charge Compression Ignition (HCCI) Engines: Key Research and Development Issues*, Society of Automotive Engineers, ISBN: 978-0768011234.

25 Vasu, S.S., Davidson, D.F., and Hanson, R. K. (2008) Jet fuel ignition delay times: Shock tube experiments over wide conditions and surrogate model predictions. *Combustion and Flame*, **152** (1–2), 125–143.

26 Lackner, M. (ed.) (2009) *Alternative Ignition Systems*, ProcessEng Engineering GmbH, Wien, ISBN: 978-3-902655-05-9.

27 Michael Johnson, D. (2010) The potential for vapour cloud explosions – Lessons from the Buncefield accident. *Journal of Loss Prevention in the Process Industries*, **23** (6), 921–927.

28 Lewis, B. and Von Elbe, G. (1987) *Combustion, Flames & Explosions of Gases*, 3rd edn, Academic Press Inc., ISBN: 978-0124467514.

29 Hattwig, M. and Steen, H. (2004) *Handbook of Explosion Prevention and Protection. An Evidence-Based Review*, 1st edn, Wiley-VCH, ISBN: 978-3527307180.

30 Lackner, M. Winter, F. and Agarwal, A.K. (eds) (2010) *Handbook of Combustion*, Wiley-VCH Verlag GmbH, ISBN: 978-3527324491.

4
Environmental Impacts

4.1
Pollutants: Formation and Impact

4.1.1
Introduction

An ideal combustion process converts fuels plus oxygen into harmless reaction products and withdraws the heat of combustion for useful work. For instance the combustion of hydrogen in oxygen only produces water. As soon as carbonaceous species are involved, CO_2 is obtained, which has become a concern in the global warming discussion (see Section 4.2). However, combustion processes also yield a wide range of side products which are completely undesirable [1]. These byproducts are termed *pollutants*. The formation of combustion-derived pollutants and their impact is illustrated in Figure 4.1.

Health effects, acid rain, smog, global warming and ozone layer depletion are significant impacts of combustion-derived pollutants. The most relevant pollutants are, apart from CO_2,

- Unburnt hydrocarbons (UHC)
- Carbon monoxide (CO)
- Nitrogen oxides (NO_x)
- Sulfur oxides (SO_x)
- Dioxins
- Particulate matter (PM)
- Dust
- Soot
- Ash
- Alkali metals
- Heavy metals

Smoke is a visible pollutant from combustion processes. *Noise* and *luminosity* can also be a nuisance, for instance in the case of flares. With neighbors in close

Combustion: From Basics to Applications, First Edition. Maximilian Lackner, Árpád B. Palotás, and Franz Winter.
© 2013 Wiley-VCH Verlag GmbH & Co. KGaA. Published 2013 by Wiley-VCH Verlag GmbH & Co. KGaA.

COMBUSTION GENERATED POLLUTANT EMISSIONS

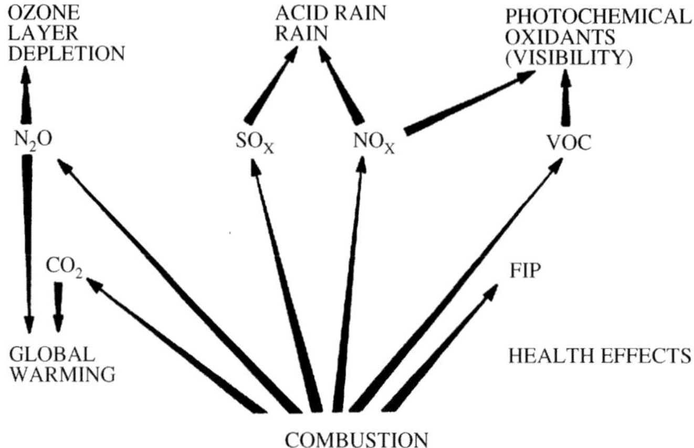

Figure 4.1 Pollutants from combustion and their consequences. VOC = volatile organic compounds; FIP = fine inorganic particles [2].

proximity, measures to dampen sound and light effects of (open) flames also have to be considered. Steam injection is one such measure.

Gaseous fuels tend to burn in the cleanest way. Liquids and solids are more complex to control. Solid fuels also produce *ash*. In power plants, the ash is divided into bottom ash and fly ash. Ash can be used for extraction of valuable materials, yet it also has to be considered a pollutant as it can contain heavy metals. Liquid and gaseous fuels tend to produce very little ash. *Waste heat* can also lead to undesired effects, for example, if the temperature of a river that is used for cooling purposes is raised excessively which leads to oxygen loss and detrimental effects on the local biota.

4.1.2
Description of Most Relevant Pollutants

4.1.2.1 Unburnt Hydrocarbons (UHC)

In order to extract the maximum possible heat from a combustion process, one strives to achieve complete fuel conversion. In the case of too high turbulence, too lean mixture, the presence of cold vessel walls/crevices, and so on, not all fuel is fully oxidized, and partially consumed fuel is expelled via the exhaust. The resulting pollutants can be traces of the fuel itself, decomposition products or flame-generated species. Particularly with natural gas engines, the parameter NMHC (non-methane hydrocarbons) is used. Because CH_4 is present in the atmosphere at 1.7 ppm, its background level is often higher than methane emissions from a combustion process. In contrast, the parameter VOC (volatile organic compounds) encompasses all hydrocarbons that are emitted unburnt (or from another, non-combustion process). VOC can stem from the fuel storage and supply system or the combustion

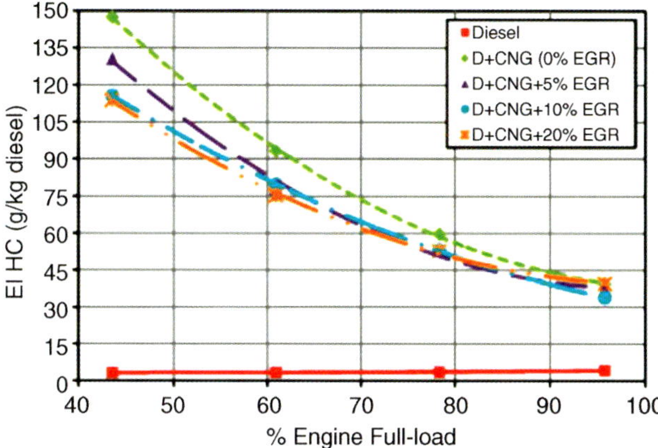

Figure 4.2 Unburned hydrocarbon emission index (EI HC) for different operating modes [4].

process itself. Another term is products of incomplete combustion (PIC) [3]. Fuel-rich flames produce polycyclic aromatic hydrocarbons (PAH), see also soot below.

In the atmosphere, unburnt hydrocarbons can lead to smog formation. Eventually, they are oxidized to CO_2 and H_2O. If hydrocarbons are spilled in a confined space, for example, a garage, they can replace oxygen and lead to suffocation, apart from the explosion risk. This is especially a risk with heavier-than-air gaseous fuels such as LPG (liquefied petroleum gas). Note: If fine particles are used in solid fuel combustion, fly ash can also carry away unconsumed fuel. Soot is unburnt carbon emitted as a solid, whereas UHC refers to liquid and gaseous species.

Efficient means for UHC reduction are flue gas recirculation (FGR) and exhaust gas recirculation (EGR), see below.

The use of natural gas as a partial supplement for liquid diesel fuel is a very promising solution for reducing pollutant emissions, particularly NO_x and PM, from conventional diesel engines. In most applications of this technique, natural gas is inducted or injected in the intake manifold to mix uniformly with air, and the homogeneous natural gas–air mixture is then introduced to the cylinder as a result of the engine suction.

These engines, referred to as dual-fuel engines, suffer from lower thermal efficiency and higher CO and UHC emissions, particularly at part load. The use of EGR is expected to partially resolve these problems and to provide further reduction in NO_x emission as well. Figure 4.2 shows the trend of UHC emission as a function of engine load for diesel fuel (D) combined with compressed natural gas (CNG).

For more information on UHC, see [5–7].

4.1.2.2 CO

All carbonaceous fuels will yield at least some CO. It is produced by incomplete combustion. The MAC value of CO is $35\,\mathrm{mg\,m^{-3}}$ ($30\,\mathrm{ml\,m^{-3}}$). The definition of the MAC value (maximum allowable concentration) is the highest allowable concentration of a substance as gas, liquid or dust suspended in air at a work place that does not induce harm to workers over a daily exposure of 8 h (1 shift).

The particular danger of CO is that it is a colorless, odorless, and tasteless gas. It is combustible (blue flame) and slightly lighter than air. CO hinders the oxygen transportation in blood as it combines with hemoglobin to produce carboxyhemoglobin. Concentrations of less than 1000 ppm (0.1%) may cause up to 50% of the human body's hemoglobin to convert to carboxyhemoglobin [8]. Common symptoms of carbon monoxide poisoning include headache, nausea, vomiting, dizziness, fatigue, and a feeling of general weakness. CO poisoning is the most common type of fatal air poisoning in many countries [9]. Modern vehicle engines emit 5–15 ppm of CO [10], while typical levels in the atmosphere are 0.1 to 0.5 ppm.

For private homes, with open stoves and natural gas burners, it is recommended to install cheap CO-sensors to detect potentially dangerous situations. In the atmosphere, CO will be oxidized to CO_2.

4.1.2.3 NO_x

Nitrogen oxides are a mix of essentially NO and NO_2. They are a major contributor to "acid rain". They increase ground level ozone concentration and smog formation. Washed out from the atmosphere, nitrification and eutrophication can occur. *Nitrification* is the process by which bacteria in soil and water oxidize ammonia and ammonium ions to form nitrites and nitrates, which can be absorbed by more complex organisms, such as the roots of green plants (nitrogen cycle). *Eutrophication* is an increase in the rate of supply of nutrients in an ecosystem. The term mainly refers to phosphates and nitrates. They promote excessive growth of algae. When these algae die and decompose, the water is depleted of available oxygen, causing the death of other organisms like fish. Apart from being air pollutants, NO_x (more specifically N_2O) contribute to global warming. NO_x from combustion processes is typically composed of ~95% NO and 5% NO_2. In the atmosphere, the NO is further oxidized to NO_2. Depending on the route of formation, one can distinguish between various nitrogen oxides, see Figure 4.3:

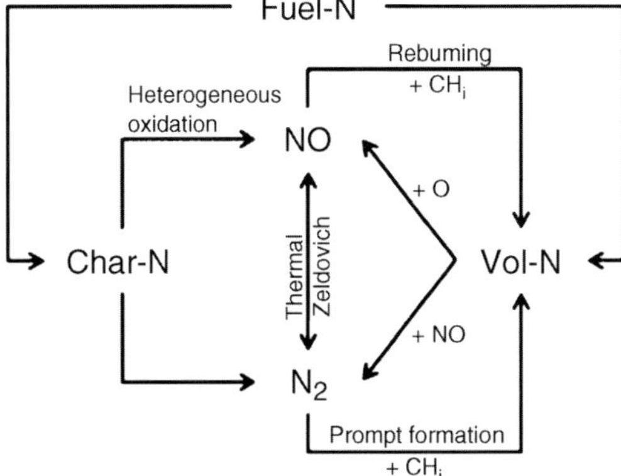

Figure 4.3 The overall mechanism of NO formation and reduction. Vol-N is an intermediate gaseous compound, for example, HCN or NH_3 [11].

Thermal NO (Zeldovich NO)

$$N_2 + O^{\bullet} \rightarrow NO + N^{\bullet} \tag{4.1}$$

$$O_2 + N^{\bullet} \rightarrow NO + O^{\bullet} \tag{4.2}$$

$$N^{\bullet} + OH^{\bullet} \rightarrow NO + H^{\bullet} \tag{4.3}$$

As the reactions above show, thermal NO is formed by an $O^{\bullet}$ radical attacking molecular nitrogen. As the triple bond in N_2 is very strong (binding energy $945\,kJ\,mol^{-1}$) this reaction is only possible at high temperatures. As the name implies, thermal NO is only formed at elevated temperatures, 2000 K and more are needed for significant amounts to be produced.

Prompt NO (Fenimore NO)

$$CH^{\bullet} + N_2 \rightarrow\rightarrow HCN + N^{\bullet} \rightarrow\rightarrow NO \tag{4.4}$$

The mechanism for Fenimore NO was postulated based on experimental results in which NO was formed to a small extent in flame regions which are too cold for thermal NO. It involves the attack of molecular nitrogen by the $CH^{\bullet}$ radical.

NO from N_2O (Laughing Gas)

$$O^{\bullet} + N_2 + M \rightarrow N_2O + M \tag{4.5}$$

$$N_2O + O^{\bullet} \rightarrow NO + NO \tag{4.6}$$

The third formation route of NO is via N_2O (laughing gas). This reaction occurs for example, in fluidized bed combustion [12]. Nitrous oxide is used as an oxidizer in rocket motors. In vehicle racing, it can be used as a fuel additive. Laughing gas itself is not combustible, however, at elevated temperatures it produces oxygen in the engine. N_2O contributes to global warming and ozone layer depletion (see Section 4.2).

NO Formation from Fuel-N

Most organic fuels, such as biomass, contain some nitrogen. This nitrogen can be converted, at least partially, to NO_x during combustion. The formation mechanism proceeds via the intermediates HCN and NH_3. Table 4.1 gives typical values of nitrogen oxides in various combustion processes.

NO can also be produced naturally during the electrical discharges of a flash of lightning.

Nitrous oxide (N_2O) is a greenhouse gas. It must not be confused with nitric oxide (formula NO) or nitrogen dioxide (formula NO_2).

A major emitter of NO_2 is the shipping industry. Marine vessels account for one third of global anthropogenic NO_x emissions, which is more than emissions from heavy trucks. Also, a substantial share of PM (particulate matter) is emitted by container ships, an area that is only now coming under environmental scrutiny.

NO_x can be reduced by combustion temperature reduction, which can be achieved by flue gas recirculation (FGR), exhaust gas recirculation (EGR) or reburning [14].

Table 4.1 Formation of NO_x [13].

Fuel	Thermal NO_x	Fuel-NO_x from liquid fuel components	Fuel-NO_x from solid fuel components (coke)	Prompt NO_x
Diesel/gasoline in internal combustion engines	90–95 ppm	—	—	5–10 ppm
Natural gas	100 ppm	—	—	—
Heavy fuel oil	40–60 ppm	60–40 ppm	—	—
Bituminous coal, dry furnace	10–30 ppm	50–70 ppm	20–30 ppm	—
Bituminous coal, slag tap furnace	40–60 ppm	30–40 ppm	10–20 ppm	—
Brown coal (lignite)	<10 ppm	>80 ppm	<10 ppm	—

Details on NO_x formation and abatement in combustion can be found in [15–17].

4.1.2.4 SO_2

SO_2 is produced from sulfur contained in the fuel. The MAC value for SO_2 is $1.3\,\mathrm{mg\,m^{-3}}$. It has a pungent, irritating smell. SO_2 can form SO_3 and subsequently H_2SO_4, which brings about heavy corrosion upon immission (acid rain). Also, sulfur dioxide emissions are a precursor to particulates in the atmosphere. S-rich fuels are wool, hair, rubber and foamed rubber, as found in many household articles. Waste incineration plants are equipped with a flue gas cleaning system that also removes SO_2 from the off gases. It can be precipitated in a scrubber to produce gypsum. Modern transportation fuels are low in sulfur content. Coal has a relatively large sulfur content.

Figure 4.4, taken from [18], shows the estimated global emissions of anthropogenic (man-made) SO_2 emissions (volcanoes are the most important natural emitters).

With fossil fuels reaching a consumption peak in the coming decades, it is assumed that anthropogenic SO_2 emissions will also decrease. Another sulfur-species which can be emitted during fuel production and distribution is H_2S. In combustion, it is converted to SO_2. Further information on combustion-derived SO_x can be found in [19,20].

4.1.2.5 Dioxins

A dioxin is a heterocyclic ring with 6 atoms where 2 carbon atoms are replaced by oxygen (1,2-dioxin or 1,4-dioxin). In this context, the term "dioxins" refers to certain dioxins and dioxin-like substances (i.e., compounds with a similar toxicity). They are highly toxic and persistant organic pollutants (POP, see also "dirty dozen"). Most well-known are the polychlorinated dibenzo-*p*-dioxins (PCDDs), or simply (but wrongly), *dioxins*. Technically, PCDDs are derivatives of dibenzo-*p*-dioxin. Out of the 75 PCDDs, seven are extremely toxic. A similar group of compounds are polychlorinated dibenzofurans (PCDFs), or simply *furans*. They are derivatives of

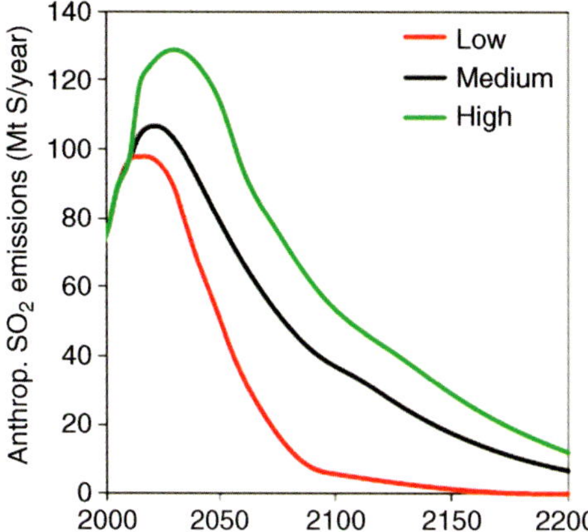

Figure 4.4 Estimation of global, global annual SO$_2$ emissions until 2200, taken from [18]. Taken from [18].

dibenzofuran. There are 135 congeners (these are derivatives differing only in the number and location of chlorine atoms). They are not dioxins as such, but ten of them have "dioxin-like" properties, as do some polychlorinated biphenyls (PCBs). Figure 4.5 shows the chemical structure of PCDD and PCDF.

PCDDs and PCDFs can be formed in combustion processes when carbon, oxygen and chlorine are present in a certain temperature window, 400–700 °C. The presence of metals such as copper can increase their yield. Figure 4.6 depicts the formation routes.

In Figure 4.7 one can see that there is a clear temperature optimum for dioxin formation.

This means that formation of dioxins is favored at relatively "poor" burning conditions of organic matter, such as municipal solid waste or hospital waste, at too low temperatures. Another important source of dioxins is metal smelting and refining. For details, see [23–25].

4.1.2.6 Particulate Matter (PM)

In terms of combustion and environmental science, the term particulate matter (PM) means any type of air pollutant that consists of solid or liquid particles suspended in the atmosphere as aerosols.

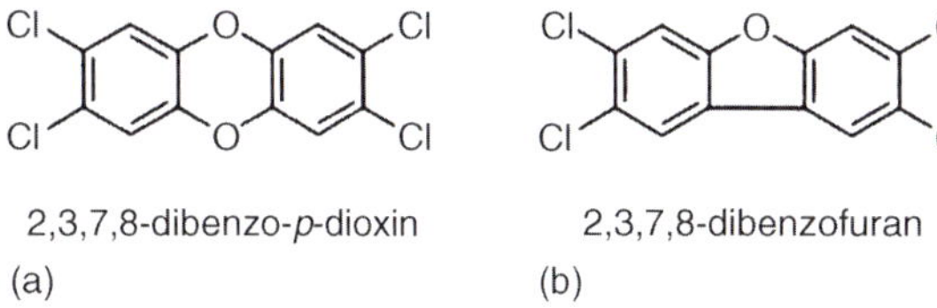

2,3,7,8-dibenzo-p-dioxin

(a)

2,3,7,8-dibenzofuran

(b)

Figure 4.5 PCDD (a) and PCDF (b), the most toxic of the 210 dioxin variants [21].

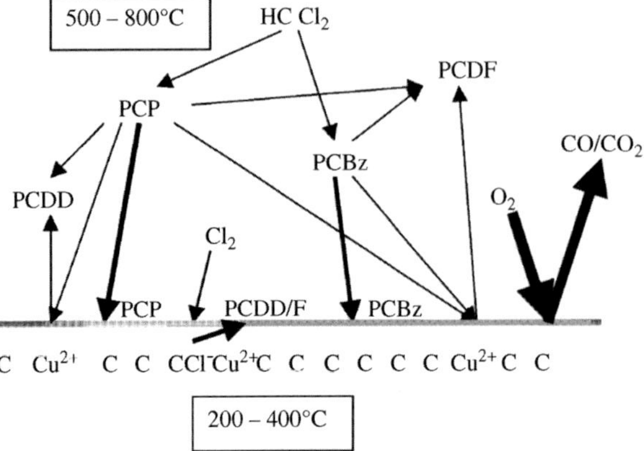

Figure 4.6 Formation routes for dioxin, taken from [22]

There are different formation routes, combustion being just one. Natural sources of particulates are volcanoes, forest and grassland fires, dust storms and sea spray, which account for the largest part of PM. Main anthropogenic sources include agriculture, hydrocarbon combustion, industrial processes, and transportation. Anthropogenic aerosols account for 10% of the total mass of aerosols in the atmosphere. Combustion-derived PM emissions result mainly from diesel engines, domestic fuels and biomass fuels [26–28]. Figure 4.8 illustrates how aerosols are generated in combustion processes.

By convention, PM is classified by size into fine particles, inhalable particles and total particulate matter. The three classes are defined based on their health effects.

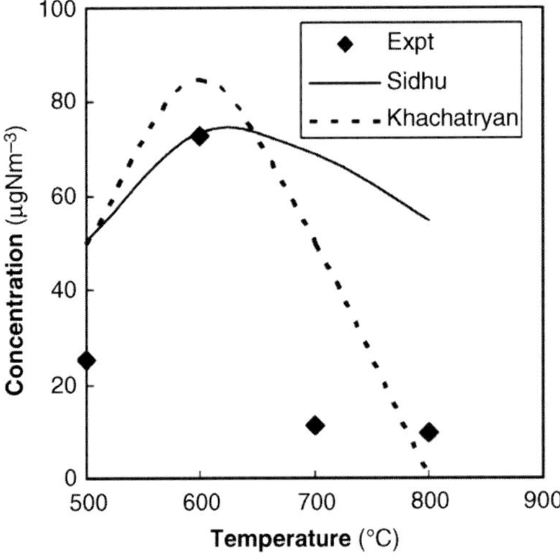

Figure 4.7 Temperature optimum for dioxin formation. Depicted are gaseous PCDD concentrations at various temperatures (♦). --- and — are models. Chart reproduced from [22].

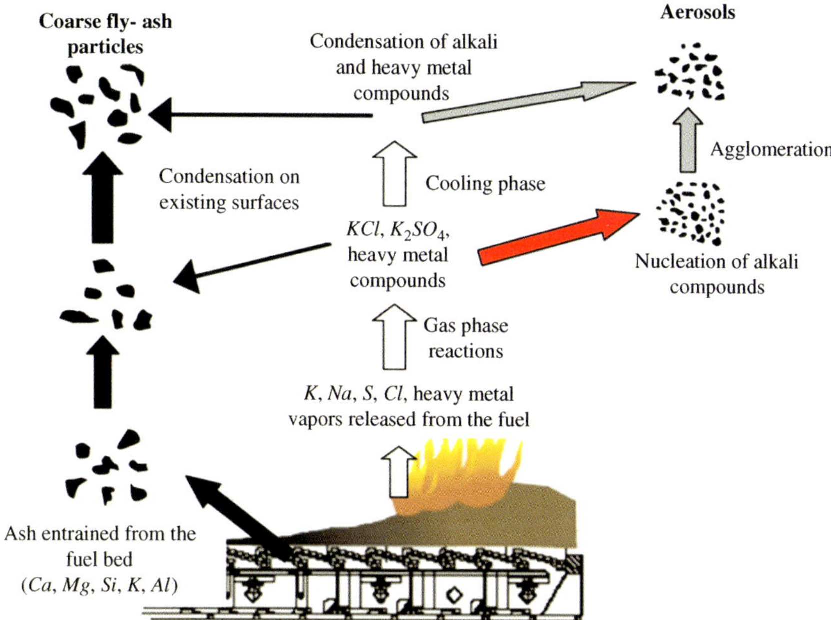

Figure 4.8 Aerosol formation during fixed bed combustion of untreated wood. Reproduced with permission from [29].

Fine particles have diameters less than 2.5 μm. They are mostly invisible to the naked eye and penetrate through most conventional filters. Nanoparticles, that is, particles of the order of 100 nm, are considered the most dangerous. PM from diesel engines is called diesel particulate matter (DPM). It can be around 100 nm in size. DPM can carry adsorbed carcinogenic compounds such as benzopyrenes.

The diameters of inhalable particles are between 2.5 and 10 μm. They still find their way through the human respiratory system and accumulate in the lungs, although they are easier to detect and filter out. Particles coarser than 10 μm are included in the total PM category and receive less attention since their tendency to settle and stay outside the respiratory system is higher [30].

There are a number of negative effects attributed to PM2.5 and PM10 emission. Most importantly, PM2.5 and PM10 have been proven to pose health risks to exposed humans. Adverse health effects include heart disease and heart rate irregularities [31], lung disease and asthma [32,33] and, via interactions with aeroallergens, allergy [34]. It has been shown that PM2.5 and PM10 fractions may reduce visibility and cause haze [35]. Modeling studies indicate that certain types of PM emission can also contribute to global warming [36]. Deposition and fouling from PM sources are esthetically unpleasant and can also change the chemical balance of surface or ground water.

The composition of PM depends mainly on its source. Typical constituents are inorganic oxides, chlorides (salt), sulfates, nitrates, organic carbon, inorganic carbon (black carbon) and biological entities, for example, bacteria, pollen, spores or viruses [37].

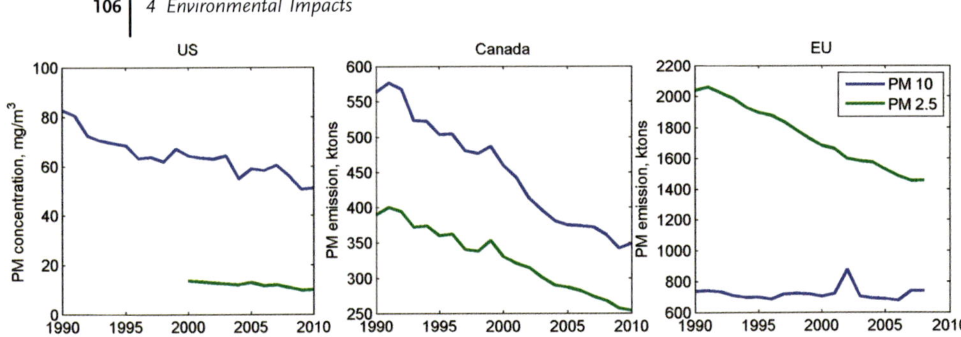

Figure 4.9 Some global trends in PM emission and concentration [30,38,39].

The trends in European and American air quality with respect to PM are promising. The global trend shows a steady decrease in PM categories over the last 20 years. Figure 4.9 depicts some global trends in PM concentrations and emissions.

The main contributors to atmospheric PM in Europe are shown in Figure 4.10 [38]. As can be seen, the contributions to PM10 and PM2.5 emissions are very similar and the indicated categories cover nearly all PM emission. The most prominent emitters are small-scale combustors and households, industry, automotive sector and agriculture.

Figure 4.11 shows the origin of PM2.5 particles in Beijing, China.

Different fuels cause different amounts of PM, see Figure 4.12.

One can see from Figure 4.12 that straw produces a high amount of PM, as compared to wood pellets. Larger furnaces tend to produce more PM than smaller ones. Table 4.2 shows the current (2012) air quality standards for PM for Europe and the US.

4.1.2.7 Soot

Soot is the solid by-product of the incomplete combustion of carbonaceous fuels, among them hydrocarbons. When discussing soot emissions, from an environmental point of view, solid carbon emissions can be grouped into two categories:

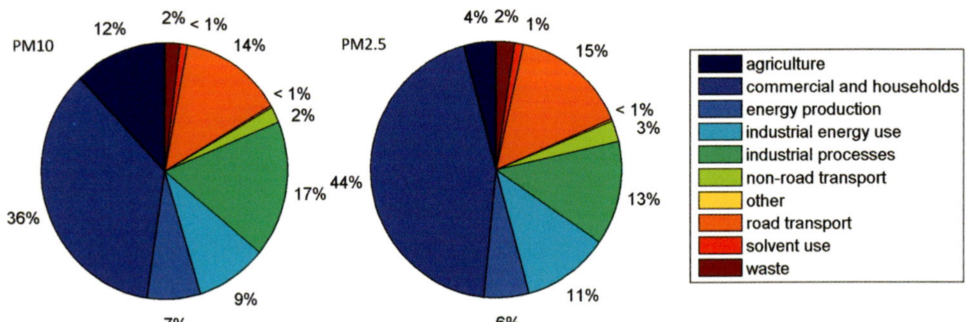

Figure 4.10 Main contributors to European PM emissions [38].

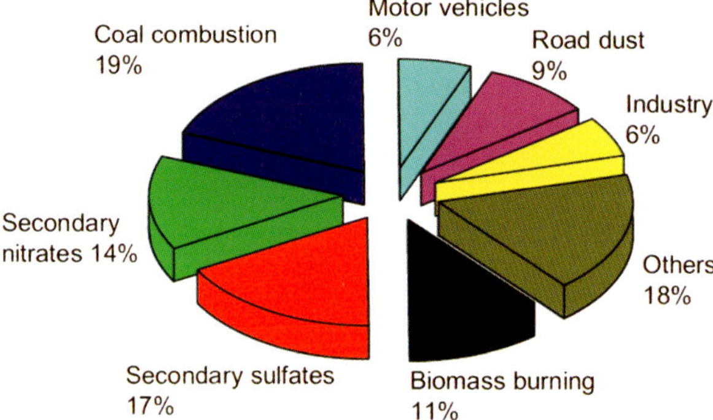

Figure 4.11 Source contribution of PM2.5 in Beijing/China [26].

organic carbon and black carbon emissions. Soot generally falls into the latter category, as soot particles mostly consist of *graphene* layers in various arrangements and structures. Graphene is a substance composed of pure carbon, with the atoms arranged in a regular hexagonal pattern similar to graphite. Black carbon is one of the few air pollutants that can absorb solar radiation and thus have a special effect on climate (Figure 4.13).

The total soot emission of a flame depends on many factors. First, not every type of fuel tends to produce sooting flames under standard conditions (methane is an example of a non-sooting fuel). The sooting tendency of a fuel can be quantified by quick laboratory tests on the basis of which various predictive models are established for surrogate fuels [41]. Secondly, the total soot output of a flame is based on competing formation and gasification mechanisms.

The most widely accepted soot formation models assume that soot particle formation is initiated by the formation of small aromatic molecules from the

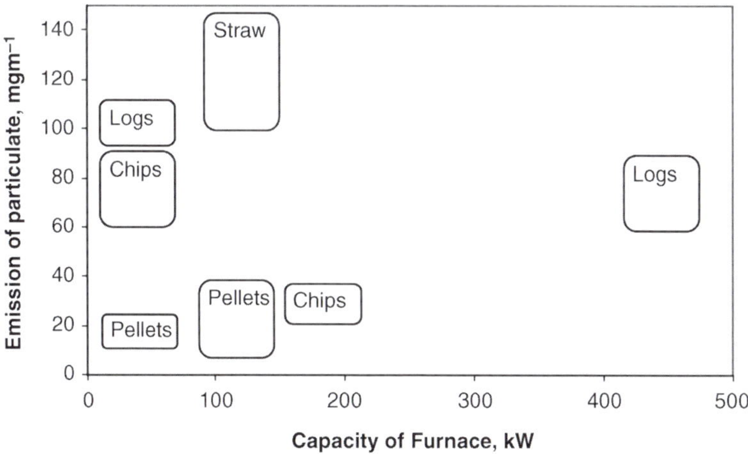

Figure 4.12 Combustion-derived PM10 plotted against appliance thermal capacity [40].

Table 4.2 Current European and American PM standards in $\mu g\,m^{-3}$ [30,38].

	Europe		United States	
	annual	24-hour	annual	24-hour
PM10	40[a]	50[b]	—	150[c]
PM2.5	25[a]	—	15[d]	35[e]

a) no annual threshold violation permitted.
b) 35 annual threshold violations permitted.
c) one annual threshold violation permitted.
d) arithmetic mean.
e) 98% percentile.

long-chain hydrocarbons that are broken up by the combustion reaction. A particularly important precursor for soot formation is believed to be acetylene [42]. From the short-chain precursors, soot formation generally proceeds via the formation of larger polycyclic aromatic hydrocarbons (PAHs). With the growth of PAHs, the first observable solid soot particles appear, with diameters of approximately 1 nm [43].

There are two accepted mechanisms that govern particle growth: collisional (coalescent) coagulation and surface growth. In the transitional regime between $Kn \rightarrow \infty$ and $Kn < 1$, coagulation is understood to be effected by van der Waals forces [44]. Kn is the Knudsen number, a dimensionless number defined as the ratio of the molecular mean free path length to a representative length scale [45]. When particles are small and collisional coagulation can play a considerable role, the van der Waals force is a function of the atomic structure and the distance between the particles. Surface growth happens in parallel with coagulation and is controlled by the density of active surface sites, with the main growth species being hydrogen [46], acetylene and PAH [47].

Figure 4.13 Burning tires generate a heavy smoke, dominated by the forming soot.

It is possible to reduce overall soot emission after it has formed in the flame by achieving near-complete oxidation. Oxidation reactions are the sinks of soot dynamics, as they convert solid combustion products to gaseous products (mostly CO_2). Oxidation is believed to happen as a result of the attack of molecular oxygen, although different oxidizing species are considered, for example, the OH radical, atomic O, H_2O and CO_2.

Soot formation and oxidation are heavily controlled by chemical kinetics. In practical flames, soot is formed and gasified in a matter of milliseconds, thus in most cases it is impossible to reach any kind of equilibrium; therefore, soot dynamics often cannot be modeled by using thermodynamic principles. A rather inaccurate, but still widely used expression to model soot oxidation is the Nagle–Strickland–Constable (NSC) relation [48]:

$$w = \left(\frac{k_A P_{O_2}}{1 + k_Z P_{O_2}} \right) x + k_B P_{O_2} (1 - x) \tag{4.7}$$

$$x = \left(1 + \frac{k_T}{k_B P_{O_2}} \right)^{-1} \tag{4.8}$$

Where P_{O_2} is the partial pressure of oxygen. k_A and k_B are the reaction rate coefficients of the two types of carbon atoms ("edge site" and "basal plane"). x is the fraction of the more reactive sites (A) of the two kinds (A and B). The NSC model assumes that oxidation happens by reactions with two distinctly different rate constants, corresponding to edge and basal plane carbon atoms. The reaction rate w is given in (g of carbon atoms) $cm^{-2} s^{-1}$. The NSC equation was developed for the oxidation of graphite in the temperature range 1273–2273 K. When studying real soot particles, one often finds that their reaction rates differ significantly from those predicted by the NSC model. These differences can be many orders of magnitude and can be explained by the characteristics of soot nanostructure.

Another important mechanism is oxidation by OH radicals. The most widely used rate expression to describe this route was developed by Fenimore and Jones [49]. Their model is based on the introduction of the collision efficiency factor to describe the frequency of removal of carbon atoms:

$$w = 1.29 \cdot 10^3 \Gamma_{OH} P_{OH} \sqrt{T} \tag{4.9}$$

Where w is the rate in $kg\,m^{-2} s^{-1}$, P_{OH} is the partial pressure in atm and Γ_{OH} is the collision efficiency factor for OH. Neoh [50] found this factor to be approximately 0.13 and relatively independent of equivalence ratio and position in the flame. OH can be a dominant oxidant at near stoichiometric and fuel-rich conditions.

Soot emissions have similar negative effects as general particulate emissions. However, in flames, soot can often play favorable roles as well. The most significant benefit of soot is the increased radiative heat output of the flame. Burning hydrocarbons and their gaseous products (H_2O and CO_2) typically radiate between 1 and 5 μm, with their total spectral emissivities often being much less than 0.5 [51]. On the other hand, soot, as a solid body, radiates in the whole range of the spectrum (although under practical conditions, most strongly in the yellow–red visible band) and its emissivity can often be close to unity [52]. A typical industrial application that utilizes the improved radiative output of sooting flames is glass production. Another favorable role is the reduction of NO_x by the carbon in soot, as observed in some studies [42].

There is a considerable interest in preventing soot emissions by the use of fuel additives. The best multifunctional additives act via multiple mechanisms, for example, they can inhibit formation and/or coagulation to enhance oxidation before leaving the reaction zone [53].

There are many methods for the diagnostics of sooting flames and the measurement of soot concentration, size distribution, emissivity and temperature. These methods are usually non-intrusive and apply optical principles. For example, two-color pyrometry was developed for the measurement of flame temperature. Other soot diagnostics include light extinction and light scattering (laser induced scattering, LIS) techniques to measure soot concentration and size distribution. The laser induced incandescence signal (LII) can also be utilized to image soot concentration and size distribution. These techniques are discussed in Chapter 5.

Offline diagnostics and analytical methods for soot include X-ray diffractometry (XRD), light microscopy and electron microscopy. Electron microscopy, particularly transmission electron microscopy (TEM), has been successfully applied to study and quantify soot structure in the past through the use of image analysis [54] for laboratory generated and/or processed samples [55], soot emitted from diesel engines [56], and samples from large scale fires [57]. Linking structural properties, obtained either by XRD or TEM, with the kinetic observations on soot formation and oxidation is a current topic of research [58]. Figure 4.14 shows typical soot structures in TEM micrographs.

4.1.2.8 Ash

Solid fuels generate ash. It can be defined as the non-aqueous residue after burning a sample or material. Ash mainly consists of inorganic matter, mostly salts and oxides. In wood, ash typically represents 0.43 to 1.82% of the mass of burned material (dry basis) [59]. In a furnace, bottom ash and fly ash are produced. Ash is obtained on the order of hundreds of millions of tons per year, see data for the USA in Figure 4.15.

Figure 4.16 below shows the typical composition of bottom ash and fly ash from different combustion processes.

There exist several ways to utilize the ash, for example, as basis for construction materials, see [61, 62]. In "urban mining" projects, researchers try to gain valuable materials such as phosphates from ash. Ash can also contain heavy metals (see below).

4.1.2.9 Alkali Metals

Alkali and alkaline earth metals emissions are mainly a corrosion issue for furnaces. Together with chlorine species, they form volatile salts that are low-melting and can block heat transfer surfaces. Alkali metal emissions are a topic of concern in biomass firing, see Figure 4.17.

Further information is given in [64–66].

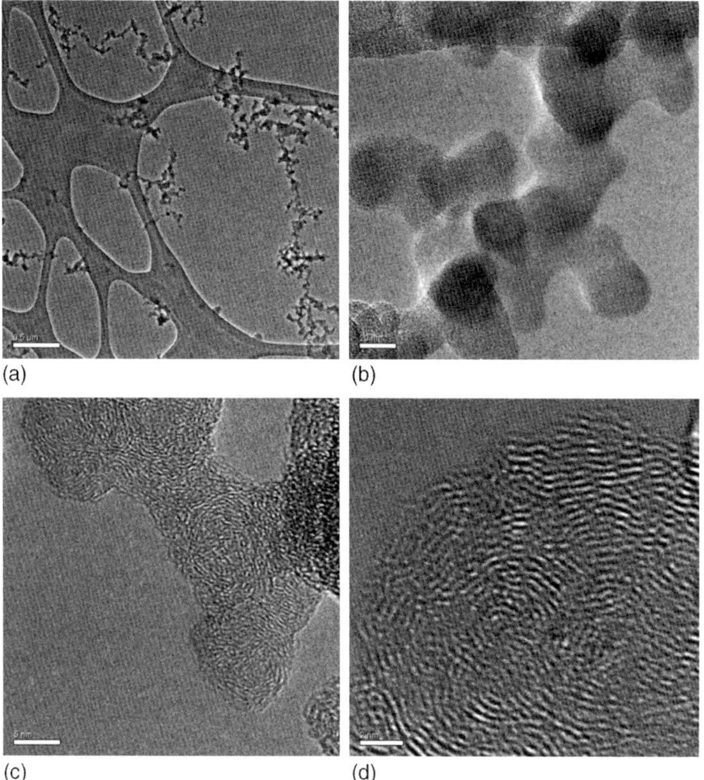

(a) (b) (c) (d)

Figure 4.14 Typical soot structures in TEM micrographs. (a) Soot aggregates are visible. Aggregates are made of coagulated soot particles. They grow to various sizes, typically up to tens of micrometers. (b) Spherical soot particles in an aggregate. Soot particles are also called primary particles or spherules, and their sizes are typically of the order of a hundred nanometers. (c) The atomic structure of a soot aggregate becomes visible if the applied magnification is high enough. Only high-resolution TEMs can achieve these magnifications. The dark and bright stripes are formed by the principle of phase contrast imaging and correspond to graphene layers. (d) At ultrahigh magnifications (around 1 million times), crystallographic properties of the graphene layers can be measured. For example, the interlayer spacing between each graphene sheet is approximately the same as that of graphite – 0.335 nm.

4.1.2.10 Heavy Metals

Common heavy metals in combustion residues are vanadium (petroleum), mercury and cadmium. They can be emitted from waste incineration without proper flue gas cleaning. The emission of lead has decreased greatly since tetraethyl lead has been widely replaced as an anti-knock additive in automotive gasoline. Figure 4.18 classifies various elements according to their mobility.

In that study, coal, ash, and limestone samples from a fluidized bed combustion (FBC) plant, a pulverized coal combustion (PC) plant, and a cyclone (CYC) plant in Illinois/US were analyzed to determine the combustion behavior of mineral matter,

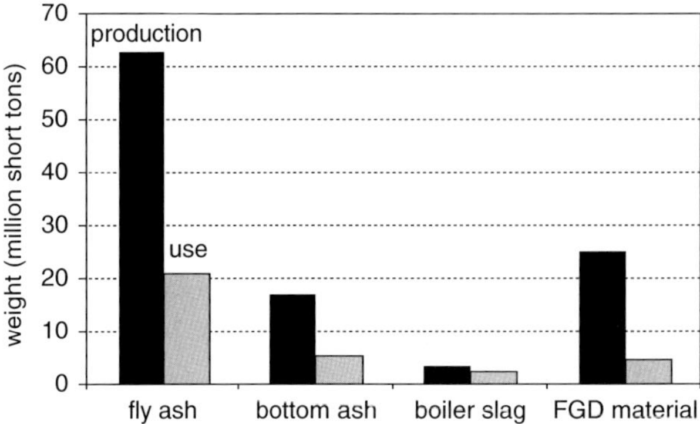

Figure 4.15 US production and use of coal combustion residues for the year 1999 (plotted using data from ACAA, the American Coal Ash Association). FGD = flue gas desulfurization [60].

and to propose beneficial uses for the power plant ashes. Mercury is a particularly critical heavy metal in combustion. Figure 4.19 shows possible pathways in a combustion process.

One can immobilize Hg in an alkali-activated fly-ash matrix. Figure 4.20 depicts the capture rate of mercury as a function of the sulfur content of the fuel.

For details on heavy metal emissions, see [69–73].

4.1.3
Concepts for Pollutant Reduction

In order to minimize adverse effects of combustion, several technologies have been developed to reduce pollutants. These can be grouped in three main categories:

- Pre-combustion (e.g., S-reduced fuel [74])
- During combustion (e.g., mixing, temperature profile, lambda)
- Post-combustion (e.g., flue gas cleaning, DENOX by SNCR and SCR [75])

This section focuses on power plants.

Staging is a concept to lower combustion temperatures, thereby decreasing NO_x emissions during combustion. Air staging and fuel staging are depicted in Figures 4.21 and 4.22.

An example of post-combustion pollutant reduction is the installation of dust filters. Figure 4.23 depicts the collection efficiency of three technologies: fabric filters, electrostatic precipitators and cyclones, as a function of particle size.

In industrial installations, several techniques for flue gas cleaning are typically combined, see the example in Figure 4.24. This figure shows a combination of different technologies for emission reduction: air staging, solids separation, dry scrubber (adsorbent injection), wet scrubbers, low-dust SCR (selective catalytic reduction of NO_x).

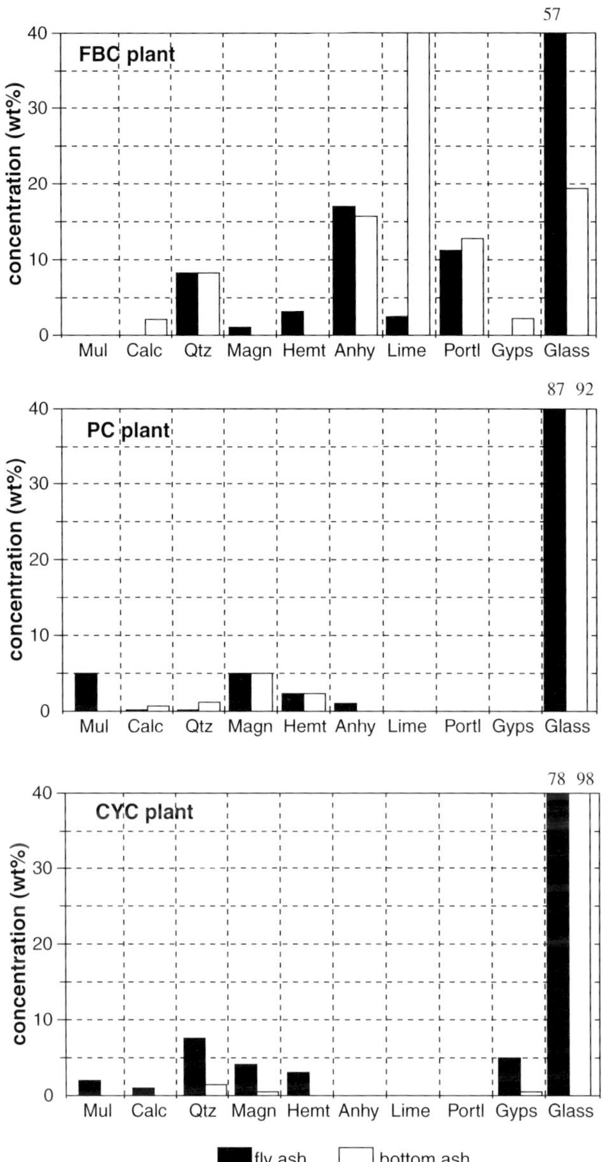

Figure 4.16 Mineralogical compositions of ashes from the FBC, PC, and CYC plants. Key: Mul, mullite; Calc, calcite; Qtz, quartz; Magn, magnetite; Hemt, hematite; Anhyd, anhydrite; Lime, lime, Portl, portlandite; Gyps, gypsym; Glass, amorphous phase. FBC = fluidized bed combustion; PC = pulverized coal combustion; CYC = cyclone [60].

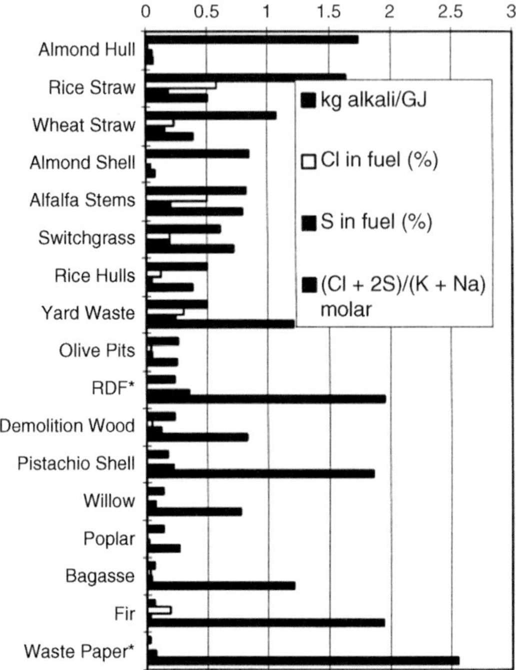

Figure 4.17 Alkali index, chlorine and sulfur concentrations, and chloride–sulfate ratios for biomass [63].

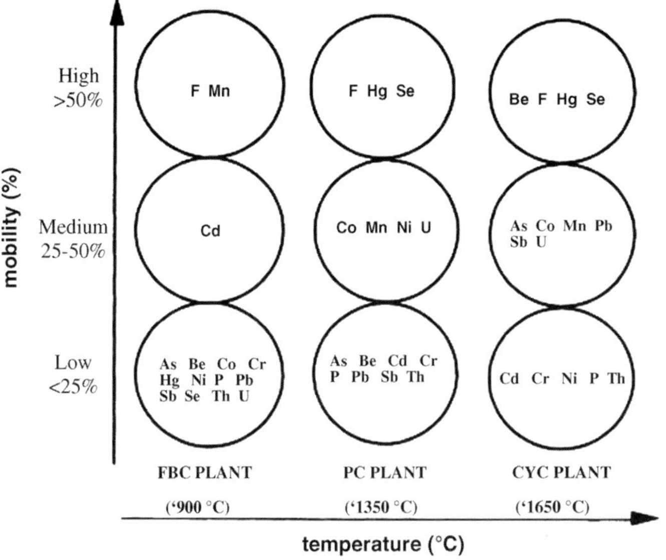

Figure 4.18 Combustion mobilities of 15 HAP elements at the FBC, PC, and CYC plants. HAP = hazardous air pollutants.

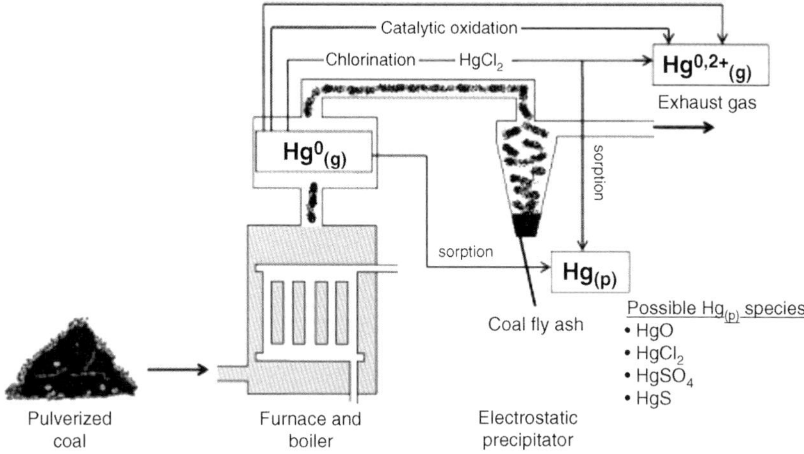

Figure 4.19 Possible fates of mercury during coal combustion [67].

To capture SO_2, limestone can be injected into fluidized beds [78]. The produced gypsum can be purified and utilized. Sorbents can also be deployed to capture heavy metals [79].

Figure 4.25 shows a photograph of an electrostatic precipitator (ESP) and two scrubbers for the flue gas treatment of a thermal power plant.

For a review of pollutant control in automotive and aero engines, see [81–83].

Another pollutant from combustion is *carbon dioxide* (CO_2), where a major research interest is CO_2 sequestration (carbon capture and storage), see Section 4.2 or [84] for details.

A review on air pollution in general is presented in [1,85].

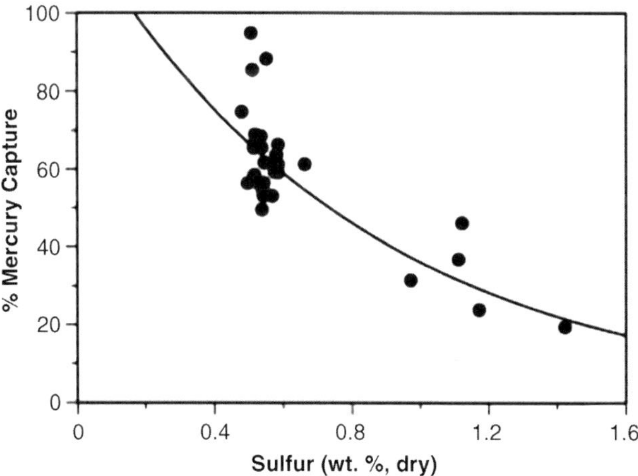

Figure 4.20 Decreasing Hg capture with increasing coal S. Data points show weekly averages observed for two units equipped with CESP emission controls where fly-ash C exceeds 5% (average 11%). CESP = cold-side electrostatic precipitator [68].

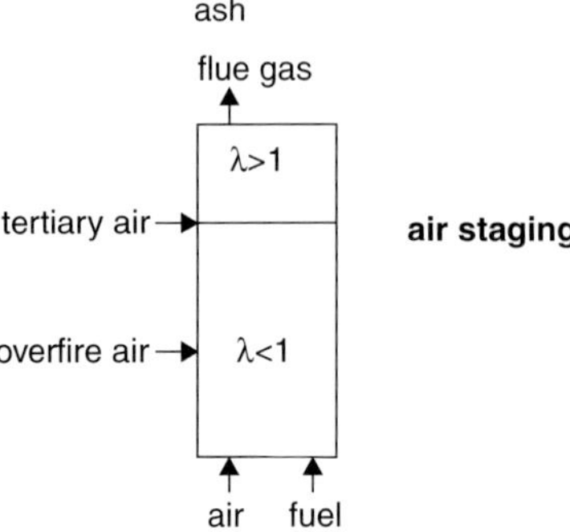

Figure 4.21 Scheme of air staging [76].

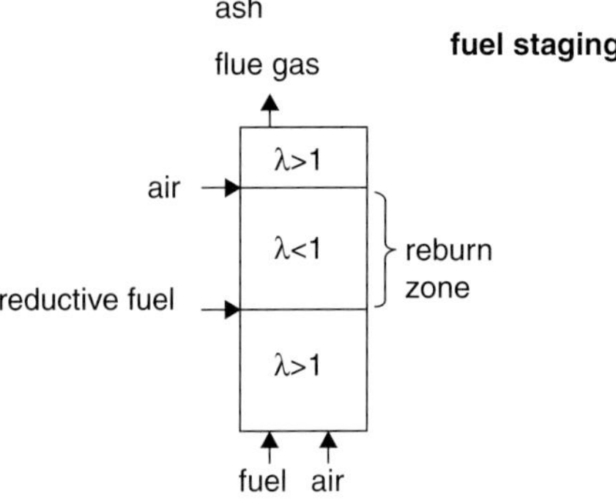

Figure 4.22 Scheme of fuel staging [76].

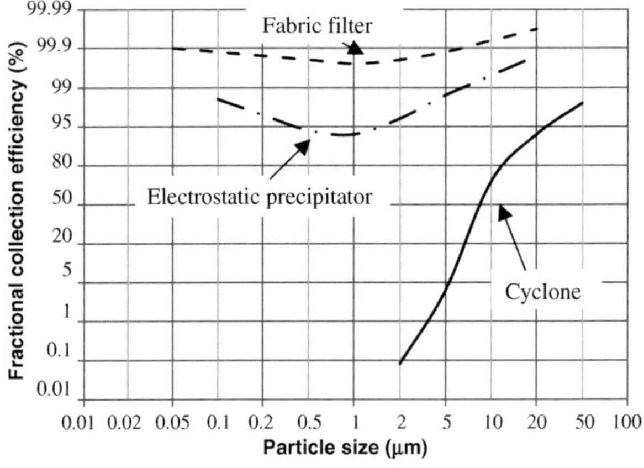

Figure 4.23 Collector efficiency of conventional gas cleaning technologies [29].

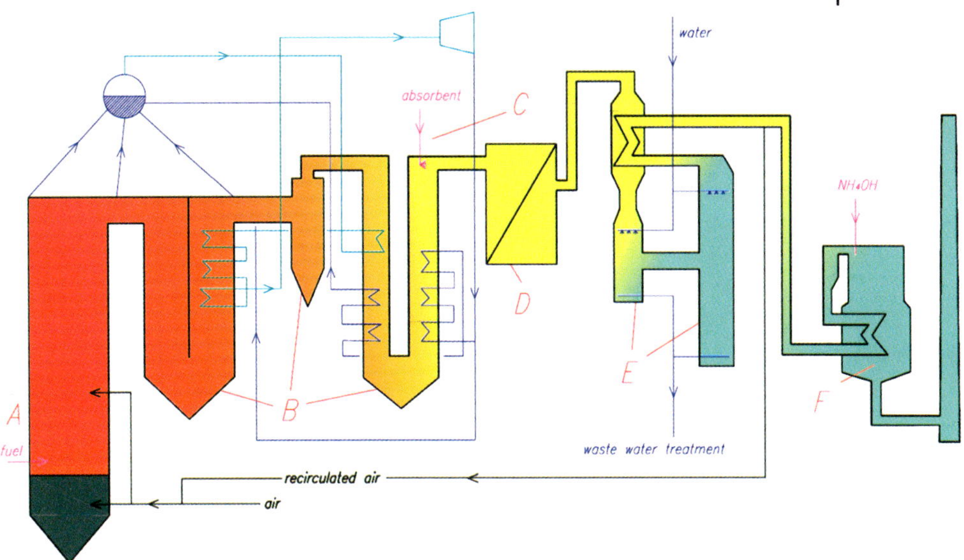

Figure 4.24 Basic set-up of the flue gas cleaning system in the waste-to-energy industry. A: fluidized bed combustor; B: gravitation and/ or centrifugal separators; C: dry flue gas cleaning; D: baghouse filter or electrostatic precipitator; E: wet scrubbers; F: selective catalytic reduction (SCR) [77].

4.1.4
Summary

In this section, the most important pollutants from combustion processes have been described, with a glimpse on technologies for their reduction. Pollutant reduction, apart from safety and optimization of energy efficiency, is the single most important aim of combustion engineers. Existing combustion processes can be upgraded to become cleaner, and proper design can optimize new installations.

Figure 4.25 Flue gas treatment in a Danish power plant (Mâbjergværket). Adiox™ is a process for dioxin removal, see [80] for details.

4.1.5
Web Resources

There are good resources on the internet to learn more about combustion-derived pollution. A few are listed here for the interested reader:

Summaries of EU legislation	http://europa.eu/legislation_summaries/environment/air_pollution/l28028_en.htm
US Environmental Protection Agency	http://www.epa.gov/iaq/combust.html
California Air Resources Board	http://www.arb.ca.gov/research/indoor/combustion.htm
Natural gas boiler burners	http://www.cleanboiler.org/Eff_Improve/Primer/Boiler_Combustion.asp
International Energy Association	http://ieacombustion.net/

4.2
Combustion and Climate Change

4.2.1
Introduction

Man has used fire for various purposes for the last 500 000 years. However, during the Industrial Revolution (1750–1850), the consumption of fossil fuels soared, leading to a strong increase in emission of carbon dioxide and other combustion byproducts into the atmosphere. Today, the rate at which carbon is being released into the atmosphere is by three orders of magnitude larger than the rate at which it was fixed millions of years ago, and the level of CO_2 has climbed from 280 ppm to 397 ppm, see Figure 4.26:

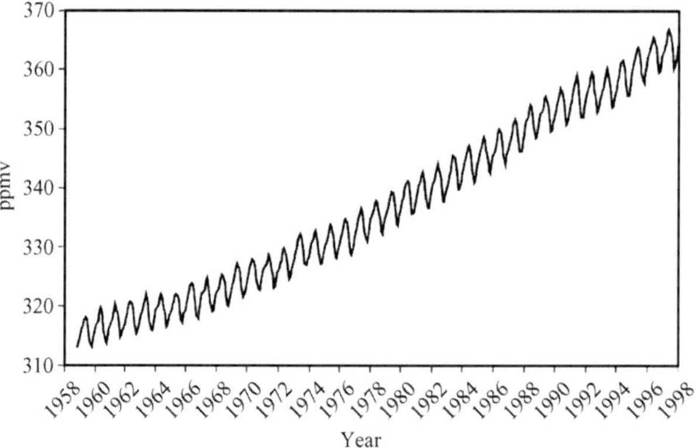

Figure 4.26 Monthly measured average CO_2 concentration from Mauna Loa, Hawaii. The seasonal variations can be explained by the uptake and production of CO_2 by the terrestrial biosphere [86].

Table 4.3 Comparison of the global-warming potential (GWP) of the six most relevant greenhouse gases (GHG).

	Greenhouse Gas	GWP	Sources
CO_2	Carbon Dioxide	1	Fossil fuel combustion, forest clearing, cement production
CH_4	Methane	70	Landfills, production and distribution of natural gas and petroleum, fermentation from the digestive system of livestock, rice cultivation, fossil fuel combustion
N_2O	Nitrous Oxide	290	Fossil fuel combustion, fertilizers
HCF	Hydrofluorocarbons	12 000 (HFC-23)	Refrigeration gases, aluminum smelting, semiconductor manufacturing
PFC	Perfluorocarbons	3800 (HFC-134a)	Aluminum production, semiconductor industry
SF_6	Sulfur hexafluoride	16 000	Electrical transmissions and distribution systems, circuit breakers, magnesium production

CO_2 is, leaving aside water vapor (H_2O) and ozone (O_3), one of six greenhouse gases (GHG). GHG can absorb (and emit) infrared radiation, hence trapping heat in the atmosphere. The global-warming potential (GWP) is a relative measure of how much heat a GHG traps in the atmosphere. The calculations in Table 4.3 are for a 20-year time horizon.

The natural greenhouse effect is larger than the man-made (anthropogenic) one. However, man's influence on the climate can cause a dangerous imbalance [87].

The accumulation of gases with GWP in the atmosphere leads to a temperature increase. Figure 4.27 shows the calculated temperature increase over 200 years from

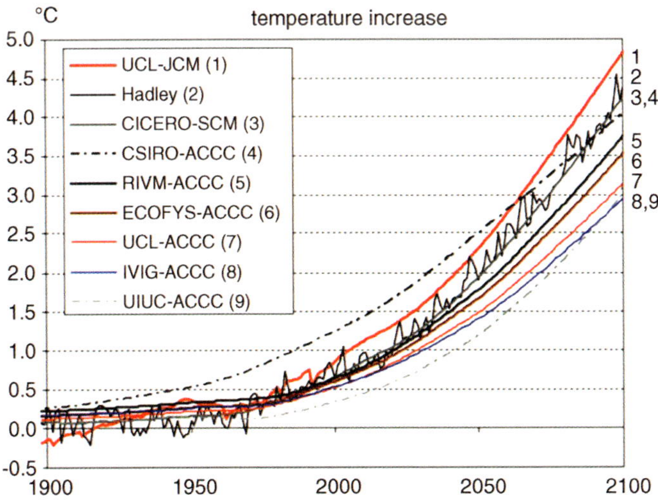

Figure 4.27 Comparison of the temperature increase calculated by nine different models [88].

nine different models [88]. One can see that the models predict a global temperature increase between 3 and 5 °C by the year 2100. Combustion processes are at the center of the discussion about man-made climate change. "Climate change" is the preferred term for global warming. The burning of fossil fuels releases climate-active gases into the atmosphere, causing an imbalance with the natural climate. The "natural greenhouse effect" is mainly caused by water vapor in the atmosphere. Without this effect, the temperature on Earth would be -18 °C instead of 15 °C. Of the anthropogenic greenhouse effect, 63% can be traced back to CO_2, most of which is released from the combustion of fossil fuels [89]. This section presents several key aspects of combustion and climate change.

4.2.2
Primary Energy Production

Figure 4.28 shows the global breakdown of energy consumption [90]. The primary energy measures the thermal energy equivalent in the fuel that was used to produce a useful form of energy. For nuclear energy, an average efficiency of 33% was assumed. For geothermal energy, an efficiency of 10% was used. In the case where energy is obtained directly in the form of electricity, such as in hydropower, wind and photovoltaic plants, the energy equivalent of electricity is used. Further data can be accessed via the IEA website [91]. The IEA Energy Outlook 2012 also presents a projection of changes in power generation from 2010–2035, where one can see a strong increase in renewables, see Figure 4.29. Renewables account for half the new capacity in power generation on a global scale [92].

4.2.3
Combustion and Global Warming by Sectors

Combustion processes release the climate-active CO_2. Also, the waste heat from a combustion process dissipates into the environment. For 1 kWh of electricity, approx. 2 kWh of waste heat and 1 kg of CO_2 are released into the environment. Due to the long residence time of CO_2 in the atmosphere on the order of 100 years,

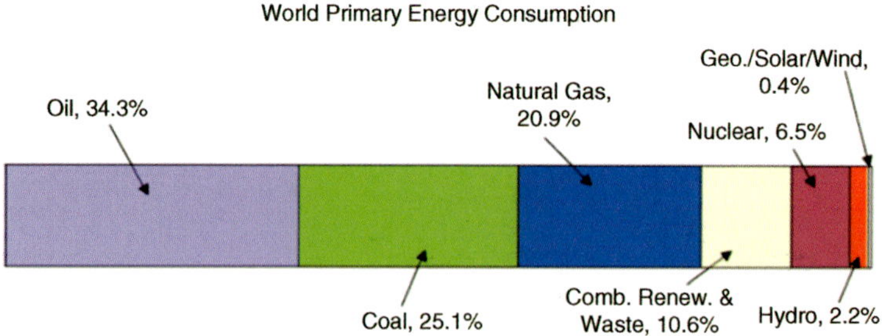

Figure 4.28 Breakdown of the World's primary energy consumption (2004). In total, 11 059 MToe (million tons of oil equivalent) were used [90].

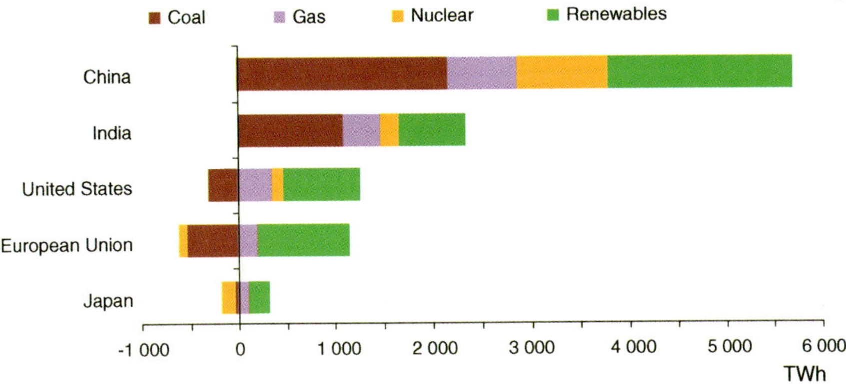

Figure 4.29 Anticipated change in power generation from 2010 to 2035: Renewables gain significance [92].

the effect of the GHG is approx. 25 times stronger than the effect of the waste heat, accounting for 92% of the atmospheric "heat up" caused by the electricity production studied [93]. Other indirect effects of combustion processes on global warming are reviewed in [88].

Major sectors where carbonaceous fuels are burned are

- Energy production
- Industry
- Transportation
- Households

Figure 4.30 shows the breakdown of GHG emissions by sector for the US in 2010. For details on the power industry, see [94]. The effect of vehicle emissions on global warming are reviewed in [95].

4.2.4
Mitigation of Global Warming in the Context of Combustion

Actions to reduce the impact of combustion processes on our climate can be grouped as follows:

- Energy efficiency measures
- Reduction of CO_2 emissions
- Use of renewable fuels
- Other measures against climate change

4.2.4.1 Energy Efficiency
The by far most cost-effective and easy-to-implement measure against excessive CO_2 emissions is an improvement in energy efficiency [87,97,98]. Figure 4.31 shows the

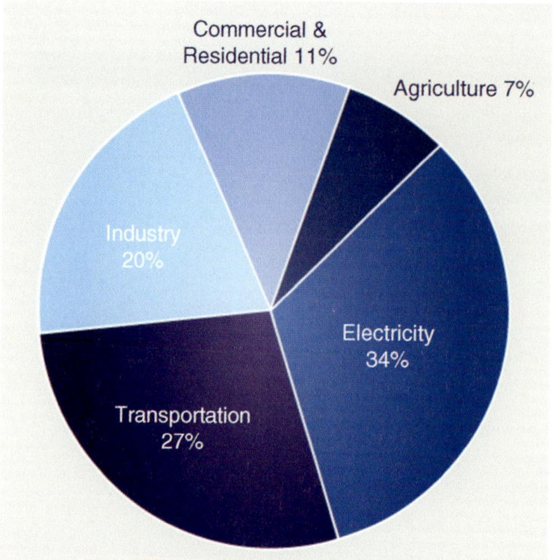

Figure 4.30 Total emissions of greenhouse gases (GHG) in the US in 2010: 6822 Million Metric Tons of CO_2 equivalent. Land use, land-use change, and forestry in the United States is a net sink and offset approximately 15% of these GHG emissions [96].

unused potential of energy efficiency measures. As can be seen industry has already used ∼40% of its energy efficiency improvement potential. The least progress has been made in buildings where, for example, thermal insulation can bring substantial energy and cost savings. Also in transportation and power generation, huge potentials for energy efficiency measures are available.

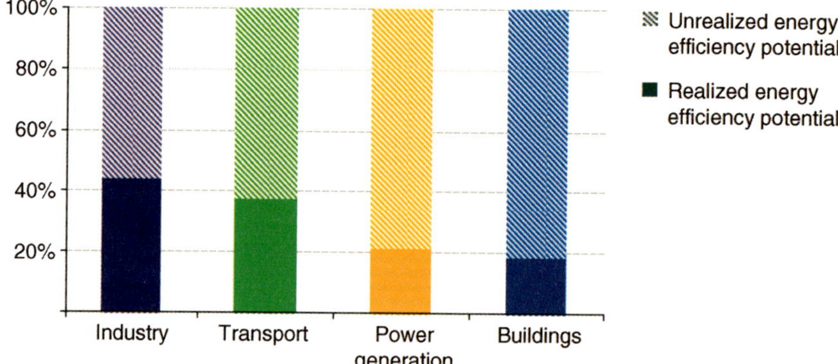

Figure 4.31 Energy efficiency potential by different sectors. According to the IEA Energy Outlook 2012, two thirds of the potential will remain untapped in the period to 2035 [92].

Table 4.4 CO_2 emissions for various fuels.

Fuel	Specific Carbon Content [kg$_C$/kg$_{fuel}$]	Specific Energy Content [kWh/kg$_{fuel}$]	Specific CO_2 Emission [kg$_{CO_2}$/kg$_{fuel}$]	Specific CO_2 Emission [kg$_{CO_2}$/kWh]
Bituminous coal	0.75	7.5	2.3	0.37
Gasoline	0.9	12.5	3.3	0.27
Diesel	0.86	11.8	3.2	0.24
Natural Gas (Methane)	0.75	12	2.8	0.23

Note: The heat loss, 55 to 75%, in power generation is not included in the numbers.

4.2.4.2 Reduction of CO_2 Emissions

By switching to a fuel with less carbon content, the specific emissions of CO_2 will be lower as Table 4.4 illustrates. So, in essence, natural gas, which mostly consists of CH_4, is a cleaner fossil fuel than coal. For a full assessment, though, one has to take more factors into account: CH_4 is a stronger greenhouse gas than CO_2, so a production of natural gas with significant exploration, production and distribution losses of CH_4 can reverse the energy balance.

The use of renewable fuels does – in theory – not put any net CO_2 into the atmosphere, because renewable fuels are seen to be in a continuous cycle of CO_2 fixation and subsequent combustion. "In theory" is stated because for biofuel production, storage and distribution, normally at least some fossil energy (with net CO_2 release into the atmosphere) is used. Renewable fuels are also called "biofuels" or "biomass". There is a wide array of biomass fuels, from agricultural residues such as straw to dedicated energy crops and wood and biogas.

Hydrogen seems, at first sight, to be the ideal energy carrier, because its combustion does not produce any CO_2, or SO_x, and in pure oxygen (oxyfuel combustion) not even NO_x. However, it has to be borne in mind that H_2 is only a means of energy storage and has to be produced in some way.

For the use of biomass combustion to mitigate global warming, see [99].

4.2.4.3 Use of Renewable Fuels

Wood has been an important fuel since man's taming of flames. During the Industrial Revolution, the share of fossil fuels increased, due to their high energy density, good availability and low cost.

In the last decades, fossil fuels have gone up in price, as increasingly remote and deep reservoirs have to be exploited. Take crude oil, for instance. Today, deep sea drilling and secondary and tertiary oil recovery techniques are being used. The question of when fossil fuels will be depleted is difficult to answer. On the one hand, the extent of known reserves bears inaccuracies. On the other hand, new reserves can still be discovered, and new techniques can exploit resources that were previously considered useless. For instance, the surge in world market price for crude oil has made oil recovery from oil sands and deep water reservoirs profitable. Another potentially interesting natural gas reservoir is methane hydrates and clathrates. However, there is no suitable technology available yet to recover them safely, environmentally friendly and economically.

Figure 4.32 shows the ultimate recoverable reserves (URR) for the three most important fossil fuels – oil, natural gas and coal – as estimated by different agencies and corporations [100]. URR is the sum of past cumulated consumption plus proven reserves of the respective fuel.

The point in time when "peak oil" is reached is not known yet. Figure 4.33 shows an estimate from five different sources of the development of fossil fuel usage over the next 200 years [100].

All five scenarios assume a peak before 2050 and a gradual phasing out of fossil fuels. It is considered to be a no-brainer that renewable fuels will gain importance. Renewable fuels include traditional biomass such as wood. Other *solid* renewable fuels are several so-called *"energy crops"*. These are fast-growing plants that are harvested for the purpose of combustion. Typical energy crops are switch grass, miscanthus (elephant grass), willow and poplar, plus whole-crops such as maize and millet, which can be turned into silage and subsequently into biogas. Another type of renewable fuel is "leftovers" from agricultural processes, such as straw, grape seeds and bark.

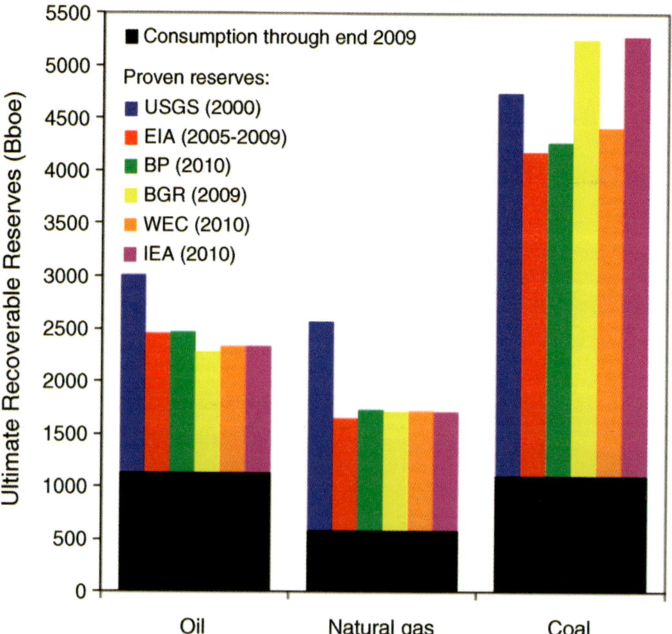

Figure 4.32 Ultimate recoverable reserves (URR) for oil, natural gas and coal according to six different estimations. One can see that, constant consumption assumed, coal will last longer than natural gas and crude oil [100].

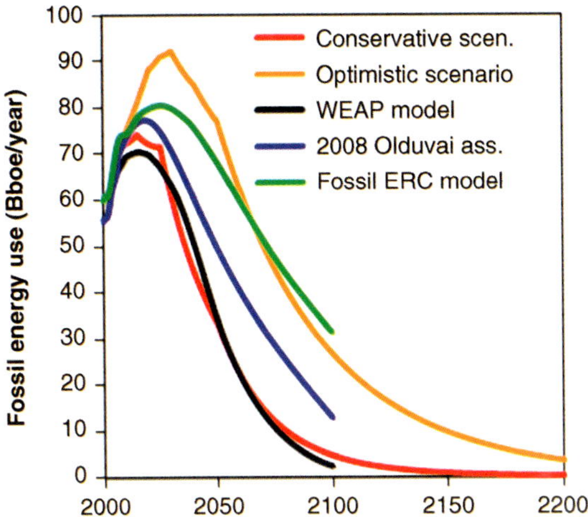

Figure 4.33 Estimated use of fossil fuels over the next two centuries [100].

Liquid renewable fuels can be produced from various sources of biomass. They are termed "biofuels". Examples are ethanol and biodiesel, plus several oils. Processes that convert biomass to liquid fuels are called BTL (biomass to liquid). An overview of routes to biofuels is shown in Figure 4.34.

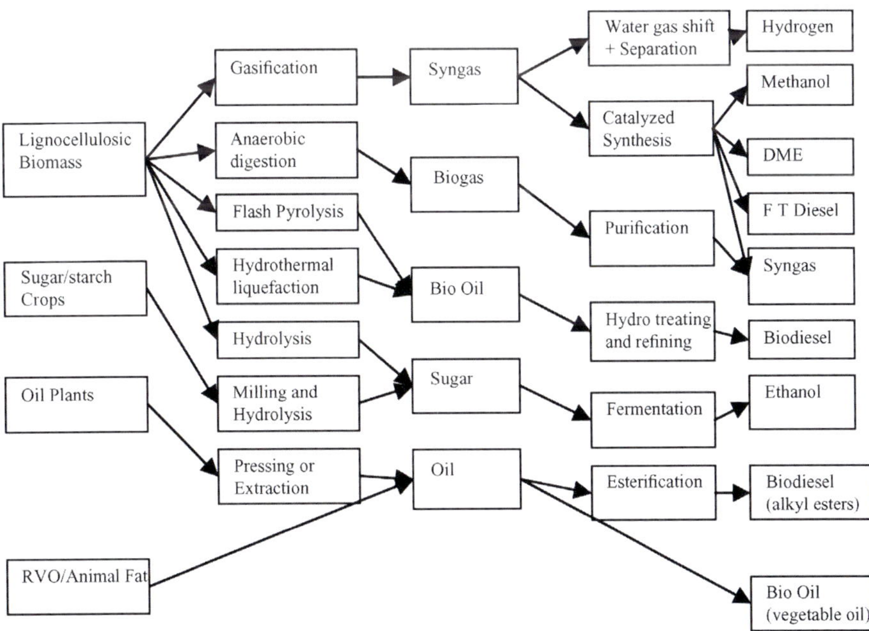

Figure 4.34 Overview of conversion routes to biofuels. Reproduced with permission from [101].

An example of a *gaseous* renewable fuel is biogas, which is obtained from the fermentation of organic matter (waste or energy crops).

Another terrestrial type of biomass with potential for increased use is bamboo. Aquatic biomass (algae and their products) are also expected to gain importance as biofuels because on $1\,m^2$ of sea, more biomass can be grown than on $1\,m^2$ of conventional soil.

Note: Materials burned in waste incineration plants are often included in "renewable fuels".

4.2.4.4 Other Measures Against Climate Change

In this category, the reduction of N_2O and CH_4 emissions can be mentioned as an example, as these, too, are potent greenhouse-active gases. They are not only emitted from combustion and associated processes, but also from land use activities [102–104]. Nitrous oxide emissions are produced in fluidized bed combustion [105]. CH_4 emissions occur mainly from the agriculture, energy and waste sectors [106]. For instance, landfill gas can contain significant amounts of CH_4. Other sources for fugitive emissions are the operations of natural gas exploration, production, storage and distribution.

4.2.5
Carbon Sequestration

Removing the carbon from fossil fuels is another possible route to "greener" combustion. Researchers are currently exploring several routes for carbon removal from combustion processes. These can be subsumed under the term *"carbon sequestration"* or *"carbon capture and storage"* (CCS). CO_2 is a "bad" feedstock for the chemical industry, as it is not reactive and has a low energy content, however, concepts do exist [107] (it is, however, a good supercritical solvent in some processes). Concepts foresee pre-combustion carbon stripping ("decarbonization") [108] and post-combustion carbon removal as CO_2, see for example, [109,110]. Figure 4.35 shows how CO_2 can be removed from a combustion process. Once the CO_2 has been removed, it needs to be purified, see Figure 4.36.

Among the new technologies to enrich CO_2 in the waste gas stream, the two below deserve mention:

- Chemical looping combustion (CLC) [112,113]
- Oxyfuel combustion [114]

Chemical looping combustion is depicted schematically in Figure 4.37. This process typically employs a dual fluidized bed system where a metal oxide serves as the bed material providing the oxygen for combustion in the fuel reactor. The reduced metal is then transferred to the second bed (air reactor)

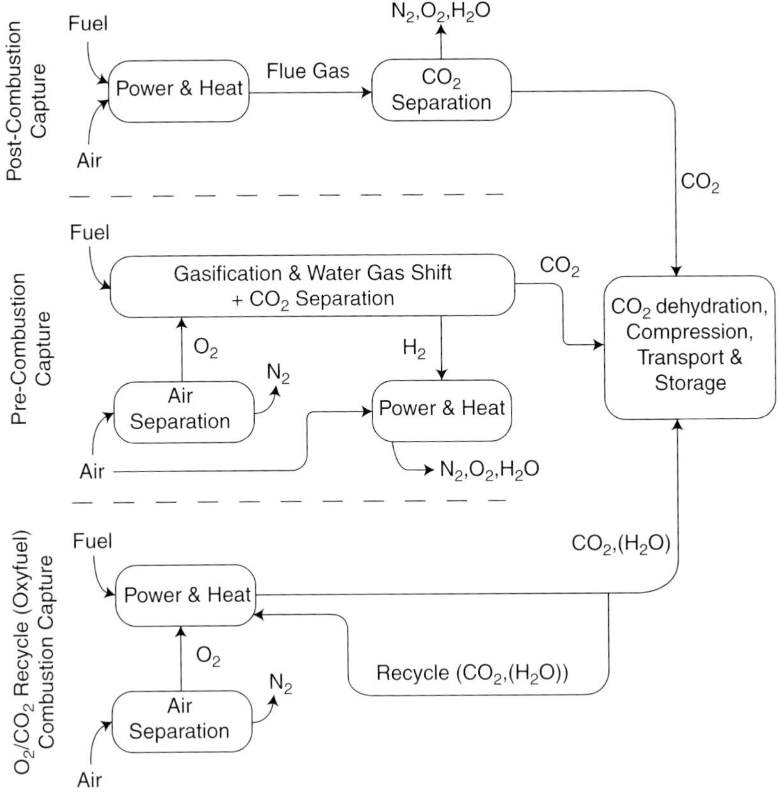

Figure 4.35 Possible, overall plant configurations for the three main categories of carbon capture technologies [111].

and re-oxidized before being sent back to the fuel reactor, completing the loop. Isolation of the fuel from the air avoids the formation of (thermal) nitrogen oxides (NO_x). Also, the flue gas is primarily composed of carbon dioxide and water vapor.

Oxyfuel combustion is shown schematically in Figure 4.38. As the name implies, a fuel is burned in pure oxygen. Since no diluting inert gases (air drags 78% of N_2 into a combustor) are present, higher combustion temperatures, and hence higher efficiencies, can be reached. Similarly to chemical looping combustion, the flue gas stream is highly enriched in CO_2. A common example of oxyfuel combustion is a torch with oxygen and acetylene.

For CO_2 sequestration, several ideas have been generated. The most straightforward one is carbon fixation in biomass, for example, by reforestation. Storage of supercritical CO_2 at the bottom of the ocean, in sea water, saline mines,

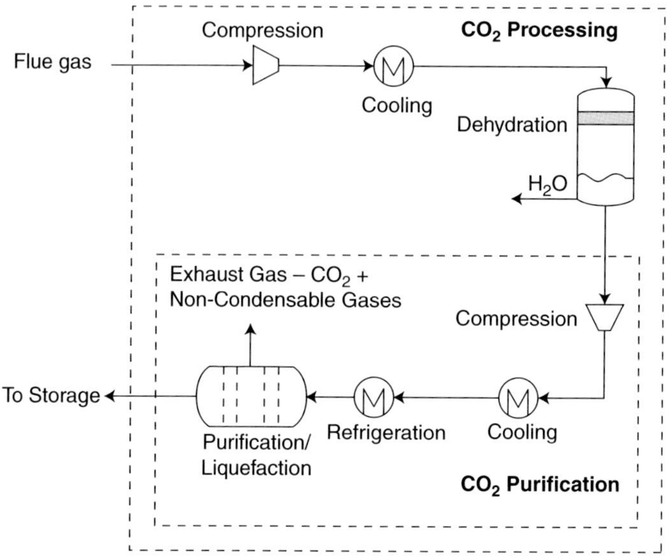

Figure 4.36 Possible process scheme for CO_2 processing and purification, that is, removal of non-condensable gas species, in an oxy-fuel power plant [111].

depleted oil and gas fields and other sinks has also been envisioned, see for example, Figure 4.39.

A recommendable textbook on climate change mitigation is [87].

4.2.6
Web Resources

There are good resources on the internet to learn more about combustion and climate change. A few of them are listed here for the interested reader:

Intergovernmental Panel on Climate Change	http://www.ipcc.ch/
European Commission, Climate Action	http://ec.europa.eu/clima/policies/brief/causes/index_en.htm
Teachers' domain	http://www.teachersdomain.org/resource/envh10.health.lp58a/
United Nations Framework Convention on Climate Change	http://unfccc.int/

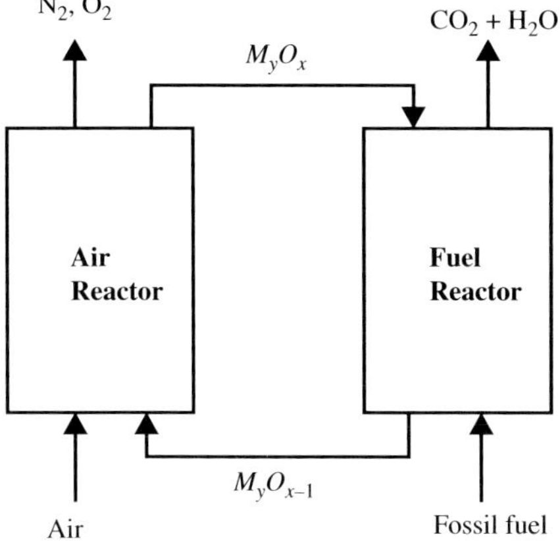

Figure 4.37 Schematic representation of the chemical-looping combustion (CLC) process. Reproduced with permission from [112].

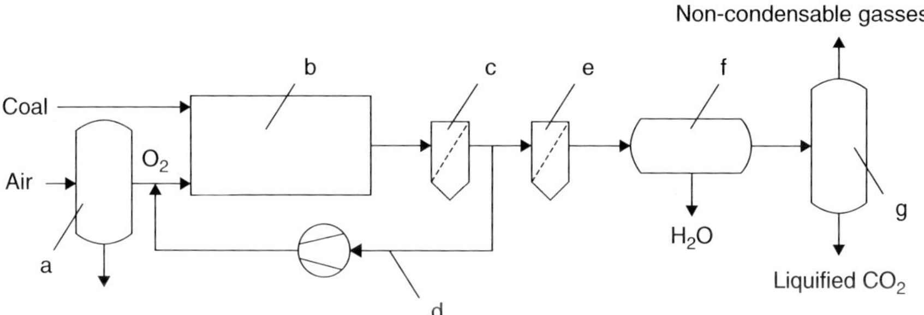

Figure 4.38 Fundamental principle of the oxyfuel combustion process. a: air separation unit; b: furnace and heat transfer sections, that is, boiler; c and e: flue gas cleaning; d: flue gas recycle; f: condenser; g: CO$_2$ compression and purification [23].

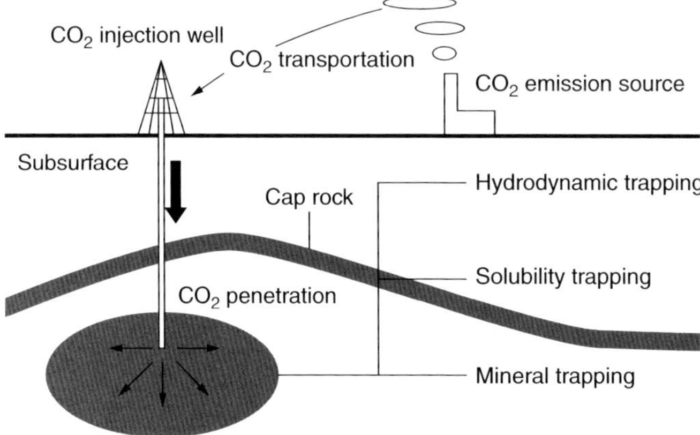

Figure 4.39 Schematic illustration of CO$_2$ sequestration [115].

References

1 Vovelle, C. (2008) *Pollutants from Combustion Formation and Impact on Atmospheric Chemistry*, Springer, reprint, ISBN: 978-0792361350.

2 Beér, J.M. (2000) Combustion technology developments in power generation in response to environmental challenges. *Progress in Energy and Combustion Science*, **26** (4–6), 301–327.

3 Daniels, S.L. (1989) Products of incomplete combustion (O_x, CO_x, HO_x, NO_x, SO_xRO_x, MO_x andPO_x). *Journal of Hazardous Materials*, **22** (2), 161–173.

4 Abdelaal, M.M. and Hegab, A.H. (2012) Combustion and emission characteristics of a natural gas-fueled diesel engine with EGR. *Energy Conversion and Management*, **64**, 301–312.

5 Guo, H., Zou, S.C., Tsai, W.Y., Chan, L.Y., and Blake, D.R. (2011) Emission characteristics of nonmethane hydrocarbons from private cars and taxis at different driving speeds in Hong Kong. *Atmospheric Environment*, **45** (16), 2711–2721.

6 Payri, F., Bermúdez, V.R., Tormos, B., and Linares, W.G. (2009) Hydrocarbon emissions speciation in diesel and biodiesel exhausts. *Atmospheric Environment*, **43** (6), 1273–1279.

7 Alkidas, A.C. (1999) Combustion-chamber crevices: the major source of engine-out hydrocarbon emissions under fully warmed conditions. *Progress in Energy and Combustion Science*, **25** (3), 253–273.

8 Tikuisis, P., Kane, D.M., McLellan, T.M., Buick, F., and Fairburn, S.M. (1992) Rate of formation of carboxyhemoglobin in exercising humans exposed to carbon monoxide. *Journal of Applied Physiology*, **72** (4), 1311–1319.

9 Omaye, S.T. (2002) Metabolic modulation of carbon monoxide toxicity. *Toxicology*, **180** (2), 139–150. doi: 10.1016/S0300-483X(02)00387-6, PMID 12324190.

10 Green, W. An Introduction to Indoor Air Quality: Carbon Monoxide (CO), United States Environmental Protection Agency, Retrieved 16 December 2008.

11 Toftegaard, M.B., Brix, J., Jensen, P.A., Glarborg, P., and Jensen, A.D. (2010) Oxy-fuel combustion of solid fuels. *Progress in Energy and Combustion Science*, **36** (5), 581–625.

12 Tsupari, E., Monni, S., Tormonen, K., Pellikka, T., and Syri, S. (2007) Estimation of annual CH_4 and N_2O emissions from fluidised bed combustion: An advanced measurement-based method and its application to Finland. *International Journal of Greenhouse Gas Control*, **1** (3), 289–297.

13 Schrod, M., Semel, J., and Steiner, R. (1985) Verfahren zur minderung von NOx-emissionen in rauchgasen, *Chemie Ingenieur Technik*, **57** (9), 717–727.

14 Smoot, L.D., Hill, S.C., and Xu, H. (1998) NOx control through reburning. *Progress in Energy and Combustion Science*, **24** (5), 385–408.

15 Kent Hoekman, S. and Robbins, C. (2012) Review of the effects of biodiesel on NOx emissions. *Fuel Processing Technology*, **96**, 237–249.

16 Kremer, H. and Schulz, W. (1988) Influence of temperature on the formation of NOx during pulverized coal combustion. *Symposium (International) on Combustion*, **21** (1), 1217–1222.

17 Beér, J.M. (1994) Minimizing NOx emissions from stationary combustion; reaction engineering methodology. *Chemical Engineering Science*, **49** (24), Part A 4067–4083.

18 Chiari, L. and Zecca, A. (2011) Constraints of fossil fuels depletion on global warming projections. *Energy Policy*, **39** (9), 5026–5034.

19 Williams, A., Jones, J.M., Ma, L., and Pourkashanian, M. (2012) Pollutants from the combustion of solid biomass fuels. *Progress in Energy and Combustion Science*, **38** (2), 113–137.

20 Stanger, R. and Wall, T. (2011) Sulphur impacts during pulverised coal combustion in oxy-fuel technology for carbon capture and storage. *Progress in Energy and Combustion Science*, **37** (1), 69–88.

21 C&EN, Chemical & Engineering News (July 8 2010) U.K. Airborne Dioxin Levels Plummeted in the 1990s, doi: 10.1021/cen070710083016.

22 Stanmore, B.R. (2004) The formation of dioxins in combustion systems. *Combustion and Flame*, **136** (3), 398–427.

23 Lavric, E.D., Konnov, A.A., and De Ruyck, J. (2004) Dioxin levels in wood combustion—a review. *Biomass and Bioenergy*, **26** (2), 115–145.

24 Tame, N.W., Dlugogorski, B.Z., and Kennedy, E.M. (2007) Formation of dioxins and furans during combustion of treated wood. *Progress in Energy and Combustion Science*, **33** (4), 384–408.

25 Kulkarni, P.S., Crespo, J.G., and Afonso, C.A.M. (2008) Dioxins sources and current remediation technologies — A review. *Environment International*, **34** (1), 139–153.

26 Yao, Q., Li, S.-Q., Xu, H.-W., Zhuo, J.-K., and Song, Q. (2009) Studies on formation and control of combustion particulate matter in China: A review. *Energy*, **34** (9), 1296–1309.

27 Ninomiya, Y., Zhang, L., Sato, A., and Dong, Z. (2004) Influence of coal particle size on particulate matter emission and its chemical species produced during coal combustion. *Fuel Processing Technology*, **85** (8–10), 1065–1088.

28 Villeneuve, J., Palacios, J.H., Savoie, P., and Godbout, S. (2012) A critical review of emission standards and regulations regarding biomass combustion in small scale units (<3 MW). *Bioresource Technology*, **111**, 1–11.

29 Ghafghazi, S., Sowlati, T., Sokhansanj, S., Bi, X., and Melin, S. (2011) Particulate matter emissions from combustion of wood in district heating applications. *Renewable and Sustainable Energy Reviews*, **15** (6), 3019–3028.

30 US EPA (Environmental Protection Agency) (2013) www.epa.gov (accessed May 30, 2013).

31 Schwartz, J. (1999) Air pollution and hospital admissions for heart disease in eight US countries. *Epidemiology*, **10** (1), 17–22.

32 Li, N., Hao, M., Phalen, R.F., Hinds, W.C., and Nel, A.E. (2003) Particle air pollutants and asthma – A paradigm for the role of oxidative stress in PM-induced adverse health effects. *Clinical Immunology*, **109**, 250–265.

33 Li, N., Xia, T., and Nel, A.E. (2008) The role of oxidative stress in ambient particulate matter-induced diseases and its implications in the toxicity of engineered nanoparticles. *Free Radical Biology and Medicine*, **44**, 1689–1699.

34 Behrendta, H., Beckerb, W.M., Friedrichsa, K.H., Darsowa, U., and Tomingasa, R. (1992) Interaction between aeroallergens and airborne particulate matter. *International Archives of Allergy and Immunology*, **99**, 425–428.

35 Kim, Y.J., Kim, K.W., Kim, S.D., Lee, B.K., and Han, J.S. (2006) Fine particulate matter characteristics and its impact on visibility impairment at two urban sites in Korea: Seoul and Incheon. *Atmospheric Environment*, **40**, S593–S605.

36 Jacobson, M.Z. Control of fossil-fuel particulate black carbon and organic matter, possibly the most effective method of slowing global warming. *Journal of Geophysical Research*, **107** (D19), 16-1–16-22.

37 NASA (2013) www.nasa.gov (accessed May 30, 2013).

38 European Environment Agency (2013) www.eea.europa.eu (accessed May 30, 2013).

39 Environment Canada (2013) www.ec.gc.ca (accessed May 30, 2013).

40 Williams, A., Jones, J.M., Ma, L., and Pourkashanian, M. (2012) Pollutants from the combustion of solid biomass fuels. *Progress in Energy and Combustion Science*, **38** (2), 113–137.

41 Yan, S., Eddings, E.G., Palotás, Á.B., Pugmire, R.J., and Sarofim, A.F. (2005) Prediction of sooting tendency for hydrocarbon liquids in diffusion flames. *Energy and Fuels*, **19** (6), 2408–2415.

42 Haynes, B.S. and Wagner, H.G. (1981) Soot formation. *Progress in Energy and Combustion Science*, **7** (4), 229–273.

43 Kennedy, I.M. (1997) Models of soot formation and oxidation. *Progress in Energy and Combustion Science*, **23**, 95–132.

44 Harris, S.J. and Kennedy, I.M. (1988) The coagulation of soot particles with van der

Waals forces. *Combustion Science and Technology*, **59**, 443–454.

45 Zlokarnik, M. (2006) *Scale-up in Chemical Engineering*, 2nd edn, Wiley-VCH Verlag GmbH & Co. KGaA, ISBN: 978-3527314218.

46 Frenklach, M. and Wang, H. (1991) Detailed modeling of soot particle nucleation and growth. *Symposium (International) on Combustion*, **23** (1), 1559–1566.

47 Howard, J.B. (1991) Carbon addition and oxidation reactions in heterogeneous combustion and soot formation. *Symposium (International) on Combustion*, **23** (1), 1107–1127.

48 Nagle, J. and Strickland-Constable, R.F. (1962) Oxidation of carbon between 1000–2000C in *Proceedings of the Fifth Conference on Carbon*, Pergamon Press, pp. 154–164.

49 Fenimore, C.P. and Jones, G.W. (1967) Oxidation of soot by hydroxyl radicals. *Journal of Physical Chemistry*, **71**, 593–597.

50 Neoh, K.H., Howard, J.B., and Sarofim, A. F. (1981) *Particulate Carbon: Formation during Combustion*, (G.W.S.D.C. Siegla ed.): Plenum, New York, pp. 261–277.

51 Quinn Brewster, M. (1992) *Thermal Radiative Transfer and Properties*, John Wiley & Sons Inc., New York.

52 Lowes, T.M. and Newall, A.J. (1971) The emissivities of flame soot dispersions. *Combustion and Flame*, **16** (2), 191–194.

53 Marsh, N.D., Preciado, I., Eddings, E.G., Sarofim, A.F., Palotas, A.B., and Robertson, J.D. (2007) Evaluation of organometallic fuel additives for soot suppression. *Combustion Science and Technology*, **179** (5), 987–1001.

54 Palotás, Á.B., Rainey, L.C., Sarofim, A.F., and Vander Sande, J.B. (1996) Soot morphology: an application of image analysis in high–resolution transmission electron microscopy. *Microscopy Research and Technique*, **33**, 266–278.

55 Palotás, Á.B., Rainey, L.C., Sarofim, A.F., Vander Sande, J.B., and Ciambelli, P. (1996) Effect of oxidation on the microstructure of carbon blacks. *Energy & Fuels*, **10**, 254–259.

56 Rainey, L.C., Palotás, Á.B., Sarofim, A.F., and Vander Sande, J.B. (1996) Application of high resolution electron microscopy for the characterization and source assignment of diesel particulates. *Applied Occupational and Environmental Hygiene*, **11** (7), 777–781.

57 Shaddix, C.R., Palotás, Á.B., Megaridis, C.M., Choi, M.Y., and Yang, N.Y.C. (2005) Soot graphitic order in laminar diffusion flames and a large-scale JP-8 pool fire. *International Journal of Heat and Mass Transfer*, **48** (17), 3604–3614.

58 Tóth, P., Palotás, Á.B., Lighty, J., and Echavarria, C.A. (2012) Quantitative differentiation of poorly ordered soot nanostructures: a semi-empirical approach. *Fuel*, **99**, 1–6.

59 Misra, M.K., Ragland, K.W., and Baker, A. J. (1993) Wood ash composition as a function of furnace temperature. *Biomass and Bioenergy*, **4** (2), 103.

60 Demir, I., Hughes, R.E., and DeMaris, P.J. (2001) Formation and use of coal combustion residues from three types of power plants burning Illinois coals. *Fuel*, **80** (11), 1659–1673.

61 Reijnders, L. (2005) Disposal, uses and treatments of combustion ashes: a review. *Resources, Conservation and Recycling*, **43** (3), 313–336.

62 Insam, H. and Knapp, B.A. (2011) *Recycling of Biomass Ashes*, 1st edn, Springer Berlin Heidelberg, ISBN: 978-3642193538.

63 Jenkins, B.M., Baxter, L.L., Miles, T.R. Jr., and Miles, T.R. (1998) Combustion properties of biomass. *Fuel Processing Technology*, **54** (1–3), 17–46.

64 Escobar, I., Oleschko, H., Wolf, K.-J., and Müller, M. (2008) Alkali removal from hot flue gas by solid sorbents in pressurized pulverized coal combustion. *Powder Technology*, **180** (1–2), 51–56.

65 Khan, A.A., de Jong, W., Jansens, P.J., and Spliethoff, H. (2009) Biomass combustion in fluidized bed boilers: Potential problems and remedies. *Fuel Processing Technology*, **90** (1), 21–50.

66 Bryers, R.W. (1996) Fireside slagging, fouling, and high-temperature corrosion of heat-transfer surface due to impurities in steam-raising fuels. *Progress in Energy and Combustion Science*, **22** (1), 29–120.

67 Donatello, S., Fernández-Jiménez, A., and Palomo, A. (2012) An assessment of Mercury immobilisation in alkali activated flyash (AAFA) cements. *Journal of Hazardous Materials*, **213–214** (30), 207–215.

68 Kolker, A., Senior, C.L., and Quick, J.C. (2006) Mercury in coal and the impact of coal quality on mercury emissions from combustion systems. *Applied Geochemistry*, **21** (11), 1821–1836.

69 Valavanidis, A., Iliopoulos, N., Gotsis, G., and Fiotakis, K. (2008) Persistent free radicals, heavy metals and PAHs generated in particulate soot emissions and residue ash from controlled combustion of common types of plastic. *Journal of Hazardous Materials*, **156** (1–3), 277–284.

70 Xu, M., Yan, R., Zheng, C., Qiao, Y., Han, J., and Sheng, C. (2004) Status of trace element emission in a coal combustion process: a review. *Fuel Processing Technology*, **85** (2–3), 215–237.

71 Yoo, J.-I., Kim, K.-H., Jang, H.-N., Seo, Y.-C., Seok, K.-S., Hong, J.-H., and Jang, M. (2002) Emission characteristics of particulate matter and heavy metals from small incinerators and boilers. *Atmospheric Environment*, **36** (32), 5057–5066.

72 Pulles, T., van der Gon, H.D., Appelman, W., and Verheul, M. (2012) Emission factors for heavy metals from diesel and petrol used in European vehicles. *Atmospheric Environment*, **61**, 641–651.

73 Shendrikar, A.D. and Ensor, D.S. (1986) Critical review: measurement of mercury combustion aerosols in emissions from stationary sources. *Waste Management & Research*, **4** (1), 75–93.

74 Alhamed, Y.A. and Bamufleh, H.S. (2009) Sulfur removal from model diesel fuel using granular activated carbon from dates' stones activated by $ZnCl_2$. *Fuel*, **88** (1), 87–94.

75 Nguyen, T.D.B., Lim, Y.-I., Eom, W.-H., Kim, S.-J., and Yoo, K.-S. (2010) Experiment and CFD simulation of hybrid SNCR–SCR using urea solution in a pilot-scale reactor. *Computers & Chemical Engineering*, **34** (10), 1580–1589.

76 Gohlke, O., Weber, T., Seguin, P., and Labore, Y. (2010) A new process for NO_x reduction in combustion systems for the generation of energy from waste. *Waste Management*, **30** (7), 1348–1354.

77 Purgar, A. and Winter, F. (2012) Concepts of emission reduction in fluidized bed combustion of biomass. Energetika & Biomass (E&B 2012), Czech Technical University, Faculty of Mechanical Engineering, Prague, April 25–27.

78 Anthony, E.J., Bulewicz, E.M., and Jia, L. (2007) Reactivation of limestone sorbents in FBC for SO_2 capture. *Progress in Energy and Combustion Science*, **33** (2), 171–210.

79 Chen, J.-C., Wey, M.-Y., and Ou, W.-Y. (1999) Capture of heavy metals by sorbents in incineration flue gas. *Science of The Total Environment*, **228** (1), 67–77.

80 Andersson, S., Kreisz, S., and Hunsinger, H. (2005) Dioxin removal: Adiox for wet scrubbers and dry absorbers. *Filtration & Separation*, **42** (10), 22–25.

81 Hochgreb, S. (1998) *Combustion-Related Emissions in SI Engines, in Handbook of Air Pollution From Internal Combustion Engines*, (ed. E. Sher), Elsevier, pp. 118–170.

82 Henein, N.A. (1976) Analysis of pollutant formation and control and fuel economy in diesel engines. *Progress in Energy and Combustion Science*, **1** (4), 165–207.

83 Wulff, A. and Hourmouziadis, J. (1997) Technology review of aeroengine pollutant emissions. *Aerospace Science and Technology*, **1** (8), 557–572.

84 Chen, W.Y. Seiner, J. Suzuki, T., and Lackner, M. (eds) (2012) *Handbook of Climate Change Mitigation*, Springer, ISBN: 978-1441979902.

85 Lewtas, J. (2007) Air pollution combustion emissions: Characterization of causative agents and mechanisms associated with cancer, reproductive, and cardiovascular effects. *Mutation Research/Reviews in Mutation Research*, **636** (1–3), 95–133.

86 den Elzen, M., Fuglestvedt, J., Höhne, N., Trudinger, C., Lowe, J., Matthews, B., Romstad, B., de Campos, C.P., and Andronova, N. (2005) Analysing countries' contribution to climate change: scientific and policy-related choices. *Environmental Science & Policy*, **8** (6), 614–636.

87 Chen, W.Y. Seiner, J. Suzuki, T., and Lackner, M. (eds) (2012) *Handbook of Climate Change Mitigation*, Springer, ISBN: 978-1441979902.

88 Wuebbles, D.J. and Jain, A.K. (2001) Concerns about climate change and the role of fossil fuel use. *Fuel Processing Technology*, **71** (1–3), 99–119.

89 http://ec.europa.eu/clima/policies/brief/causes/index_en.htm (accessed May 30, 2013).

90 Ghoniem, A.F. (2011) Needs, resources and climate change: Clean and efficient conversion technologies. *Progress in Energy and Combustion Science*, **37** (1), 15–51.

91 International Energy Association (IEA) (2012) http://www.iea.org (accessed May 30, 2013)

92 http://www.worldenergyoutlook.org/pressmedia/recentpresentations/PresentationWEO2012launch.pdf (accessed May 30, 2013).

93 Zevenhoven, R. and Beyene, A. (2011) The relative contribution of waste heat from power plants to global warming. *Energy*, **36** (6), 3754–3762.

94 Akorede, M.F., Hizam, H., AbKadir, M.Z.A., Aris, I., and Buba, S.D. (2012) Mitigating the anthropogenic global warming in the electric power industry. *Renewable and Sustainable Energy Reviews*, **16** (5), 2747–2761.

95 Wade, J., Holman, C., and Fergusson, M. (1994) Passenger car global warming potential: Current and projected levels in the UK. *Energy Policy*, **22** (6), 509–522.

96 http://www.epa.gov/climatechange/ghgemissions/sources.html (accessed May 30, 2013).

97 Keepin, B. and Kats, G. (1988) Greenhouse warming: Comparative analysis of nuclear and efficiency abatement strategies. *Energy Policy*, **16** (6), 538–561.

98 Kats, G.H. (1990) Slowing global warming and sustaining development: The promise of energy efficiency. *Energy Policy*, **18** (1), 25–33.

99 Zheng, Y.H., Li, Z.F., Feng, S.F., Lucas, M., Wu, G.L., Li, Y., Li, C.H., and Jiang, G.M. (2010) Biomass energy utilization in rural areas may contribute to alleviating energy crisis and global warming: A case study in a typical agro-village of Shandong, China. *Renewable and Sustainable Energy Reviews*, **14** (9), 3132–3139.

100 Chiari, L. and Zecca, A. (2011) Constraints of fossil fuels depletion on global warming projections. *Energy Policy*, **39** (9), 5026–5034.

101 Agarwal, A.K. (2007) Biofuels (alcohols and biodiesel) applications as fuels for internal combustion engines. *Progress in Energy and Combustion Science*, **33** (3), 233–271.

102 Nol, L., Verburg, P.H., and Eddy, J. (2012) Trends in future N_2O emissions due to land use change. *Journal of Environmental Management*, **94** (1), 78–90.

103 Kammann, C., Müller, C., Grünhage, L., and Jäger, H.-J. (2008) Elevated CO_2 stimulates N_2O emissions in permanent grassland. *Soil Biology and Biochemistry*, **40** (9), 2194–2205.

104 van Beek, C.L., Meerburg, B.G., Schils, R.L.M., Verhagen, J., and Kuikman, P.J. (2010) Feeding the world's increasing population while limiting climate change impacts: linking N_2O and CH_4 emissions from agriculture to population growth. *Environmental Science & Policy*, **13** (2), 89–96.

105 Tsupari, E., Monni, S., Tormonen, K., Pellikka, T., and Syri, S. (2007) Estimation of annual CH_4 and N_2O emissions from fluidised bed combustion: An advanced measurement-based method and its application to Finland. *International Journal of Greenhouse Gas Control*, **1** (3), 289–297.

106 Yusuf, R.O., Noor, Z.Z., Abba, A.H., Abu Hassan, M.A., and Din, M.F.M. (2012) Methane emission by sectors: A comprehensive review of emission sources and mitigation methods. *Renewable and Sustainable Energy Reviews*, **16** (7), 5059–5070.

107 Aresta, M. and Dibenedetto, A. (2007) Utilisation of CO_2 as a chemical feedstock: opportunities and challenges. *Dalton Transactions*, 2975–2992.

108 bbasi, T. and Abbasi, S.A. (2011) Decarbonization of fossil fuels as a strategy to control global warming.

Renewable and Sustainable Energy Reviews, **15** (4), 1828–1834.

109 Peng, Y., Zhao, B., and Li, L. (2012) Advance in post-combustion CO_2 capture with alkaline solution: a brief review. *Energy Procedia*, **14**, 1515–1522.

110 Lee, Z.H., Lee, K.T., Bhatia, S., and Mohamed, A.R. (2012) Post-combustion carbon dioxide capture: Evolution towards utilization of nanomaterials. *Renewable and Sustainable Energy Reviews*, **16** (5), 2599–2609.

111 Toftegaard, M.B., Brix, J., Jensen, P.A., Glarborg, P., and Jensen, A.D. (2010) Oxy-fuel combustion of solid fuels. *Progress in Energy and Combustion Science*, **36** (5), 581–625.

112 Hossain, M.M. and de Lasa, H.I. (2008) Chemical-looping combustion (CLC) for inherent CO_2 separations—a review. *Chemical Engineering Science*, **63** (18), 4433–4451.

113 Wall, T., Liu, Y., Spero, C., Elliott, L., Khare, S., Rathnam, R., Zeenathal, F., Moghtaderi, B., Buhre, B., Sheng, C., Gupta, R., Yamada, T., Makino, K., and Yu, J. (2009) An overview on oxyfuel coal combustion—State of the art research and technology development. *Chemical Engineering Research and Design*, **87** (8), 1003–1016.

114 Hadjipaschalis, I., Kourtis, G., and Poullikkas, A. (2009) Assessment of oxyfuel power generation technologies. *Renewable and Sustainable Energy Reviews*, **13** (9), 2637–2644.

115 Sasaki, K., Fujii, T., Niibori, Y., Ito, T., and Hashid, T. (2008) Numerical simulation of supercritical CO_2 injection into subsurface rock masses. *Energy Conversion and Management*, **49** (1), 54–61.

5
Measurement Methods

5.1
Introduction

Combustion systems are characterized by complex interactions of flow and transport processes with a large number of elementary chemical reactions. In order to gain a better insight into combustion processes, critical parameters need to be measured. Flames are challenging environments for experimentalists since they are characterized by one or several of the following conditions:

- High pressure
- High temperature
- Multiple phases
- Unstable species
- Transient processes

For instance, there might be fuel droplets, soot and ash particles blocking probing light beams, or thermal background radiation saturating a detector.

Typical time scales and length scales to resolve combustion phenomena are listed in Table 5.1 [1].

Parameters of interest are mostly temperature, pressure, flow, particle size, and species concentrations. Speciation, that is, the determination of the valence state of metal species (ions), can be an additional piece of information to be sought [2]. Measurements can be done on open flames, in a special laboratory apparatus, a pilot plant, or a "real" combustor that might be equipped with dedicated optical access ports. In broad terms, one can distinguish between *in situ* and *ex situ* techniques. Optical methods are particularly advantageous, because they are not invasive. To choose the right technique for the measurement job at hand, these criteria have to be considered:

- Qualitative versus quantitative (relative or absolute)
- Integrating versus spatially resolved
- 1D versus 2D and 3D
- Time resolution

Table 5.1 Temporal and spatial resolution desirable for combustion research.

Area	Time scale [μs]	Length scale [m]
Laminar flames	$10^5 - 10^6$	$10^{-4} - 10^{-2}$
Turbulent flames	$10^{-2} - 10^2$	$10^{-5} - 10^{-2}$
Fire research	$10^2 - 10^3$	$10^{-5} - 10^{-2}$
Aircraft turbine	$10^3 - 10^4$	$10^{-4} - 10^{-1}$
Gas turbine	$10^{-1} - 10^1$	$10^{-5} - 10^{-4}$

- Experimental (and financial) effort
- One-time lab measurement versus continuous monitoring of a process in the field

An example of a qualitative method is P-LIF (planar laser-induced fluorescence), which provides spatially resolved 2D images. On the other hand, tunable diode laser absorption spectroscopy delivers an averaged signal over the line-of-sight, which is quantitative. Schlieren photography is an example of a very cost-effective technique, while DFWM (degenerate four wave mixing) needs costly equipment.

Modeling and experiments are closely intertwined. Any model needs experimental verification, and vice versa. This means that future progress in (computer) simulations will not render combustion-measurements obsolete [3].

Optical measurement methods play an important role in combustion diagnostics, because they can be used to obtain results *in situ*, that is, right at the spot. *In situ* measurements provide the following advantages:

- Timely result, fast response time
- Less likelihood of measurement mistakes, as no sample needs to be manipulated
- Less perturbation of the probed system (see Figure 5.1)

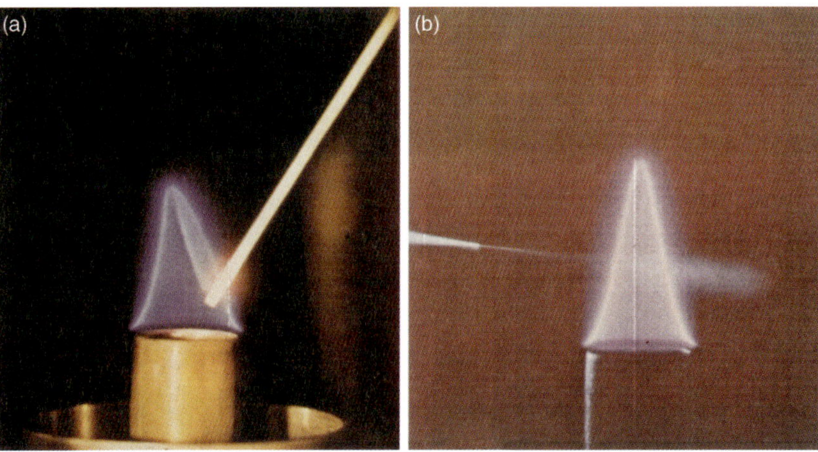

Figure 5.1 Perturbation by probes versus *in situ* measurements. Courtesy of Prof. Dr. Alfred Leipertz.

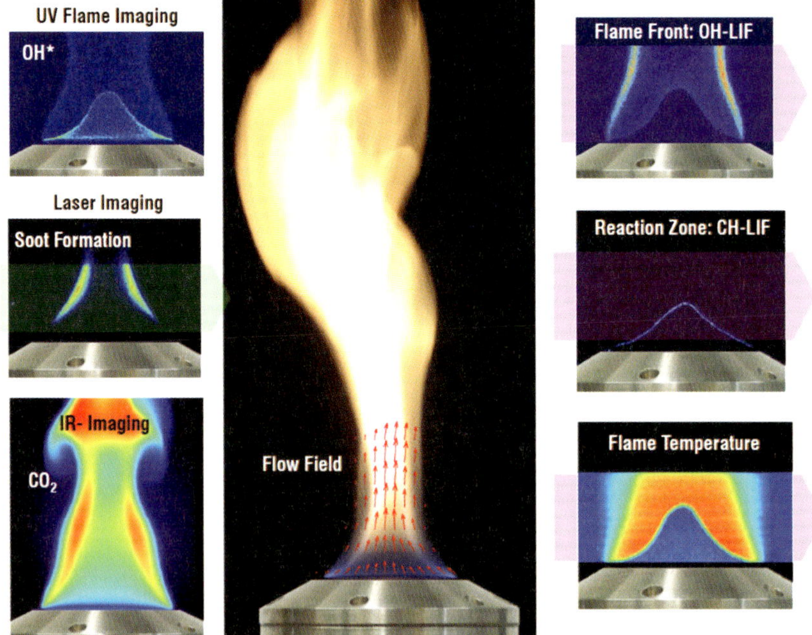

Figure 5.2 Flame characterization by optical techniques (schematic) [4].

In Figure 5.1a, one can see a magnesia stick placed in a flame, whereas in Figure 5.1b, a laser beam passes through a similar flame. The magnesia stick leads to strong changes in the shape and properties of the flame in its vicinity, causing experimental measurement errors.

Today, a wide array of optical techniques is available at the disposal of researchers for combustion measurements. For many of them there are commercial instruments available which are easy to use, see Figure 5.2 for a schematic illustration.

This chapter briefly introduces the most commonly used techniques.

5.2
In Situ **versus** *Ex Situ* **Measurements**

As stated above, a measurement can be done "at the spot", that is, directly in a flame, or outside under less harsh conditions after a suitable sample has been drawn and possibly conditioned. Figure 5.3 shows the evolution of CO from a burning wood particle during its two characteristic stages, devolatilization and char combustion [5]. As can be seen, the *in situ* and *ex situ* measurements give completely different results. Whereas the *in situ* technique shows a higher CO concentration during the first phase, devolatilization, the *ex situ* technique underpredicts the CO in this step, whereas it yields significantly higher readings for CO during char combustion, the second phase. Note the different scales (vol% vs. ppm). The explanation for the faulty *ex situ* measurement is that the CO reacts to CO_2 before it reaches the

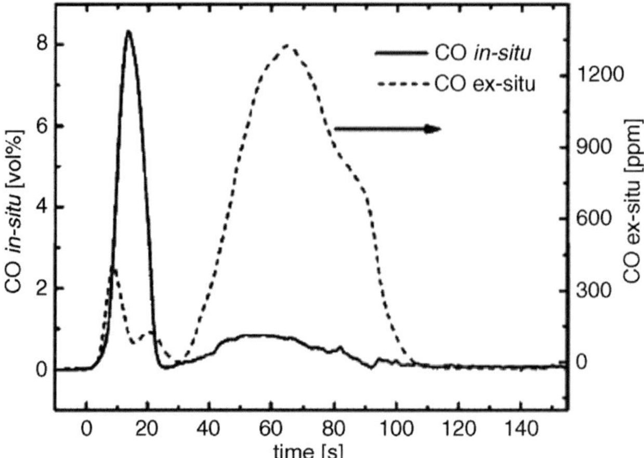

Figure 5.3 Comparison of the CO concentration histories measured *in situ*, that is, directly over the surface of the particle studied, and *ex situ* by an NDIR-spectrometer coupled to the exhaust gas duct. One can see marked differences. See text for details [5].

sampling point and analyzer. So, whenever bulky analyzers that rely on drawn samples can be avoided, the measurement results will be closer to reality.

In diagnostics, one has to distinguish between spectroscopy and spectrometry. *Spectroscopy* studies the interaction of matter with light, whereas *spectrometry* derives quantitative data from a spectrum. Hence, one can say that spectroscopy is a theoretical approach or technique, and spectrometry the practical application that gives results. Sometimes, the terms are used interchangeably, though.

5.3
Fuel Characterization

Techniques to characterize fuel samples are classic analytical ones and particularly developed ones, where the respective norms and standards offer detailed information (e.g., on how to measure the moisture content, the pour point and cloud point of a certain oil). Billions of tons of coal and other fuels are mined annually, and most of them have to be characterized prior to use. In this chapter, a selection was made to present common techniques for fuel characterization.

5.3.1
Proximate and Ultimate Analysis

According to the standard ASTM[1] D121 and D3172, proximate analysis separates the constituents of a solid fuel sample into four groups:

---

1) ASTM International (American Society for Testing and Materials International) is an international standards organization that develops and publishes voluntary consensus on technical standards for a wide range of materials, products, systems, and services, comparable to DIN and ISO.

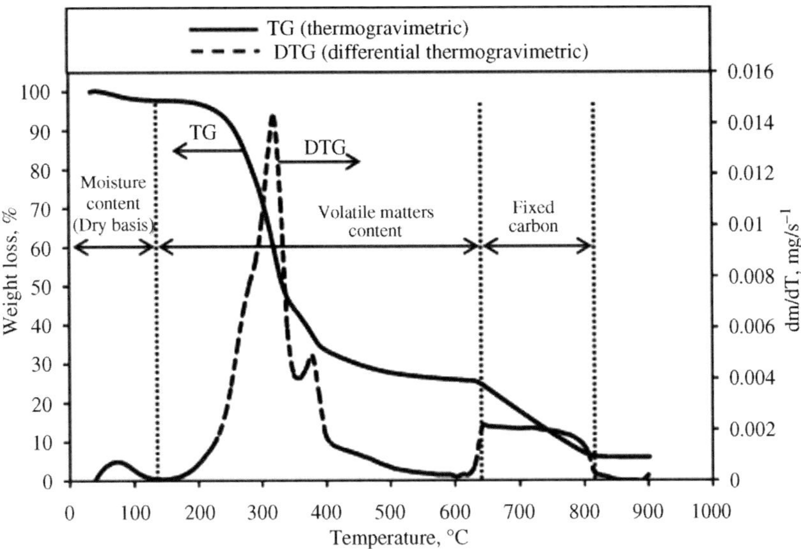

Figure 5.4 Proximate analysis of EFB using thermogravimetry. EFB (empty fruit bunches) are a waste material from the palm oil industry [7].

- moisture
- volatile matter (gases and vapors released during pyrolysis)
- fixed carbon (the nonvolatile fraction of the fuel)
- ash (the inorganic residue that remains after combustion)

As per ASTM standard D3176, ultimate analysis is the determination of a fuel's composition of the elements C, H, N, S and (by difference) O. For proximate and ultimate analysis, the reference is important, for example, "as received" (ar) or "dry ash free" (daf). The results from proximate and ultimate analysis allow a judgment on the quality of a fuel and an estimation of its heating value [6].

Figure 5.4 shows how proximate analysis can be done by thermogravimetry.

5.3.2
Thermal Analysis (TGA/DSC)

In thermal analysis, the properties of samples as a function of temperature are studied. The two most common techniques are TGA (thermogravimetric analysis) and DSC (differential scanning calorimetry). TGA and DSC are widely used to characterize polymers. In combustion science, they can reveal information for a sample on oxidation, thermal decomposition, pyrolysis, combustion kinetics, flammability and melting behavior (also of the ash). In TGA, a small sample of the material under investigation is placed on an arm of a recording microbalance (a so-called thermobalance) inside a special furnace. The furnace temperature is controlled, for example, as a ramp. The weight-change rate is determined. By contrast,

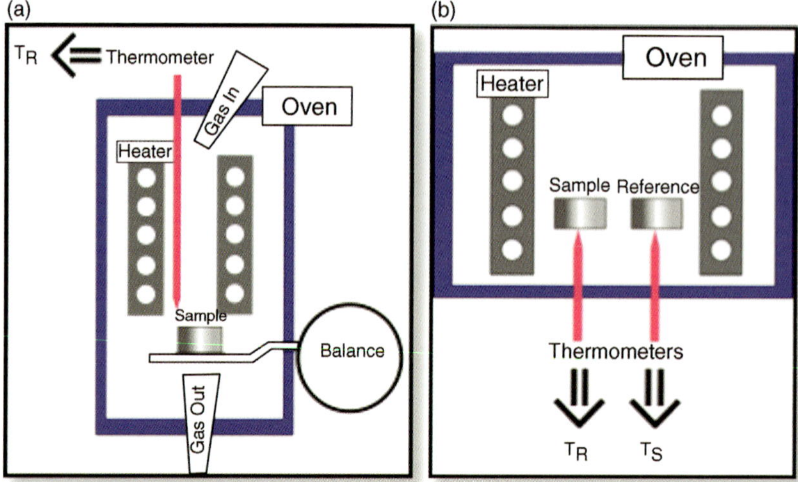

Figure 5.5 Experimental set-up for TGA (a) and DSC (b) measurements. Reproduced with permission from Prof. Dr.Thorsteinn Adalsteinsson, Department of Chemistry & Biochemistry, Santa Clara University.

in DSC, the difference in the amount of heat required to increase the temperature of a sample compared to a reference is measured as a function of temperature. The two techniques are depicted in Figure 5.5.

The simultaneous study of DSC and TGA in one instrument is referred to as simultaneous thermal analysis (STA). A typical TGA/DSC curve is shown in Figure 5.6 [8].

Examples where the technologies are used are pyrolysis and combustion kinetics [9], co-combustion characteristics [10], determination of specific heats of fuels [11], and determination of material flammability [12]. For further details, see [13].

5.3.3
Ash Melting

For solid fuel combustion systems, the properties of the ash are very important. A critical parameter is the melting behavior, both for bottom ash and for fly ash. There exist several methods to characterize ash melting, among them the *ash fusion temperature* (AFT). The literature describes a softening, hemispherical and fluid AFT [14]. The hemispherical AFT is the temperature at which controlled heating under reducing conditions causes sufficient softening to change a pyramid-shaped ash pellet to a hemispherical shaped one. Here is a selection of ash fusibility standards:

- USA: ASTM D 1857, ASTM E 953
- Australia: AS 1038.15-1995

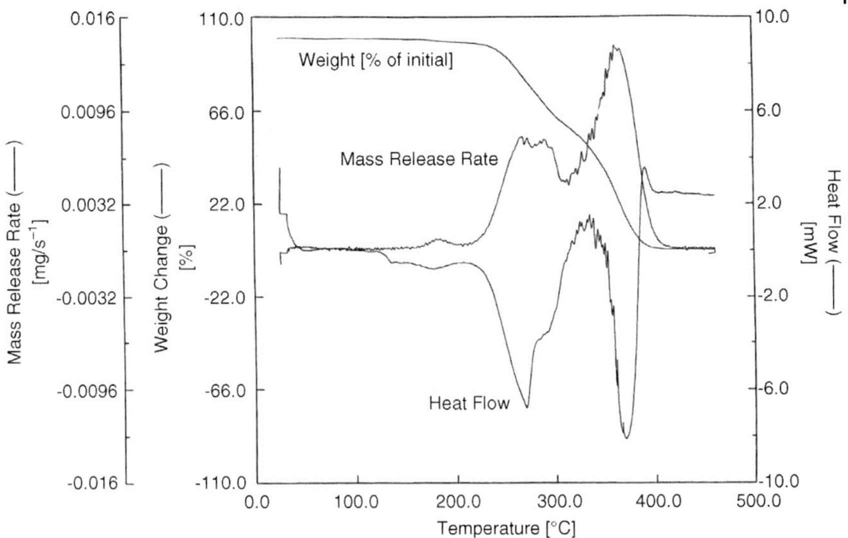

Figure 5.6 TGA/DSC curve of the polymer PMMA. The weight-curve is from TGA, the heat flow from DSC [8].

- Great Britain: BS ISO 540:2008
- Germany: DIN 51730
- Russia: GOST 2057
- France: AFNOR NF M03-048

The measurement of AFT according to the ASTM is shown in Figure 5.7 and according to the DIN standard in Figure 5.8.

Additional equipment to study ash melting is the *heating microscope* and the *Bunte-Baum dual furnace instrument* [15]. For further information on ash characterization, see [16–18].

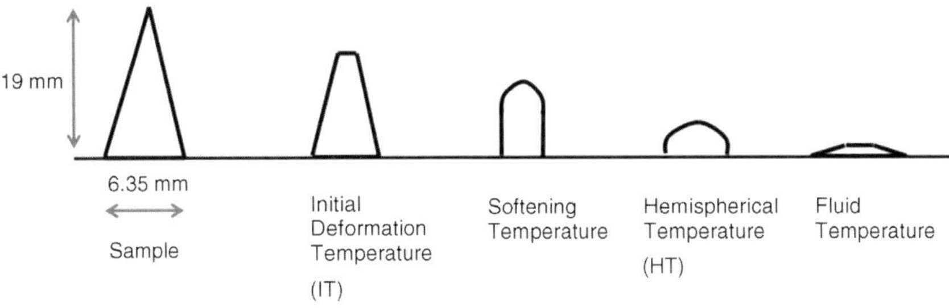

Figure 5.7 A sketch for the ASTM standard, using conical or pyramid-shaped samples for AFT determination.

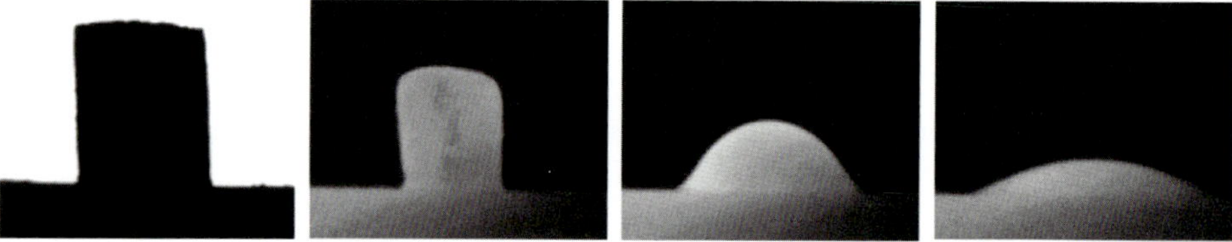

Figure 5.8 A series of photographs to demonstrate the use of cylindrical samples according to the DIN standard for AFT determination. The various critical temperatures are demonstrated. Sample: Coal ash.

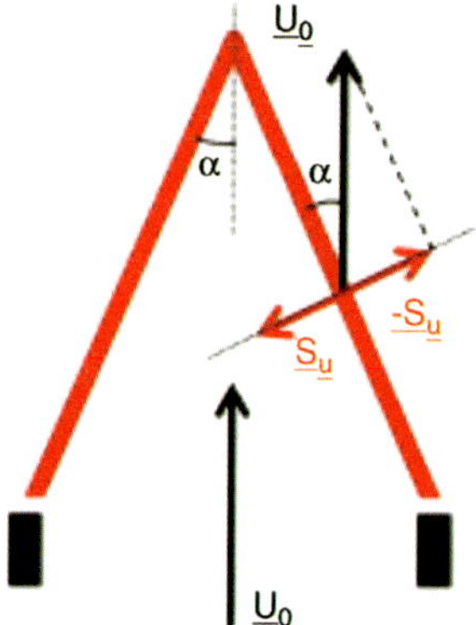

Figure 5.9 Determination of the laminar flame speed in premixed gases on a Bunsen burner [20].

5.3.4
Laminar Flame Speed

The laminar flame speed is, next to the calorific value, one of the most important parameters for gaseous fuels. It can be determined on a Bunsen burner from the flame's cone angle [19], as depicted in Figure 5.9.

The velocity of the unburned mixture U_0 at the nozzle exit can be considered to be uniform, and the laminar flame speed S_u can be expressed as the normal flow velocity component with respect to the flame surface, that is, $S_u = U_0 \sin \alpha$ [21], with α being half of the cone angle of the flame, as shown in Figure 5.9.

Other analytical techniques are used to characterize fuels, too, among them *optical microscopy, electron microscopy* and *X-ray diffraction* [22].

5.4
Investigation of Combustion Processes

In contrast to fuel characterization, which can be performed in a laboratory under ambient conditions without time criticality, the study of a "live" combustion process is significantly more demanding. In this section, non-optical and optical methods will be presented. Again, only the most widespread techniques are covered within the scope of this chapter.

An overview of combustion diagnostic techniques is given in [23–27].

5.4.1
Selection of Non-optical Methods

5.4.1.1 Suction Probe Coupled with GC/MS

A suction probe is a cooled metallic tube which allows the extraction of gas samples. Reactive species are quenched in the tube and transferred to an external analyzer. Figure 5.10 shows a suction tube with its quartz nozzle in a flame.

Commonly, a combination of gas chromatograph/mass spectrometer (GC/MS) is used. The technique can be deployed to measure species concentrations. For hydrocarbons, a GC can be coupled with an FID (flame ionization detector). Two examples from the recent literature where GC/MS was successfully used are the determination of toxic products released in the combustion of pesticides [28] and the creation/verification of chemical kinetic models for the combustion of hydrocarbon fuels [29]. A variant of GC/MS is irm-GC/MS (isotope ratio monitoring).

Sampling errors by suction probes are discussed in [31]. When sampling particles from a flow, isokinetic sampling [32] is mandatory for representative results. An example of a spectrum obtained by Py/GC/MS (pyrolysis gas chromatography mass spectrometry) is given in Figure 5.11. The MS-spectrum shows the relative

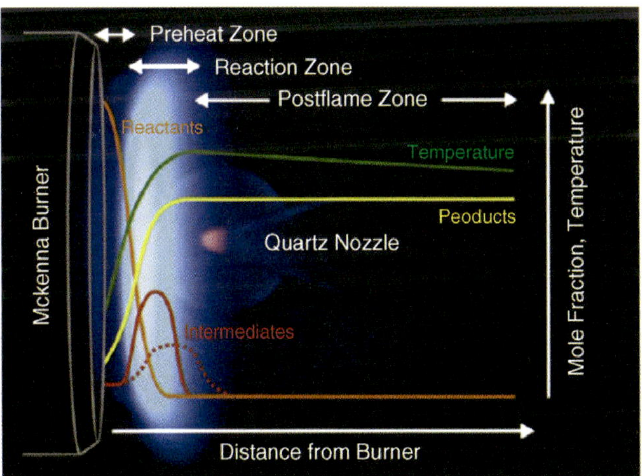

Figure 5.10 Photograph and schematic structure of a premixed, laminar, low-pressure flat flame. Temperature and mole fractions of reactants, products, and intermediate species are given as functions of distance from the burner. In the photograph, a widespread reaction (luminous) flame zone and the quartz nozzle used for molecular-beam sampling are also seen. Reprinted with permission from [30].

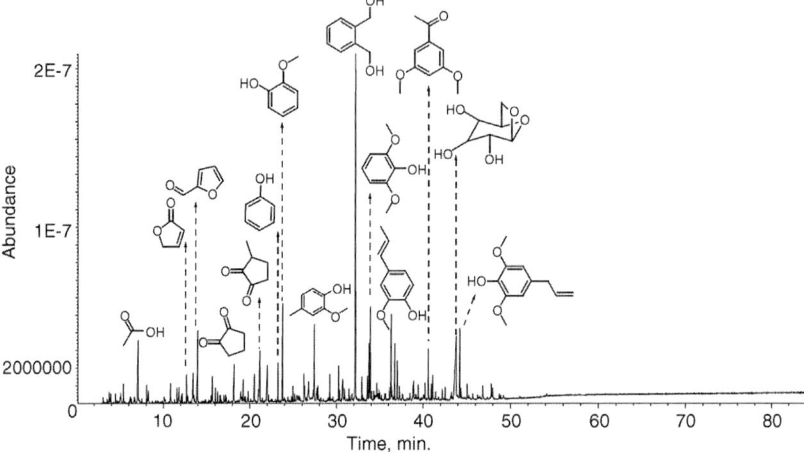

Figure 5.11 Py-GC-MS of miscanthus. Miscanthus is a tall perennial bamboo-like grass used as an energy crop. Reproduced with permission from [33].

abundance of fragments with the same mass/charge ratio, which can be assigned in a second step.

Suction probes can also be used to sample solids, as shown in Figure 5.12.

5.4.1.2 **Hot Wire Anemometry**

The purpose of anemometers is to measure wind speeds as common in weather stations. They consist of a thin wire that can be hot or cold. Hot wire anemometers

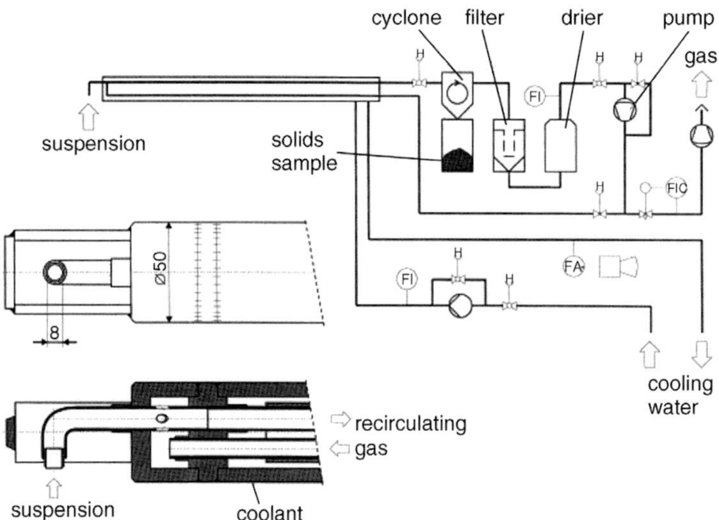

Figure 5.12 Suction probe used in a CFB combustor (CFB = circulating fluidized bed). Reproduced with permission from [34].

use a very fine wire (of the order of several µm, typically made of tungsten) that is electrically heated a few degrees above the ambient air temperature. Any air flow will cool the wire, which results in a change in electrical resistance, so the flow speed can be calculated.

Other techniques used to measure flows are laser Doppler anemometry (LDA) and particle image velocimetry (PIV), see later. For details, see [35–37].

5.4.1.3 Thermocouple

A thermocouple consists of a wire made from 2 different materials (metal alloys). Around the contact point, a temperature-dependent voltage can be measured. Thermocouples are used for temperature sensors in many areas of combustion. Table 5.2 shows the most common types of thermocouples. Apart from giving absolute temperature measurements up to approximately 2300 °C, thermocouples can be used to determine temperature gradients dT/dt. They are cost-effective and easy to use. For details, see [38,39]. For the measurement of small temperature differences with high accuracy, other technologies such as thermistors, silicon bandgap temperature sensors and resistance temperature detectors are available. An optical alternative to thermocouples is a *pyrometer* (radiation thermometry) [40], which can also measure higher temperatures, but generally with less accuracy.

5.4.1.4 Gas Potentiometric Sensors

Gas potentiometry works on the principle of solid electrolytes and can be used to determine species concentrations. One also speaks about ion-selective electrodes (ISE). The lambda probe in the exhaust pipe of automobiles is such an example. It measures the residual oxygen in the tail pipe so that combustion can be carried out under optimum conditions (so-called lambda window, see Figure 5.13).

The classic lambda sensor is made from zirconium dioxide (zirconia) and based on a solid-state electrochemical fuel cell called the Nernst cell. The output voltage corresponds to the quantity of oxygen in the exhaust gas relative to that in the atmosphere (200 mV for "lean" and 800 mV for "rich").

Table 5.2 Thermocouples. Values for continuous operation. Short term, the range is larger.

Type	Temperature range [°C]		
K	0	—	1100
J	0	—	750
N	0	—	110
R	0	—	1600
S	0	—	1600
B	200	—	1700
T	-185	—	300
E	0	—	800

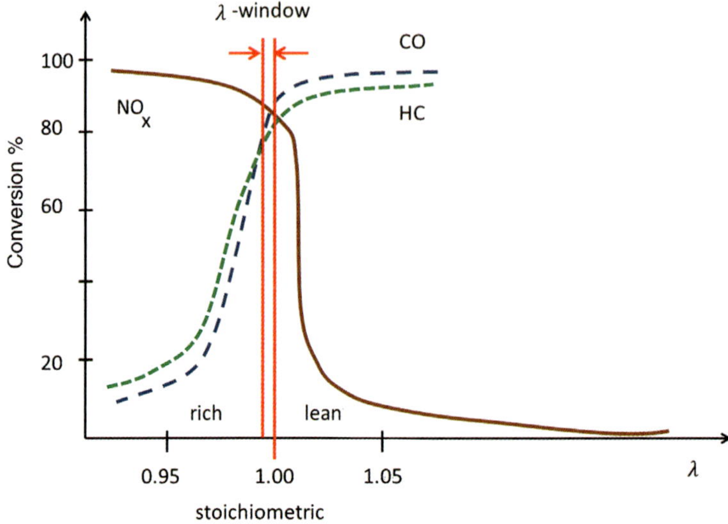

Figure 5.13 Three-way-catalysts demand that an Otto engine is operated around lambda $= 1$.

5.4.1.5 Paramagnetic Analyzer for O$_2$

The oxygen molecule is a biradical. Due to its paramagnetic behavior, the concentration of O_2 in gases can be measured accurately [41]: oxygen is attracted into a strong magnetic field, whereas most other gases are not. The two glass spheres which are suspended on a torsion wire are filled with nitrogen. In an oxygen-laden gas flowing through the cell, these N_2-filled spheres will be pushed outside the magnetic field, and a light beam directed on the central mirror can be used to measure the deflection, that is, the concentration of O_2 in the sample gas (see Figure 5.14).

Alternatively, the oxygen concentration can be determined quantitatively using fuel cells, which is also more cost-effective.

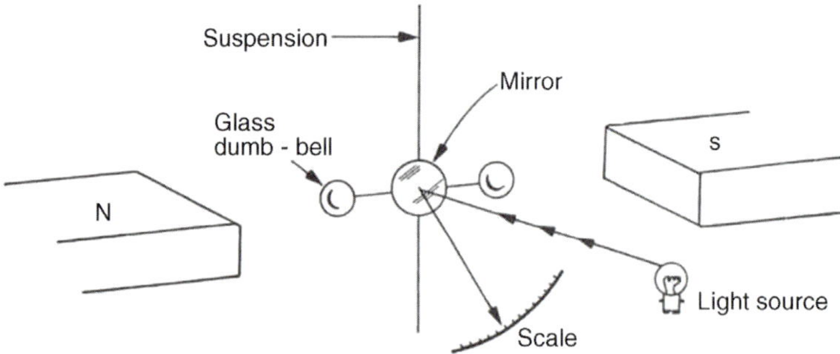

Figure 5.14 Principle of the paramagnetic oxygen analyzer [42].

5.4.2
Selection of Optical Techniques

Light and matter can interact in three ways:

- Absorption
- Emission
- Scattering

Depending on the size of particles compared to the wavelength, scattering takes places in three regimes: Rayleigh, Mie and optical.

Laser light is particularly advantageous for spectroscopy because it is mono-chromatic, spatially and temporally coherent, intense, and readily available [43]. "Laser" stands for "light amplification by stimulated emission of radiation". For details, see [44]. In spectroscopy, spectrally pure lasers are often needed. Such lasers can be of the type ECDL, DFB, DBR and VCSEL.

- ECDL   external cavity diode laser
- DFB   distributed feedback
- DBR   distributed Bragg reflector
- VCSEL   vertical-cavity surface-emitting laser

These diode lasers contain special elements that can filter out and amplify a single mode, see for example, [45].

- Flame sensors
  Since flames emit UV-light, special UV sensors can be used to record flame emissions. Missing flame emissions indicate misfiring and flame extinction in individual cylinders of large, stationary internal combustion engines installed in the field. For examples, see [46–48].
- High speed cameras
  High speed cameras may be used with a variety of techniques. One might be interested in a single image, taken at a precise point in time with short exposure time, or a time-resolved sequence, for example, covering an ignition phenomenon over several degrees of crank angle in an engine. First, straightforward images in close sequence can be obtained. By selecting a target wavelength, *chemilum-inescence* imaging can be carried out. *Schlieren photography*, a method to visualize density gradients in transparent media, can use high speed cameras, too. High speed cameras are also often used in conjunction with LIF (see later). For examples on high speed imaging in combustion, see for example, [49–51].

5.4.2.1 Chemiluminescence

Chemiluminescence (chemoluminescence) is the emission of light as the result of a chemical reaction with little or no heat production (compare fluorescence, where the excited electronic state is produced from a light absorption process). A common example of chemiluminescence is the luminol test, where blood is indicated by

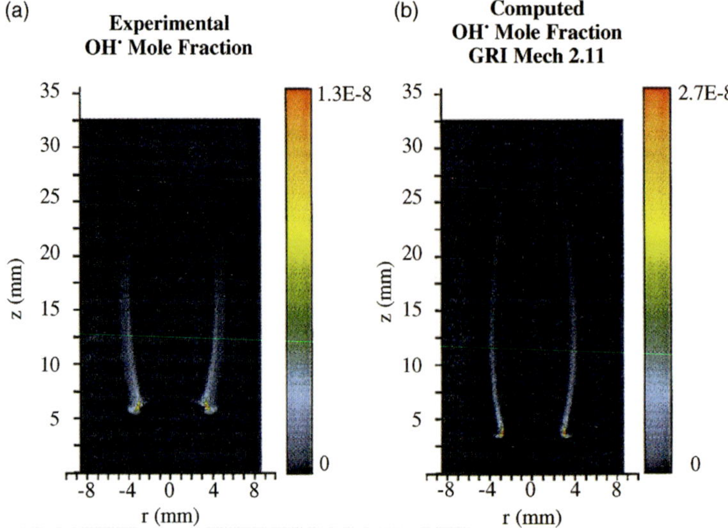

Figure 5.15 Chemiluminescence of the OH$^\bullet$ radical (a) and comparison to modeling results (b) [52].

luminescence upon contact with iron in hemoglobin. Chemiluminescence in living organisms, for example, as observed in the firefly, is called bioluminescence. In combustion research, chemiluminescence of the OH$^\bullet$ radical (peaking at 307.8 nm) can be used for diagnostic purposes. Chemiluminescence gives nice images as shown in Figure 5.15.

For further reading on chemiluminescence see [53].

5.4.2.2 Schlieren Photography

This method is experimentally very easy to realize. It is an integrating technique, delivering averaged information over the entire sampled volume. The visualized density gradients can stem from pressure and/or temperature differences. For instance, Schlieren photography can be used to study shock waves or flame kernel formation. The simple experimental set-up is shown in Figure 5.16.

Note: instead of a fast camera, a small time resolution can also be achieved by using a "fast" light source. However lasers can lead to interference patterns on Schlieren images. Figure 5.17 shows a sequence of four images obtained from a laser ignition test.

Like Schlieren photography, shadowgraphy can visualize non-uniformities in transparent media. It is simpler to set up. Recommendable textbooks on Schlieren and shadowgraph photography are [54,55].

5.4.2.3 Non-Dispersive Infrared Spectrometer

A common *ex situ* gas analyzer that is fed by a cooled suction probe from the combustion process under investigation is a non-dispersive infrared (NDIR)

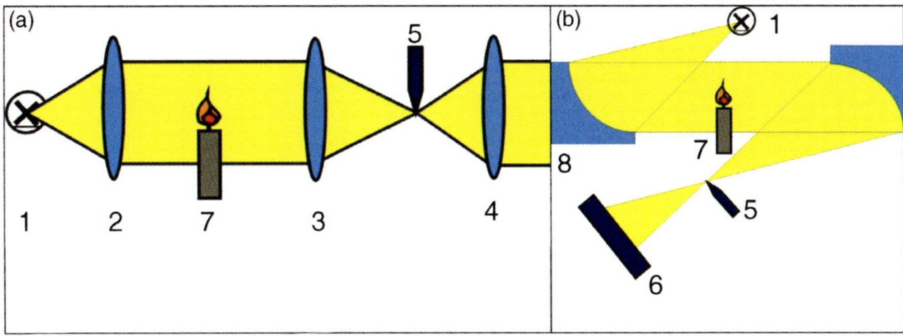

Figure 5.16 Set-up for Schlieren photography. Schlieren photography can be done with lenses (a) or mirrors (b); 1: light source; 2, 3, 4: lenses; 5: knife blade in the focal point; 6: Screen (CCD camera); 7: study object; 8, 9: parabolic mirrors.

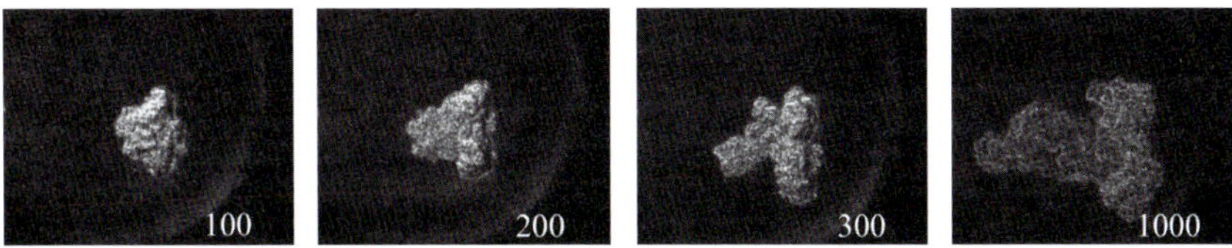

Figure 5.17 Laser ignition of an H_2/air mixture observed by Schlieren photography. Experimental conditions: Pressure $= 25$ bar, air/fuel ratio $\lambda = 6.0$.

spectrometer in which the entire available infrared light is used for the absorption measurement as opposed to techniques where one wavelength is selected, for example, in FTIR or TDLS (see below). NDIR can be used to measure the concentration of infrared-active gases such as CO and CO_2. For details, see [56].

5.4.2.4 Fourier Transform Infrared Spectrometer

Fourier transform infrared (FTIR) spectroscopy is a common and effective analytical tool for the identification or screening of unknown samples in the gaseous, liquid and solid states. Unlike NDIR instruments, where a filter selects only a narrow wavelength range matched to the target analyte, FTIR spectrometers acquire a broadband spectrum or, to be more precise, an interferogram, which is then mathematically converted back into a spectrum by performing a Fourier transform.

The predecessors of FTIR spectrometers were the dispersive instruments: grating monochromator or spectrograph. FTIR spectrometers are often simply called FTIRs. FTIRs are based on the Michelson interferometer. As such, they have the following advantages over "classic" dispersive instruments:

- **Multiplex advantage** (Fellgett advantage): All source wavelengths are measured simultaneously. Therefore, a complete spectrum can be collected very rapidly.
- **Throughput advantage** (Jacquinot advantage): A higher energy throughput than with slit-limited dispersive spectrometers is possible, resulting in higher signal-to-noise-ratios in shorter time.

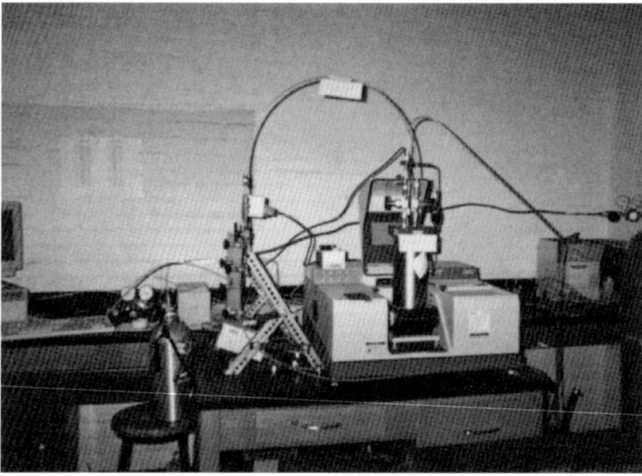

Figure 5.18 Nicolet Magna-IR 550 Fourier transform infrared spectrometer with a 10-m path length multipass gas cell [57].

- **Connes advantage:** The wavenumber scale of the interferometer is derived from a HeNe (helium neon) laser that acts as an internal reference for each scan, giving good long term stability.

Typical FTIR cells have optical path lengths of several meters. In combustion, the gas cell is typically heated to prevent the condensation of combustion products such as water and/or tar. Figures 5.18 and 5.19 depict the set-up of an FTIR spectrometer.

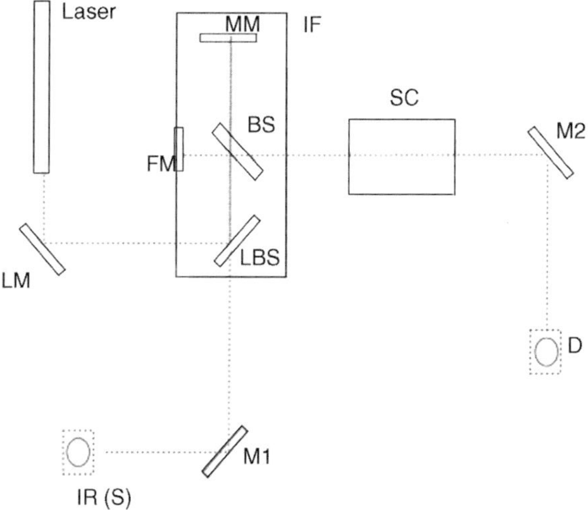

Figure 5.19 Schematic representation of the FT-IR spectrometer layout. IR(S), infrared source; LM, M1, M2, mirrors; LBS, BS, beam splitters; SC, sample compartment/cell; MM, moving mirror; FM, fixed mirror; IF, Michelson interferometer; D, mid-IR detector [57].

A challenge in combustion is often the background signal from water, which prevents trace gas measurements of other species. For details on the technology, see [58–60].

5.4.2.5 Laser-Induced Absorption Techniques

- Photo acoustic spectroscopy (PAS)
- Tunable diode laser absorption spectroscopy (TDLS, TDLAS)
- Cavity ringdown spectroscopy (CRDS)

Photo Acoustic Spectroscopy (PAS)   In PAS, the sample is excited with an intermittent light beam, often infrared radiation pulsed into the sample by means of a chopper wheel. The analyte in the sample will absorb the light. Upon dissipation of the heat, the sample expands and contracts, producing sound energy. This effect is repeated with every pulse, and an acoustic signal proportional to the concentration of the analyte in the sample cell is obtained. PAS can be applied to solid, liquid and gaseous samples. It is based on experiments by Alexander Graham Bell in 1880, when he showed that thin discs of special materials emitted an audible sound when exposed to rapidly interrupted light. Figure 5.20 shows a typical set-up.

The detection limit that can be reached by PAS is given in Table 5.3.

Major advantages of PAS over mass and NDIR spectrometers are:

- Few interfering signals (small measurement cell)
- Possible at atmospheric pressure
- No cryogenic IR detectors (which require cooling with liquid nitrogen) necessary
- Cost-effective, as microphones are cheaper than infrared detectors.

For details on PAS, see [63–66].

Tunable Diode Laser Absorption Spectroscopy (TDLS, TDLAS)   Infrared spectroscopy is a versatile tool in gas analysis. In NDIR, as introduced above, a reference cell is used to compensate for background errors. A different approach is followed in TDLS. In this

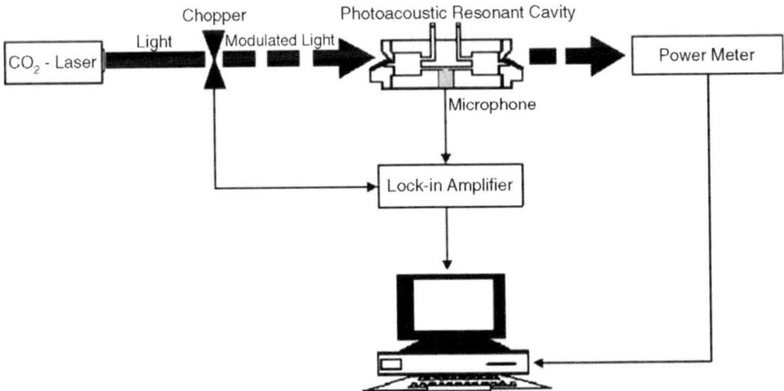

Figure 5.20   Experimental arrangement for the photoacoustic measurements [61].

Table 5.3 Detection limits in PAS [62].

Compound	Detection limit [ppm]
CH_4	$0.001 - 0.1$
CO	$0.001 - 0.1$
CO_2	$0.001 - 0.1$
HBr	$0.001 - 0.05$
HCN	$0.002 - 0.2$
HCl	$0.001 - 0.05$
HF	$0.001 - 0.08$
HI	$0.001 - 0.1$
HOCl	$0.001 - 0.1$
H_2O	$0.001 - 0.01$
H_2O_2	$0.01 - 0.05$
NH_3	$0.002 - 0.2$
NO	$0.003 - 0.3$
NO_2	$0.008 - 0.2$
N_2O	0.8
O_2	$0.05 - 5.0$
OH	$0.001 - 1.0$
H_2CO	1.0

technology, the wavelength of a diode laser that is matched to an absorption wavelength of the target molecule is scanned back and forth around the selected absorption wavelength. By doing so, background signals can be eliminated, allowing accurate and quantitative species concentration measurements of several gases relevant for combustion, such as O_2, CO, CO_2, H_2O, HCN, NO, NO_2, SO_2, HCl, HCHO and CH_4.

Figure 5.21 illustrates schematically the difference in the spectra of gaseous, liquid and solid samples.

Small molecules at low and ambient pressure will yield a spectrum where individual peaks which can be assigned to rovibronic energy levels, can be seen, whereas the interaction of probed molecules in the condensed phase will lead to more or less broad absorption bands. Figure 5.22 depicts the absorption lines of common gases in the near- and mid-infrared spectral region.

When the probing light source is matched in emission frequency to an absorption wavelength of the target species, the light attenuation upon passing the sample can be described by Lambert Beer's law, see also Figure 5.23.

$$I = I_0 e^{-klc} \tag{5.1}$$

where I is the transmitted light intensity, I_0 the initial light intensity, l the path length and c the concentration of the target analyte, with k being a constant. By rearranging Lambert Beer's law, the so-called absorbance (A) can be defined.

$$A = \ln \frac{I_0}{I} = k'lc \sim c \tag{5.2}$$

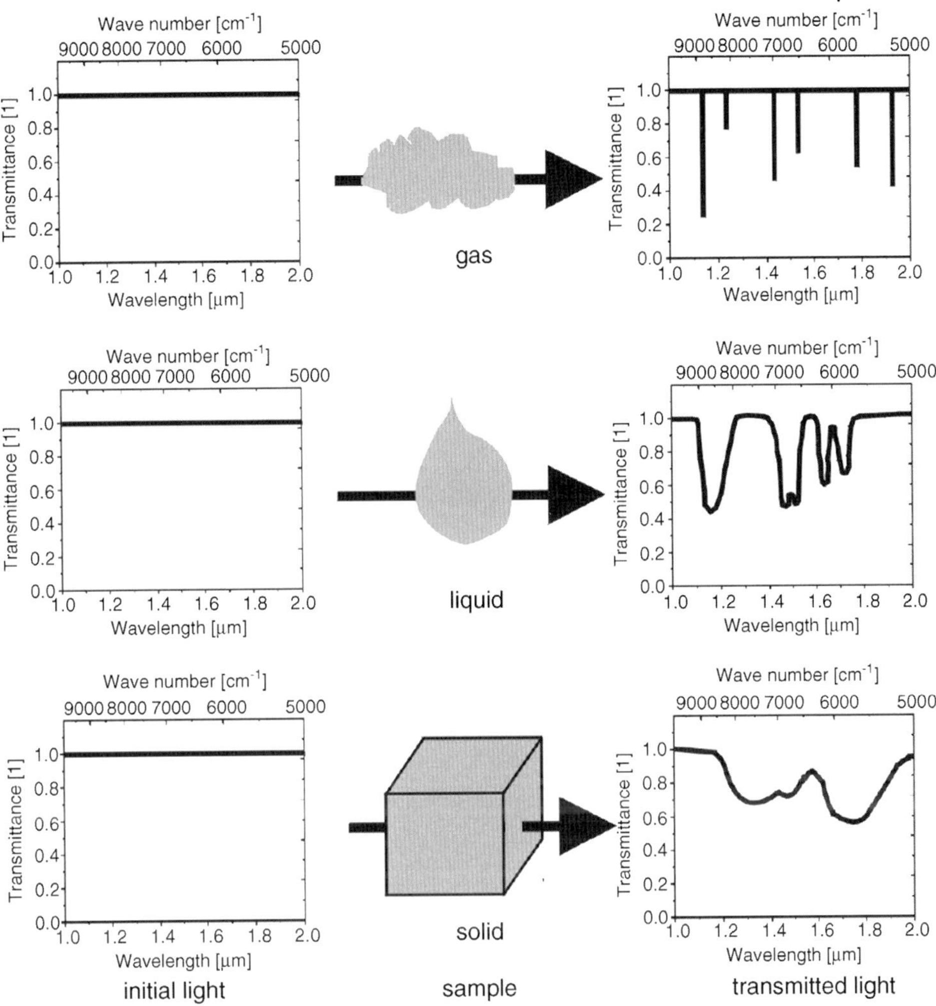

Figure 5.21 Gases of small molecules yield a "pure" spectrum with distinct absorption peaks, whereas liquids and solids tend to show absorption bands.

The absorbance is directly proportional to the number density (concentration times path length) and hence to the concentration of the analyte.

In Figure 5.24 the set-up for a TDLAS experiment to probe a flame is shown. In this configuration, the light from several diode lasers is coupled into a fiber and sent through a flame. Behind the flame, a grating is used to separate the wavelengths again. That configuration to measure with multiple lasers simultaneously is called wavelength-multiplexing. Another approach is time-multiplexing, by which different wavelengths are used consecutively.

In general, one laser (with one wavelength) is used to probe one target molecule. Sometimes, it is possible to select a laser with a wavelength range that can cover an

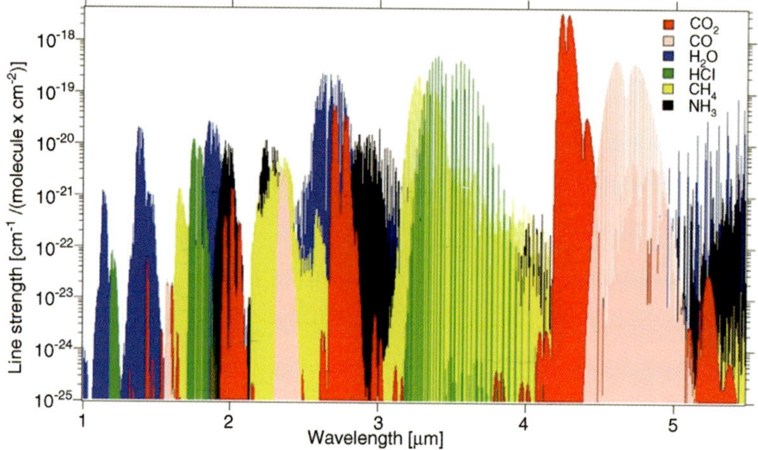

Figure 5.22 Absorption lines of several gases in the infrared spectral region, taken from the spectral database HITRAN [67].

$$I = I_0 * e^{-k*l*c} \qquad \text{(eq. 1)}$$

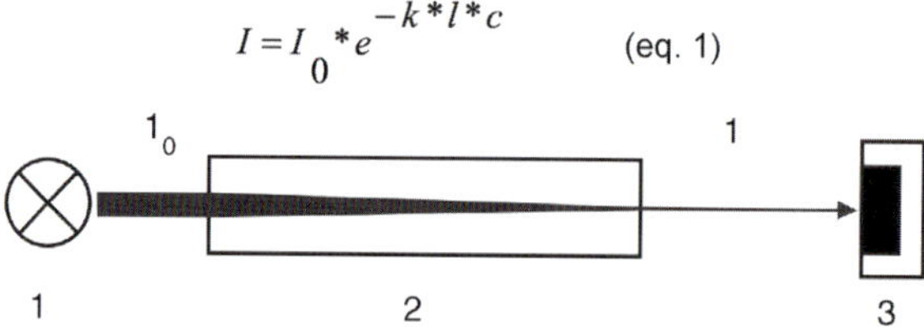

Figure 5.23 Light attenuation in TDLAS. 1: Light source (monochromatic); 2: sample volume filled with analyte; 3: detector.

absorption peak of two or more species. Apart from concentration measurements, TDLS allows temperature and temperature distribution measurements, for example, by probing several absorption lines of O_2 around 760 nm.

For details on TDLS, see [69–73]. A variant of TDLS is cavity ringdown spectroscopy (CRDS) [74,75].

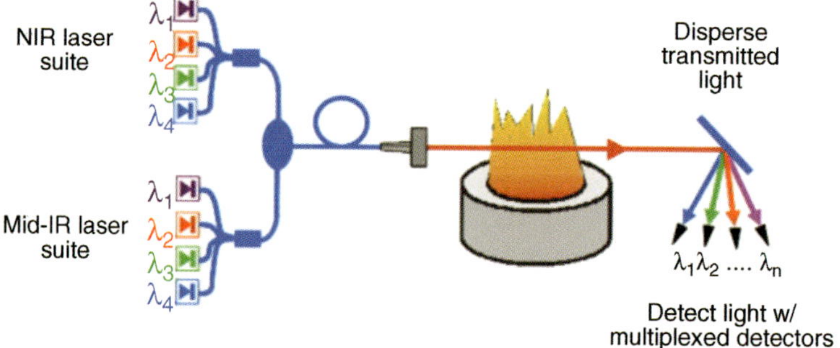

Figure 5.24 Wavelength-multiplexed diode laser absorption arrangement [68].

5.4.2.6 Laser-Induced Emission Techniques

- Laser-induced fluorescence (LIF, PLIF)
- Laser-induced incandescence (LII)

Laser-Induced Fluorescence (LIF, PLIF) LIF has become a standard technique in combustion research that can yield very helpful results. It can be performed with fuels (non-fluorescence-active fuels can be doped with a tracer such as acetone), with combustion products and, most importantly, with combustion intermediates (radicals).

LIF is a two-stage-process:

- Excitation of an atom/molecule by absorption of a photon
- Emission of a photon from the excited state.

The excitation wavelength is matched (tuned) to species of interest, compare TDLAS. LIF is species-specific, and also minority species are measurable. For instance, the $OH^{\bullet}$ radical shows the flame front, and the $CH^{\bullet}$ radical the reaction zone. Two-dimensional images obtained by P-LIF (planar LIF) are very illustrative. A drawback with LIF is that it is only semi-quantitative, because the extent of radiationless decays as mediated by oxygen is difficult to calculate.

Typical probed species/molecules are: OH, CH, NO, CN, C_2, HCO, CH_2O, HCCO, PAH (polycyclic aromatic hydrocarbons), SO and SO_2.

The strength of the fluorescence signal S_F depends on

$$S_F \propto n \cdot EL \cdot s_{abs} \Phi_{LIF} \eta_{hv} \tag{5.3}$$

S_F	fluorescence signal
n	number density of LIF species
EL	(number of laser photons)/cm^2
s_{abs}	absorption cross section
Φ_{LIF}	fluorescence efficiency $= A/(A+Q)$
η_{hv}	detection efficiency
A	fluorescence rate
Q	quenching rate

The experimental set-up for LIF is shown in Figure 5.25.
For details, see [77–80].

Laser-Induced Incandescence (LII) LII can be deployed for the analysis and characterization of soot. Examples are provided in [81,82].

The LII signal strength is given by the following formula:

$$S_{LII} \propto c_{pm} s_{abs} \eta_{hv} \tag{5.4}$$

S_{LII}	LII signal of particulate matter (soot)
c_{pm}	particulate matter (soot) volume fraction
s_{abs}	absorption cross section for particle heating
η_{hv}	detection efficiency

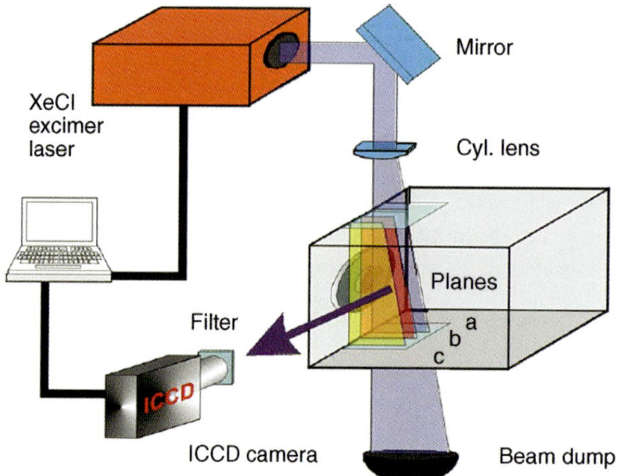

Figure 5.25 Experimental set-up to measure fuel/air mixing in a swirl burner of a gas turbine with LIF. Reproduced with permission from [76].

Figure 5.26 depicts the experimental set-up for TiRe-LII (TiRe = time-resolved). Using single laser pulses, time-resolved laser-induced incandescence (TiRe-LII) signal transients from soot particulates can be acquired during unsteady combustion events, such as in engines. Gas temperatures can be deduced from spectrally resolved soot pyrometry measurements. Using the LII model of Kock *et al.* [84], ensemble mean soot particle diameters can be determined.

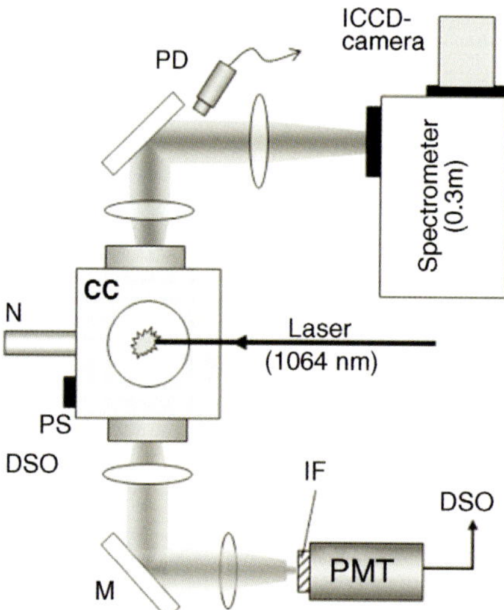

Figure 5.26 Experimental set-up for the combined TiRe-LII and soot spectral emission analysis in a high-pressure high-temperature Diesel combustion cell (CC). DSO: digital storage oscilloscope, ICCD: intensified CCD-camera, M: aluminum mirror, N: fuel injection nozzle, PD: photodiode, PS: pressure sensor, IF: interference filter, PMT: photomultiplier tube [83].

5.4.2.7 Laser-Induced Scattering Techniques

- Rayleigh scattering
- Raman spectroscopy
- Coherent anti-Stokes Raman scattering (CARS)
- Laser Doppler anemometry (LDA)
- Particle image velocimetry (PIV)

Depending on the ratio of the light wavelength and the interacted particle diameter, one can distinguish between different regimes, see Figure 5.27.

Mie scattering occurs on smoke.

Rayleigh Scattering Rayleigh scattering (Rayleigh spectroscopy) is the *elastic* scattering of light by particles much smaller than the wavelength of the electromagnetic radiation. Most photons are scattered elastically (for inelastic scattering, see Raman).

The probed particles can be individual atoms or molecules. The signal strength will depend on:

$$S_{RS} \propto n_{tot} \cdot EL \cdot s_{eff}\, \eta_{h\nu} \tag{5.5}$$

S_{RS} Rayleigh signal
n_{tot} total number density
EL (number of laser photons)/cm^2
s_{eff} effective cross section
$\eta_{h\nu}$ detection efficiency

Because the scattering is elastic, $\lambda_{Rayleigh} = \lambda_{Laser}$.
For details see [86–88].

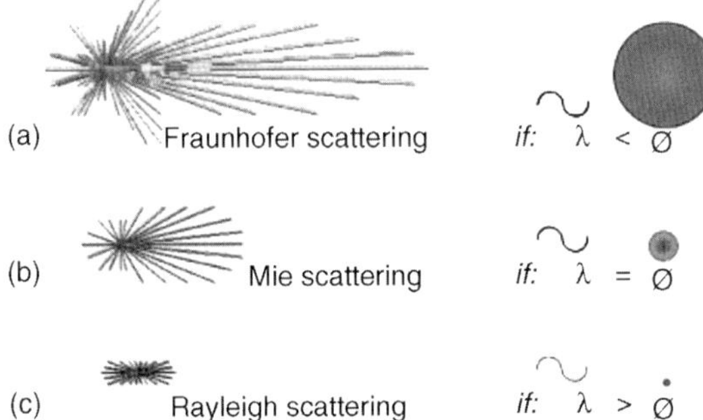

Figure 5.27 Envelopes of scattering as a function of ratio of particle size to wavelength. (a) Fraunhofer scattering (particles much larger than wavelength), (b) Mie scattering (intermediate region, e.g., particles about the size of the wavelength) and (c) Rayleigh-scattering (particles much smaller than wavelength) (λ = wavelength) [85].

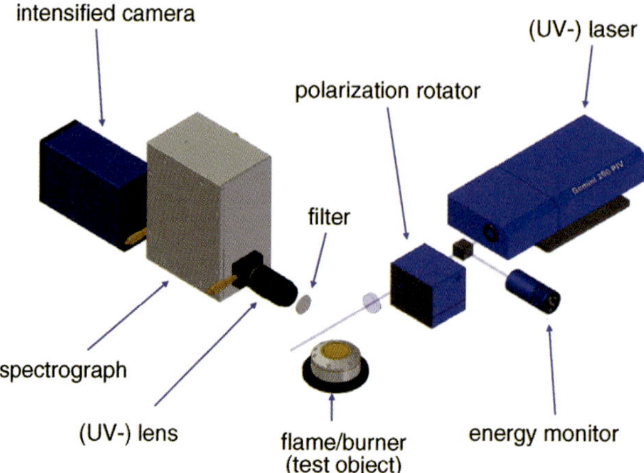

Figure 5.28 Set-up for Raman spectroscopy. Courtesy of LaVision [4].

Raman Spectroscopy Raman spectroscopy (Raman scattering) is based on vibrational spectroscopy (IR spectroscopy). It allows the study of vibrational and rotational transitions. It relies on *inelastic scattering* (=Raman scattering) of monochromatic (laser) light in the visible or near-infrared spectral region. The recorded spectrum provides information on the probed molecules from excitation of vibrations. RAMAN and IR spectroscopies can yield complementary information. Figure 5.28 shows the experimental set-up.

Raman spectroscopy is a stable and established measurement approach for combustion-relevant molecules. It allows the measurement of the major species simultaneously. Typical species/molecules are: N_2, O_2, H_2O, CO, CO_2, H_2, and CH_4.

Under good conditions, single shot results are possible (important for turbulence). The signal strength depends on

$$S_{i(li)} \cdot \propto n_i \cdot EL \cdot s_i \eta_{hv} \tag{5.6}$$

$S_{i(li)}$ Raman signal of species i at l_i
n_i number density of species i
EL (number of laser photons)/cm^2
S_i cross section
η_{hv} detection efficiency

No tunable laser is required (high energy, easy operation, no dye preparation). A common wavelength is, for example, 532 nm, where there is no need for UV optics, or 355 nm in the UV spectral region.

A big advantage of Raman is that no influence from quenching (as in LIF) is encountered.

It is well suited for high-pressure environments.

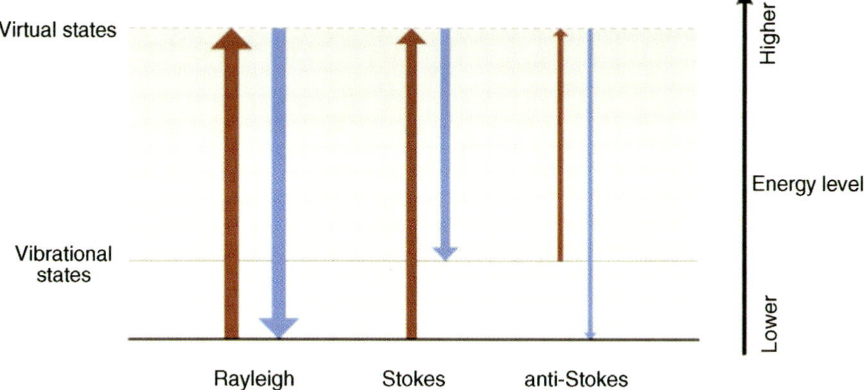

Figure 5.29 Energy diagram for Rayleigh and Raman scattering processes. Rayleigh scattering is the most intense scattering process, and Stokes scattering predominates over anti-Stokes scattering at room temperature. The thickness of the arrows indicates the relative intensity of the three processes (not to scale) [89].

The disadvantages of Raman spectroscopy are:

- Inelastic scattering gives a weak signal, superimposed strong signals can disturb (Rayleigh, LIF from PAHs).
- Only species with high enough concentration and scattering cross section (temperature dependent) are detected.

In inelastic scattering, two events are possible:

- The material absorbs energy and the emitted photon has a lower energy than the absorbed photon. This outcome is called Stokes Raman scattering.
- The material loses energy and the emitted photon has a higher energy than the absorbed photon. This outcome is called anti-Stokes Raman scattering.

Figure 5.29 shows the energy diagram for Rayleigh and Raman scattering processes.

For the wavelengths, the following relationship holds true: $\lambda_{\text{AntiStokes}} < \lambda_{\text{Laser}} < \lambda_{\text{Stokes}}$. For details, see [90–92].

Coherent Anti-Stokes Raman Scattering (CARS) Coherent anti-Stokes Raman spectroscopy, also called coherent anti-Stokes Raman scattering spectroscopy or CARS, is comparable to Raman spectroscopy and probes molecular vibrations. However, multiple photons are used to excite the target molecule, and the produced signal is a coherent electromagnetic wave. Therefore, the CARS signal is several orders of magnitude more intense than that from conventional, spontaneous Raman emission. CARS involves three laser beams and is a nonlinear process. The experimental set-up is shown in Figure 5.30.

CARS can be used for thermometry. More information can be found in [94–98].

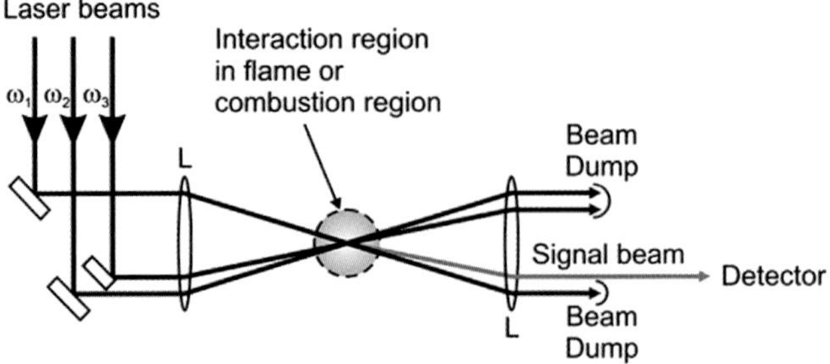

Figure 5.30 Basic optical set-up for CARS. Reproduced with permission from [93].

Laser Doppler Anemometry (LDA)   LDA is a widely used method for flow visualization and diagnostics. The technique is based on the Doppler effect. So-called seed particles are added to the fluid to be investigated and probed with a laser beam. The motion of the particles, which should follow the flow as closely as possible, will lead to a Doppler shift, which can be detected. For details on LDA see [99].

Particle Image Velocimetry (PIV)   PIV is another flowfield technique that relies on seed particles. Images are shot consecutively, and software is used to calculate velocity vectors from the particle movements between two measurements. A variant, particle tracking velocimetry (PTV), can be 2D or 3D. For details on PIV see [100–104].

5.4.3
Particle Diagnostics

Combustion-generated fine particles are encountered in smoke. Biomass and diesel engines produce relatively large fractions of fine and ultrafine particles. Smoke is an aerosol (mist) of solid particles and liquid droplets from combustion and pyrolysis processes. Leonardo Da Vinci tried to characterize smoke. He could distinguish between black smoke (containing carbon particles = soot) and white smoke (water). Today, smoke and particles are characterized by:

- Optical scattering
- Optical obscuration
- Ringelmann scale (used for measuring the apparent density of smoke)
- Sampling, for example, by in-line capturing, a filter/dilution tunnel or electrostatic precipitation.

Composition, particle size distribution and specific surface area are major parameters of interest. For a review on particle diagnostics, see [105]. See also the section on particulate matter (PM) in Chapter 4, Section 4.1.

5.4.4
Spray Diagnostics

Sprays are an important element in liquid fuel combustion. Atomization of fuels for good mixing of fuel and oxidizer can be monitored and characterized by several

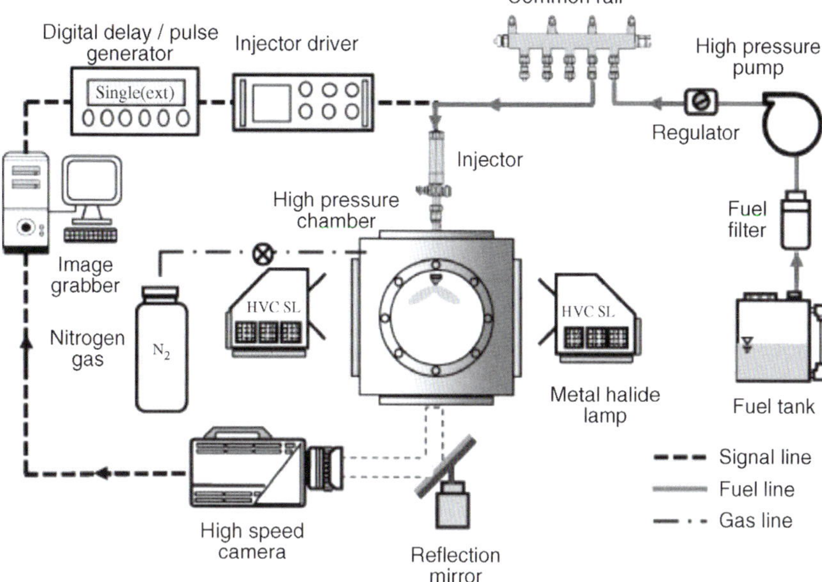

Figure 5.31 Schematic of the spray visualization system [106].

methods, of which (high speed) photography is the most straightforward one. Figure 5.31 shows the set-up of a spray visualization system.

Figure 5.32 gives an example of applied spray diagnostics by simultaneous liquid-fuel spray and natural emission images.

For details on spray diagnostics in combustion see [108].

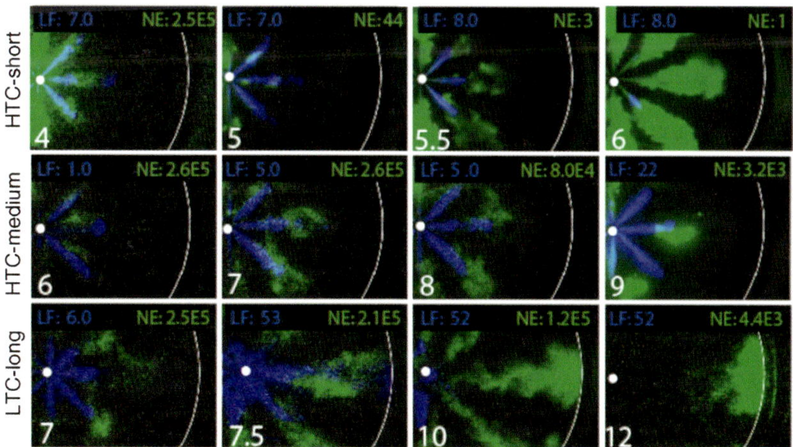

Figure 5.32 Simultaneous single-shot liquid-fuel (LF) and natural emission (NE) images for the three operating conditions. The numbers at the top of each image are the camera gains for each diagnostic, and the crank angle degrees after start of injection are at the bottom left [107]. HTC = high temperature combustion; LTC = low temperature combustion.

5.4.5
Other Techniques

Among less frequently used techniques for combustion research are: shadowgraphy, degenerate four-wave mixing (DFWM), laser induced grating spectroscopy (LIGS), polarization spectroscopy (PS), ICLAS (intracavity laser absorption spectroscopy), interferometric Mie imaging (IMI) and phase Doppler interferometry (PDI), the latter being used for droplet detection. For further reading, see also plasma spectroscopy [109]. For a review on sensors in combustion control, see [110,111].

5.4.6
Test Beds

Combustion diagnostics can be performed on a variety of "test beds", from the lab to huge industrial plants. In this section, several typical ones are presented. Since the purposes (fire research, fuel characterization, fundamental investigations, process control, etc.) are so diverse, only a fraction of common and useful installations can be covered here.

5.4.6.1 Open Flames on Laboratory Model Burners

The *Bunsen burner* is the classic burner in labs. It produces, in a safe way, a hot, sootless, non-luminous flame by premixing fuel and oxidizer in a vertical tube. Variants are the *Teclu burner* and the *Meker burner*. For flame diagnostics, typically other burners are used:

McKenna Burner (Flat Flame Burner)   The flat flame burner is a popular burner in R&D, in that it allows the investigation of homogeneous, premixed flames in a reproducible fashion. Its cross-section is depicted in Figure 5.33.

Wolfhard Parker Burner (Slot Burner)   For non-premixed flames, a slot burner (Wolfhard Parker burner) can be deployed. It can be home-made in a few steps only. Its set-up is shown in Figure 5.34.

Axisymmetric Co-annular Burner   The axisymmetric (co)annular burner is often used for flame synthesis [114]. It is similar to the Wolfhard Parker burner, see Figure 5.35.

Swirl Burner   Swirl burners produce intensive mixing of fuel and oxidizer by providing them with a "swirl" motion upon entry. They can be used to burn low-grade fuels. Figure 5.36 shows a typical set-up, and a second version is shown in Figure 5.37.

Swirl burners can be built for solid, liquid and gaseous fuels.

Low Swirl Burner   Low-swirl combustion (LSC) is an aerodynamic flame stabilization method. Originally, this type of burner was intended as a burner for basic

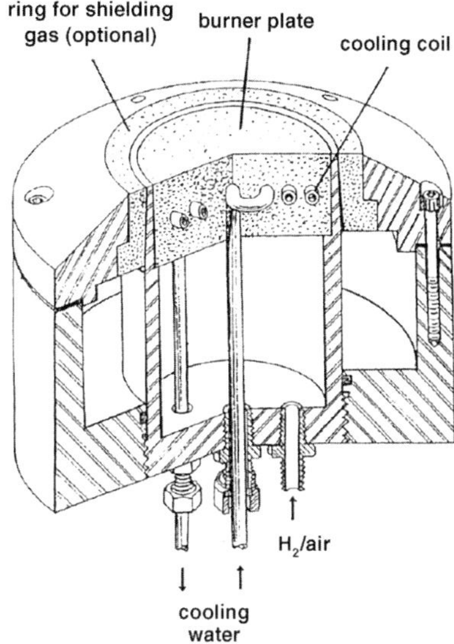

Figure 5.33 Cross-section through a McKenna flat flame burner. Reprinted with permission [112].

research on turbulent combustion by the Lawrence Berkeley National Laboratory. Low swirl burners can support ultra-lean flames that emit very low levels of NO_x, so today the technology is used for burners in industrial heaters and in gas turbines for electricity production [118]. Figure 5.38 shows a picture of a low swirl burner. Low-swirl combustion can be implemented in two ways. The original approach uses small jets to create the swirling motion. The second approach that was developed for heaters and gas turbines uses a vane-swirler. For details see [118].

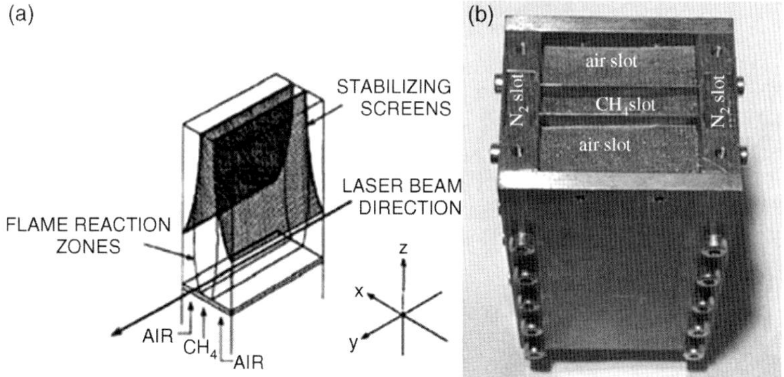

Figure 5.34 (a) Schematic of a Wolfhard–Parker Burner showing the top of the burner and placement of flame stabilizing gulls above the burner. (b) Picture of the modified burner with N_2 slots for end-flame suppression [113].

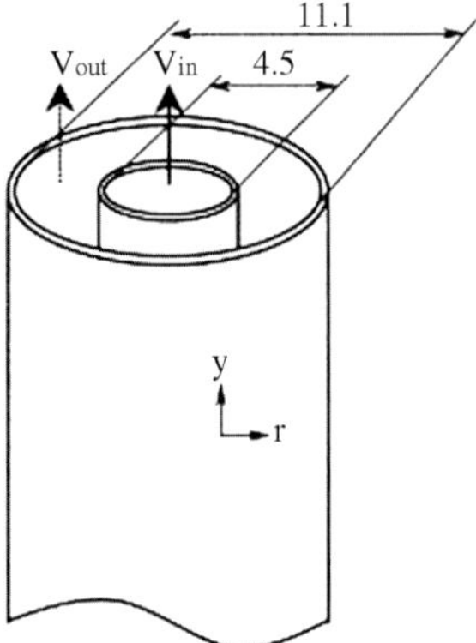

Figure 5.35 An axisymmetric co-annular burner [115].

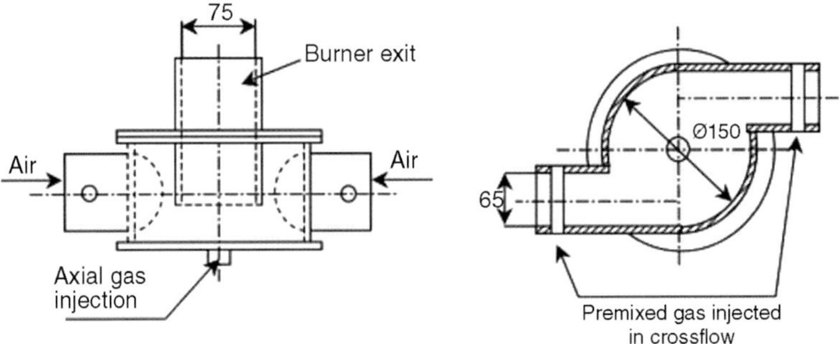

Figure 5.36 Schematic diagram of a generic swirl burner—swirl number adjustable from 0.75 upwards via use of tangential inserts. Exhaust extends backwards into swirl chamber to prevent flashback. Nominal thermal input 100 kW [116].

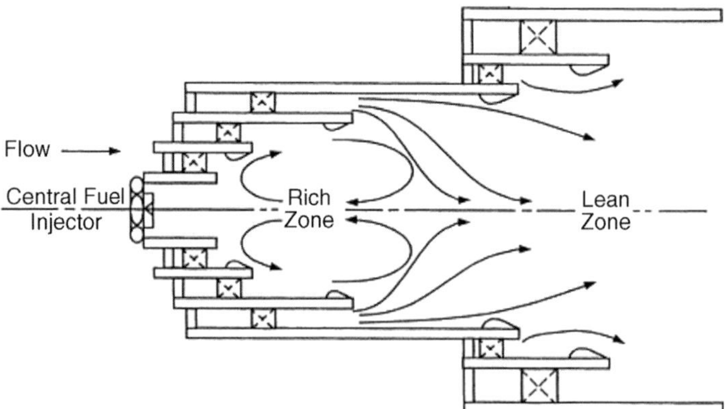

Figure 5.37 Multi-annular swirl burner (MASB) topping combustor [117].

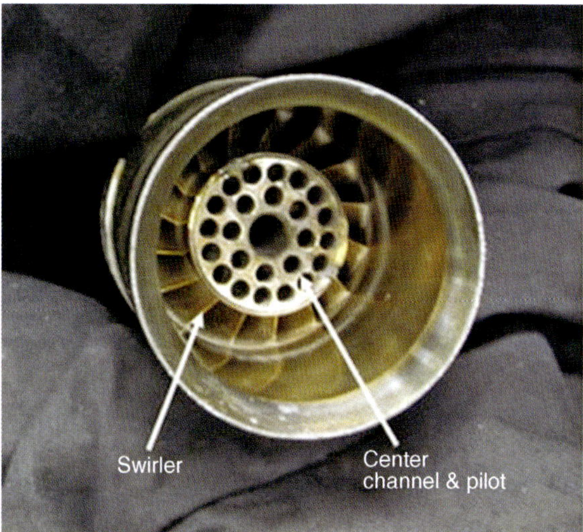

Figure 5.38 Picture of a low swirl burner [118].

Tsuji Burner A Tsuji burner is depicted in Figure 5.39. In the Tsuji burner, a gaseous fuel from a sinter-metal tube is injected into the surrounding air, which flows vertically upwards. Below the tube a stagnation point is formed. Under certain conditions, the flow field may be split into an inviscid outer flow and a boundary layer close to the surface. The Tsuji burner can be used for fundamental flame research, for example, for flame stabilization.

Opposed Tubular Burner In an opposed tubular burner, oxidizer and fuel are directed in counter-flow towards each other, see Figure 5.40. An opposed flow burner is a tool to study the effects of curvature on extinction and flame instability of non-premixed flames. Details on this type of burner can be found in [120].

There are many other special burners, such as *porous media burners* [121].

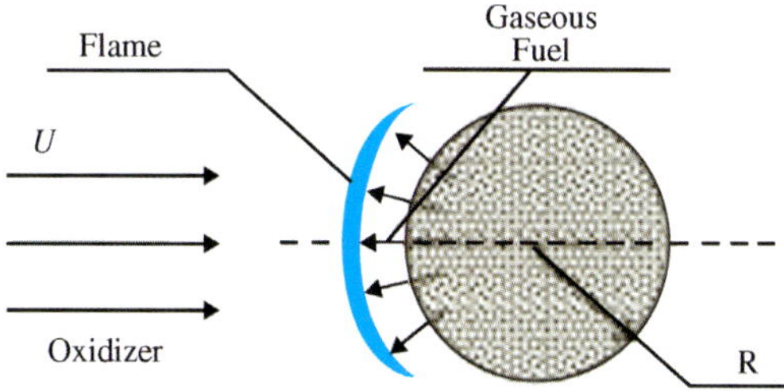

Figure 5.39 Sketch of Tsuji burner [119].

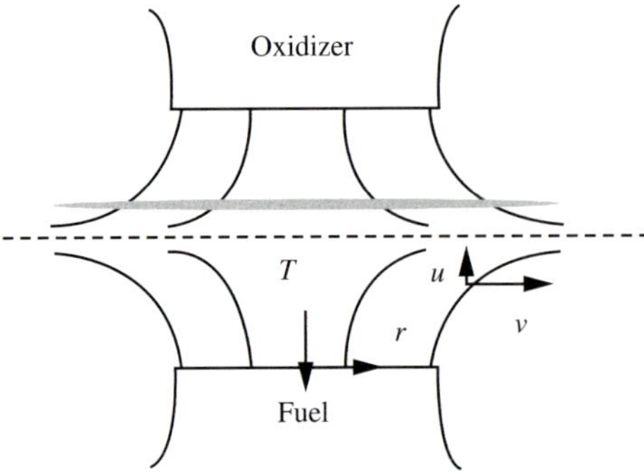

Figure 5.40 Sketch of an opposed flow burner [119].

5.4.6.2 **Combustion Bombs**

A combustion bomb is a closed vessel that can withstand the combustion/explosion pressure. Alternatively, it can be equipped with a fast pressure relief mechanism such as a bursting disk. For research, combustion bombs are typically equipped with optical access ports which allow optical (laser) diagnostics. Often, to avoid interference patterns, the windows are wedged, that is, there is a small angle between the two outer planes, typically 2–3°. Figure 5.41 shows a schematic illustration of a constant-volume combustion bomb.

5.4.6.3 **Shock Tubes**

Shock tubes allow variation of pressure (0.1–1000 bar) and temperature (500–15 000 K) in a very wide range with little experimental effort. Therefore, the shock tube technique is used to investigate gas phase reactions at high temperatures [123].

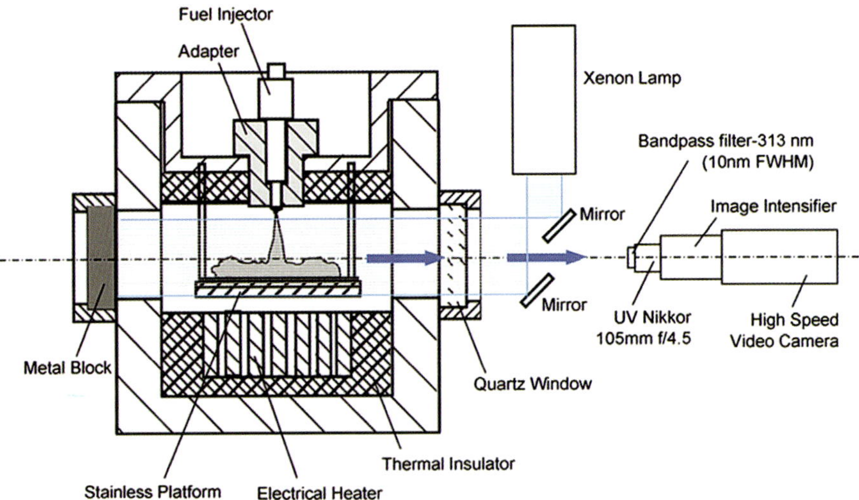

Figure 5.41 Schematic illustration of a constant-volume combustion bomb with optical set-up for direct flame and OH chemiluminescence imaging. Reprinted with permission from [122].

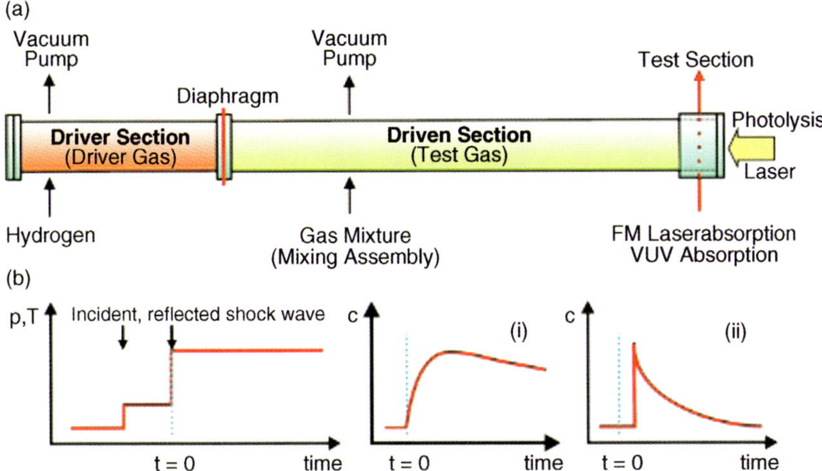

Figure 5.42 (a) Schematic set-up of a shock tube. (b) Pressure/temperature-time profile during a shock tube experiment and typical concentration–time profiles of (i) thermal and (ii) photolytic generation of the detected species [123].

A shock tube is a simple device. Essentially, it consists of a several meters long tube, which is divided by a diaphragm into 2 sections: a driver section (high pressure) and a driven section (low pressure). In the driven section, the gas under investigation can be found. The driver section is filled with a highly compressed gas, typically hydrogen or helium. When the diaphragm bursts, a shock wave will travel through the tube, compressing and heating the gas in the driven section. Shock wave reflections increase the pressure. Typical test durations are 1 μs, which is sufficient to study high-temperature gas phase reactions. An example of a shock tube is depicted schematically in Figure 5.42a.

Figure 5.42b shows the resulting pressure/temperature–time profiles. The species of interest are generated thermally by the decomposition of suitable precursor molecules (Figure 5.42b(i), or by a photolytic process (Figure 5.42b(ii)). Real-time detection of the concentration–time profiles of target species is done through optical windows by various spectroscopic absorption or emission techniques, shown in the right-hand section of Figure 5.42a. For more details, see for example, [123].

5.4.6.4 Optical Engines

To study combustion in engines, one can modify a "real" engine so that information can be derived from it by spectroscopic techniques. This is done by placing suitable windows in the combustion chamber to obtain a so-called "optical engine". Around optical engines, a test bed can be built, see Figure 5.43 for an example.

The optical engine in Figure 5.43 is equipped with a viewing window in the piston (bowl rim window), a ring window (piston crown window) and a third window in the cylinder head. Another version is shown in Figure 5.44.

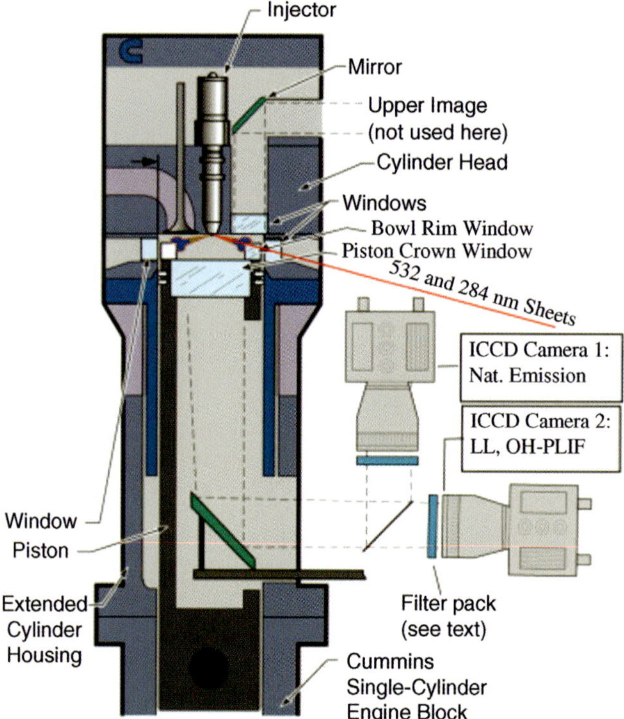

Figure 5.43 Experimental set-up for the optical engine [107].

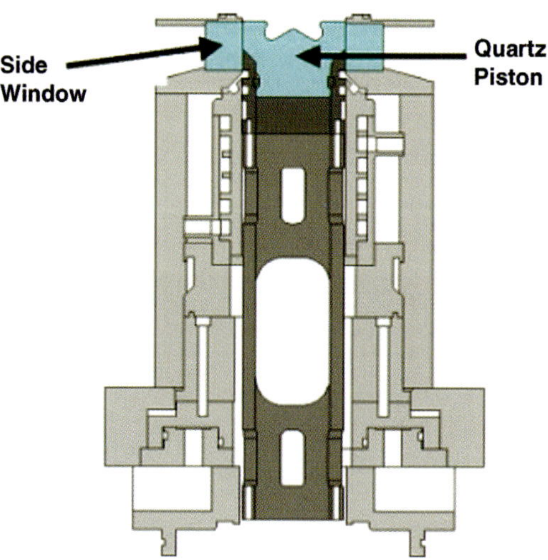

Figure 5.44 Assembly cross-section of an optical engine design with drop-liner raised [124].

5.4.6.5 Pilot Plants

In the scale-up of a combustion process, pilot plants are often used. A pilot plant is a small, fully-fledged and well instrumented installation that is built to gather data for the construction of a large, industrial plant. By first building a pilot plant, the risks of making costly or even uncorrectable construction mistakes with large plants are minimized. In size, a pilot plant lies between a bench-scale reactor and the final industrial installation. Pilot combustors are operated by numerous university institutes and industrial companies. One example was randomly picked to illustrate the principle: a grate firing system installed at the Institute of Combustion and Power Plant Technology, University of Stuttgart/Germany (Figures 5.45 and 5.46). This grate firing system consists of a horizontal moving grate for solid biomass, such as wood chips with a thermal power output of 80–240 kW. The fuel feeding from the fuel storage is managed by two stokers, and the pilot plant can be operated

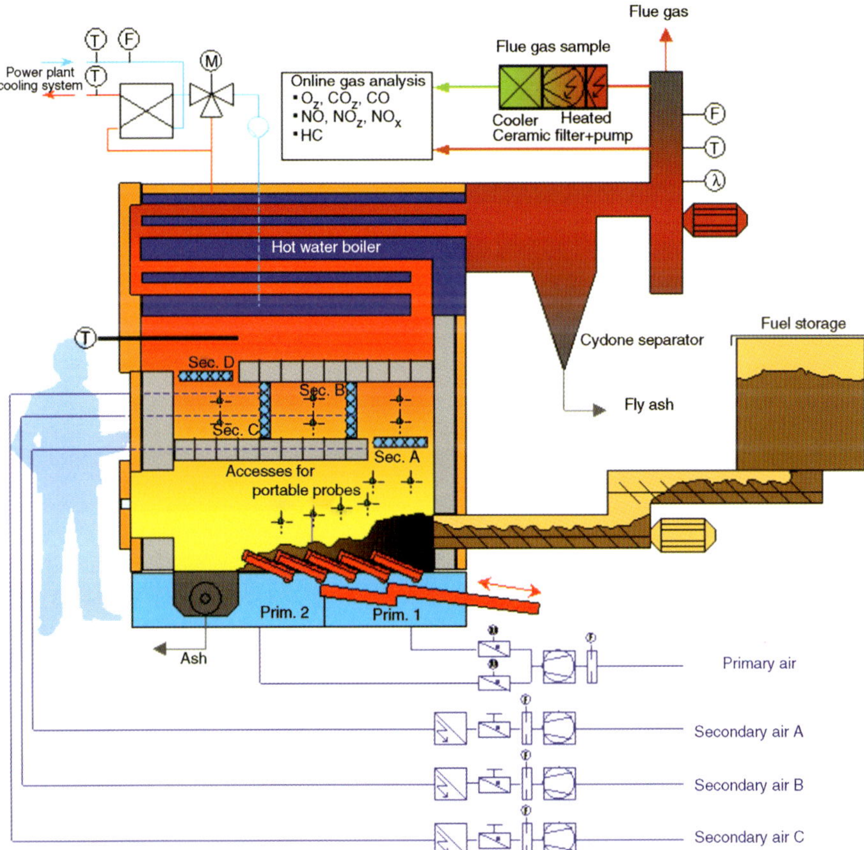

Figure 5.45 A grate firing system pilot plant installed at the Institute of Combustion and Power Plant Technology, University of Stuttgart/Germany. Note the size of the person. Pilot plants allow the gathering of ample process information [125].

Figure 5.46 A temperature probe and a cooled suction probe in the pilot plant of the Institute of Combustion and Power Plant Technology, University of Stuttgart/Germany [125].

with primary and secondary air (staged combustion for NO_x reduction). The installation is equipped with 16 ports to install suction probes. Hence, the temperature and the gas composition can be determined virtually everywhere in the combustion zone. The exhaust gas composition can be measured synchronously. The fuel mass flow of this pilot plant is $15–100\,\mathrm{kg\,h^{-1}}$ and the flue gas flow up to $700\,\mathrm{m^3\,h^{-1}}$.

Pilot plants can be built for each type of combustor in various sizes.

Based on the concept of *chemical similarity*, it is not necessary to mirror an entire combustion installation on a small scale in the lab, but instead one can pick a section of the industrial installation and recreate it in the lab for detailed study [126]. This is a cost-effective and straightforward approach useful for several applications.

Scale-up in combustion is covered in [126].

5.4.6.6 Combustors Placed on a Test Rig

Similar in concept to a pilot plant is the *test rig* set-up: A "real" combustion application is placed in a well-equipped lab for research. Figure 5.47 shows a car on a test rig.

On the test rig, ambient conditions such as temperature can be set, and the engine can be tested under various load scenarios. The test rig could also be set up in a wind tunnel, for instance.

5.4.6.7 Industrial Furnaces with Optical Access Ports

Large combustors such as power plants can be equipped with measurement equipment for process control or research purposes. Figure 5.48 shows an

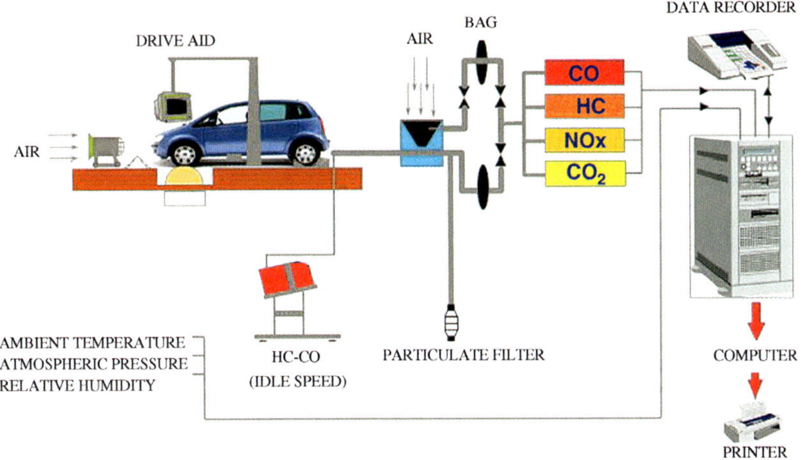

Figure 5.47 Car on a test rig to carry out emission measurements [127].

example of a TDLAS-measurement in the combustion chamber of a $600\,\mathrm{MW_{th}}$ furnace. In that example, the beam of a diode laser is sent $13\,\mathrm{m}$ through the combustion chamber of a coal-fired power plant to detect CO and H_2O species concentrations.

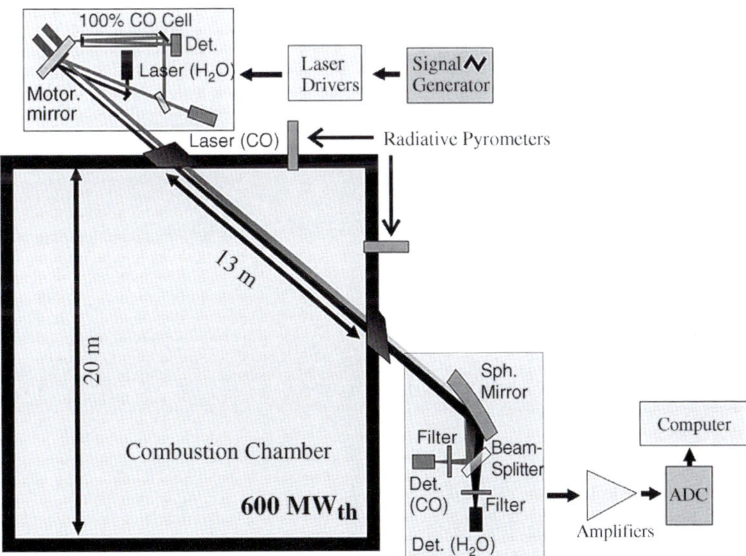

Figure 5.48 Schematics of an experimental set-up during TDLAS-based *in situ* CO/H_2O measurements at a coal-fired power plant. (Det. = detector). Reproduced with permission from [128].

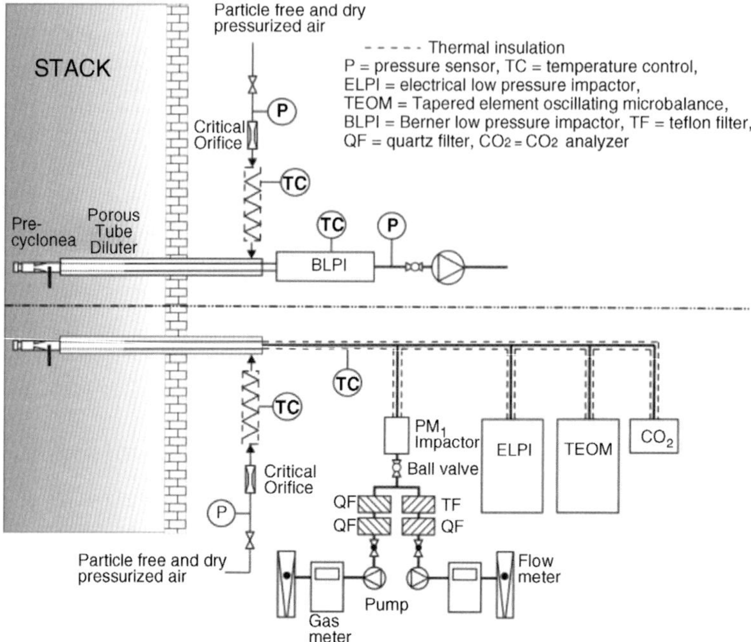

Figure 5.49 Particle emission measurement set-up in a smoke stack [129].

Another example for measurements directly in an industrial installation is shown in Figure 5.49. It shows a set-up to measure particle emissions from the flue gas stack of a power plant.

Particle emission measurement in a smoke stack can be used to monitor the performance of flue gas cleaning installations, or even to control the combustion process.

5.4.7
Advanced Combustion Control

Combustion diagnostics can be carried out to learn more about the fundamentals of a process or reaction in a laboratory environment. Also for combustors "in the field", diagnostics can be applied advantageously for control purposes. Simple examples are:

- Optimization of combustion performance in a waste incineration plant (varying properties of the fuel) by measuring the O_2 concentration in the exhaust gas.
- Reduction of emissions from Otto engines by maintaining lambda around 1.
- Minimization of ammonia slip from a DENOX plant to reduce operational costs and environmental burden.

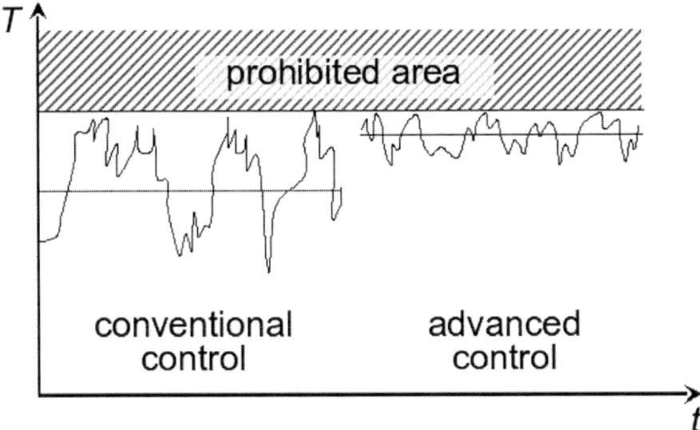

Figure 5.50 Direct economic benefit from applying advanced control. Narrower band of fluctuation allows higher average live steam temperature, which directly results in higher plant efficiency [130].

These three examples are common practice. Today, also advanced control techniques are being increasingly used for improved efficiency.

Figure 5.50 shows a typical example of controlling the steam temperature in a power plant. The higher the temperature, the better the efficiency. Using conventional control techniques, it is necessary to maintain a wider distance from the highest possible temperature to avoid damage by overshooting. If now a more advanced control technique with fewer fluctuations can be used, it will be possible to approach the optimum temperature more closely. A higher set-point of the steam temperature translates directly into better efficiency and cost savings for the plant operator.

This simple example highlights an obvious case of obtaining direct economic benefit.

The most important benefits of advanced combustion control, as can be obtained with optical diagnostics, in a power plant are:

- Reaching higher efficiency in steady state operation.
- Reduction of emissions from steady state operation.
- Making dynamic changes in a smooth and less resource-consuming way, including start-up.
- Higher equipment life time due to less thermal stress.

Details on advanced control techniques in power plants can be found in [130].

Web Resources

There are good resources on the internet to learn more about measurement methods in combustion. A few of them are presented below for the interested reader:

NASA Glenn Research Center	http://www.grc.nasa.gov/WWW/k-12/airplane/combst1.html
Symposium on Combustion Diagnostics	http://www.combustion-diagnostics.com/
R.K. Hanson research group	http://engineering.stanford.edu/profile/rkhanson
Lawrence Livermore National Laboratory	https://www-pls.llnl.gov/?url=science_and_technology-chemistry-combustion
DLR Institute of Combustion Technology	http://www.dlr.de/vt/en/

References

1 Lackner, M. Winter, F. and Agarwal, A.K. (eds) (2010) *Handbook of Combustion*, Wiley-VCH Verlag GmbH, Weinheim, ISBN: 978-3527324491.

2 Lundholm, K., Nordin, A., and Backman, R. (2007) Trace element speciation in combustion processes—Review and compilations of thermodynamic data. *Fuel Processing Technology*, **88** (11–12), 1061–1070.

3 Poinsot, T. and Veynante, D. (2012) *Theoretical and Numerical Combustion*, 3rd edn, European Centre for Research and Advanced Training in Scientific, Computation (CERFACS).

4 La Vision (2013) http://www.lavision.de/en/ (accessed May 30, 2013)

5 Lackner, M., Totschnig, G., and Winter, F. (2002) In-situ laser measurements of CO and CH_4 close to the surface of a burning single fuel particle. *Measurement Science and Technology*, **13** (10), 1545–1551.

6 Thipkhunthod, P., Meeyoo, V., Rangsunvigit, P., Kitiyanan, B., Siemanond, K., and Rirksomboon, T. (2005) Predicting the heating value of sewage sludges in Thailand from proximate and ultimate analyses. *Fuel*, **84** (7–8), 849–857.

7 Mohammed, M.A.A., Salmiaton, A., Wan Azlina, W.A.K.G., and Mohamad Amran, M.S. (2012) Gasification of oil palm empty fruit bunches: A characterization and kinetic study. *Bioresource Technology*, **110**, 628–636.

8 Westmoreland, P.R., Inguilizian, T., and Rotem, K. (2001) Flammability kinetics from TGA/DSC/GCMS, microcalorimetry and computational quantum chemistry. *Thermochimica Acta*, **367–368**, 401–405.

9 Sait, H.H., Hussain, A., Salema, A.A., and Ani, F.N. (2012) Pyrolysis and combustion kinetics of date palm biomass using thermogravimetric analysis. *Bioresource Technology*, **118**, 382–389.

10 Li, X.G., Lv, Y., Ma, B.G., Jian, S.W., and Tan, H.B. (2011) Thermogravimetric investigation on co-combustion characteristics of tobacco residue and high-ash anthracite coal. *Bioresource Technology*, **102** (20), 9783–9787.

11 Collazo, J., Pazó, J.A., Granada, E., Saavedra, Á., and Eguía, P. (2012) Determination of the specific heat of biomass materials and the combustion energy of coke by DSC analysis. *Energy*, **45** (1), 746–752.

12 Wichman, I.S. (2003) Material flammability, combustion, toxicity and fire hazard in transportation. *Progress in Energy and Combustion Science*, **29** (3), 247–299.

13 Vossoughi, S. (1986) TGA/DSC techniques as research tools for the study of the in-situ combustion process. *Thermochimica Acta*, **106**, 63–69.

14 Miziolek, A.W. (2009) *Laser Induced Breakdown Spectroscopy*, Cambridge

University Press, ISBN: 978-0521071000; Vassileva, G.C. and Vassilev, S.V. (2002) Relations between Ash-fusion temperatures and chemical and mineral composition of some bulgarian coals. *Comptes Rendus de l'Academie Bulgare des Sciences*, **55** (6), 61–66.

15 Szűcs, I., Szemmelveisz, K., and Wopera, Á. (2005) Characteristics of Slags at Biomass Combustion, Proceedings of the European Combustion Meeting 2005, ECM 2005, United Kingdom http://www.combustion.org.uk/ECM_2005/Papers/159_Szucs.pdf (accessed May 30, 2013).

16 Arvelakis, S., Folkedahl, B., Frandsen, F.J., and Hurley, J. (2008) Studying the melting behaviour of fly ash from the incineration of MSW using viscosity and heated stage XRD data. *Fuel*, **87** (10–11), 2269–2280.

17 Olanders, B. and Steenari, B.-M. (1995) Characterization of ashes from wood and straw. *Biomass and Bioenergy*, **8** (2), 105–115.

18 Gray, V.R. (1987) Prediction of ash fusion temperature from ash composition for some New Zealand coals. *Fuel*, **66** (9), 1230–1239.

19 Bouvet, N., Chauveau, C., Gökalp, I., Lee, SY., and Santoro, R.J. (2011) Characterization of syngas laminar flames using the Bunsen burner configuration. *International Journal of Hydrogen Energy*, **36** (1), 992–1005.

20 Fu, J., Tang, C., Jin, W., Thi, L.D., Huang, Z., and Zhang, Y. (2013) Study on laminar flame speed and flame structure of syngas with varied compositions using OH-PLIF and spectrograph. *International Journal of Hydrogen Energy*, **38** (3), 1636–1643.

21 Andrews, G.E. and Bradley, D. (1972) Determination of burning velocities: a critical review. *Combustion and Flame*, **18**, 133–153.

22 Boutefnouchet, H., Curfs, C., Triki, A., Boutefnouchet, A., and Vrel, D. (2012) Self-propagating high-temperature synthesis mechanisms within the Ti–C–Ni system: A time resolved X-ray diffraction study. *Powder Technology*, **217**, 443–450.

23 Ballester, J. and García-Armingol, T. (2010) Diagnostic techniques for the monitoring and control of practical flames. *Progress in Energy and Combustion Science*, **36** (4), 375–411.

24 Monkhouse, P. (2011) On-line spectroscopic and spectrometric methods for the determination of metal species in industrial processes. *Progress in Energy and Combustion Science*, **37** (2), 125–171.

25 Docquier, N. and Candel, S. (2002) Combustion control and sensors: a review. *Progress in Energy and Combustion Science*, **28** (2), 107–150.

26 Ballester, J. and García-Armingol, T. (2010) Diagnostic techniques for the monitoring and control of practical flames. *Progress in Energy and Combustion Science*, **36** (4), 375–411.

27 Lackner, M. (ed.) (2009) *Laser Diagnostics in Combustion*, ProcessEng Engineering GmbH, Wien, ISBN: 978-3-902655-03-5.

28 Chen, K., Mackie, J.C., Kennedy, E.M., and Dlugogorski, B.Z. (2012) Determination of toxic products released in combustion of pesticides. *Progress in Energy and Combustion Science*, **38** (3), 400–418.

29 Simmie, J.M. (2003) Detailed chemical kinetic models for the combustion of hydrocarbon fuels. *Progress in Energy and Combustion Science*, **29** (6), 599–634.

30 Hansen, N., Cool, T.A., Westmoreland, P.R., and Kohse-Höinghaus, K. (2009) Recent contributions of flame-sampling molecular-beam mass spectrometry to a fundamental understanding of combustion chemistry. *Progress in Energy and Combustion Science*, **35** (2), 168–191.

31 Lupant, D., Pesenti, B., and Lybaert, P. (2010) Influence of probe sampling on reacting species measurement in diluted combustion. *Experimental Thermal and Fluid Science*, **34** (5), 516–522.

32 Costa, M.A.M., CarvalhoJ Jr., J.A., Soares Neto, T.G., Anselmo, E., Lima, B.A., Kura, L.T.U., and Santos, J.C. (2012) Real-time sampling of particulate matter smaller than 2.5 μm from Amazon forest biomass combustion. *Atmospheric Environment*, **54**, 480–489.

33 Williams, A., Jones, J.M., Ma, L., and Pourkashanian, M. (2012) Pollutants from the combustion of solid biomass fuels. *Progress in Energy and Combustion Science*, **38** (2), 113–137.

34 Werther, J. (1999) Measurement techniques in fluidized beds. *Powder Technology*, **102** (1), 15–36.

35 Lomas, C.G. (2011) *Fundamentals of Hot Wire Anemometry*, Cambridge University Press, Cambridge, Reissue, ISBN: 978-0521283182.

36 Parker, K.H. and Guillon, O. (1971) Local measurements in a turbulent flame by hot-wire anemometry. *Symposium (International) on Combustion*, **13** (1), 667–674.

37 Checkel, M.D. and Thomas, A. (1994) Turbulent combustion of premixed flames in closed vessels. *Combustion and Flame*, **96** (4), 351–370.

38 Heitor, M.V. and Moreira, A.L.N. (1993) Thermocouples and sample probes for combustion studies. *Progress in Energy and Combustion Science*, **19** (3), 259–278.

39 Kerlin, T.W. (1999) *Practical Thermocouple Thermometry*, Isa-Instrument Society of America, ISBN: 978-1556176449.

40 Srinivasan, K., Sundararajan, T., Narayanan, S., Jothi, T.J.S., and Rohit Sarma, C.S.L.V. (2013) Acoustic pyrometry in flames. *Measurement*, **46** (1), 315–323.

41 Tsai, C.-L., Fann, C.-S., Wang, S.-H., and Fung, R.-F. (2001) Paramagnetic oxygen measurement using an optical-fiber microphone. *Sensors and Actuators B: Chemical*, **73** (2–3), 211–215.

42 Sheffield Department of Anaesthesia (2013) http://www.soa.group.shef.ac.uk/museum/paramagnetic_oxygen_anal.html (accessed May 30, 2013).

43 Lackner, M. (ed.) (2008) *Lasers in Chemistry – Optical Probes and Reaction Starters*, vol. **2**, Wiley-VCH Verlag GmbH, Weinheim, ISBN: 978-3527319978.

44 Siegman, A.E. (1986) *Lasers*, University Science Books, ISBN: 978-0935702118.

45 Weldon, V., O'Gorman, J., Pérez-Camacho, J.J., McDonald, D., Hegarty, J., Connolly, J.C., Morris, N.A., Martinelli, R.U., and Abeles, J.H. (1997) Laser diode based oxygen sensing: A comparison of VCSEL and DFB laser diodes emitting in the 762 nm region. *Infrared Physics & Technology*, **38** (6), 325–329.

46 Brown, D.M., Downey, E., Kretchmer, J., Michon, G., Shu, E., and Schneider, D. (1998) SiC flame sensors for gas turbine control systems. *Solid-State Electronics*, **42** (5), 755–760.

47 Ballester, J. and García-Armingol, T. (2010) Diagnostic techniques for the monitoring and control of practical flames. *Progress in Energy and Combustion Science*, **36** (4), 375–411.

48 Riedel, F. and Stössel, Z. (1993) A fail-safe sensor for flame detection. *Sensors and Actuators A: Physical*, **37-38**, 534–539.

49 Jakob, M., Hülser, T., Janssen, A., Adomeit, P., Pischinger, S., and Grünefeld, G. (2012) Simultaneous high-speed visualization of soot luminosity and OH^* chemiluminescence of alternative-fuel combustion in a HSDI diesel engine under realistic operating conditions. *Combustion and Flame*, **159** (7), 2516–2529.

50 Kasahara, J., Horii, T., Endo, T., and Fujiwara, T. (1996) Experimental observation of unsteady H_2-O_2 combustion phenomena around hypersonic projectiles using a multiframe camera. *Symposium (International) on Combustion*, **26** (2), 2903–2908.

51 Zhang, J., Jing, W., and Fang, T. (2012) High speed imaging of OH^* chemiluminescence and natural luminosity of low temperature diesel spray combustion. *Fuel*, **99**, 226–234.

52 Walsh, K.T., Long, M.B., Tanoff, M.A., and Smooke, M.D. (1998) Experimental and computational study of CH, CH^*, and OH^* in an axisymmetric laminar diffusion flame, Twenty-Seventh Symposium (International) on Combustion/The Combustion Institute, pp. 615–623.

53 Mercer, C. (2003) *Optical Metrology for Fluids, Combustion and Solids*, Springer, ISBN: 978-1402074073.

54 Settles, G.S. (2001) *Schlieren and Shadowgraph Techniques: Visualizing Phenomena in Transparent Media*, 2nd edn, Springer, Berlin Heidelberg, ISBN: 978-3540661559.

55 Ray, S.F. (1999) *Scientific Photography and Applied Imaging*, Focal Press, ISBN: 978-0240513232.

56 Park, J.S. and Yi, S.H. (2010) Temperature compensated NDIR CH_4 gas sensor with focused beam structure. *Procedia Engineering*, **5**, 1248–1251.

57 Haridass, C., Aw-Musse, A., Misra, P., and Jordan, J. (2000) Fourier Transform infrared (FT-IR) spectroscopy of trace molecular species of importance for the elucidation of atmospheric phenomena. *Computers & Electrical Engineering*, **26** (1), 4.

58 Griffiths, P.R., De Haseth, J.A., and Winefordner, J.D. (2007) *Fourier Transform Infrared Spectrometry*, 2nd edn, JohnWiley & Sons Inc., ISBN: 978-0471194040.

59 Wartewig, S. (2003) *IR and Raman Spectroscopy: Fundamental Processing*, Wiley-VCH Verlag GmbH, Weinheim, ISBN: 978-3527302451.

60 Smith, B.C. (1995) *Fundamentals of Fourier Transform Infrared Spectroscopy*, CRC Press Inc., ISBN: 978-0849324611.

61 Schramm, D.U., Sthel, M.S., da Silva, M.G., Carneiro, L.O., Junior, A.J.S., Souza, A.P., and Vargas, H. (2003) Application of laser photoacousticspectroscopy for the analysis of gas samples emitted by diesel engines. *Infrared Physics & Technology*, **44** (4), 263–269.

62 http://analyticjournal.de/anwendungsbereiche_gase/pastech_spurengase_laser.html (2013). (accessed May 30, 2013)

63 Stenberg, J., Åmand, L.-E., Hernberg, R., and Leckner, B. (1998) Measurements of gas concentrations in a fluidized bed combustor using laser-induced photoacoustic spectroscopy and Zirconia cell probes. *Combustion and Flame*, **113** (4), 477–486.

64 Wang, Q., Ma, Z., and He, Z. (2011) A novel photoacoustic spectroscopy system using diphragm based fiber Fabry-Perot sensor. *Procedia Engineering*, **15**, 5395–5399.

65 Ajtai, T., Filep, Á., Schnaiter, M., Linke, C., Vragel, M., Bozóki, Z., Szabó, G., and Leisner, T. (2010) A novel multi-wavelength photoacoustic spectrometer for the measurement of the UV–vis-NIR spectral absorption coefficient of atmospheric aerosols. *Journal of Aerosol Science*, **41** (11), 1020–1029.

66 Wang, Q., Wang, J., Li, L., and Yu, Q. (2011) An all-optical photoacoustic spectrometer for trace gas detection. *Sensors and Actuators B: Chemical*, **153** (1), 214–218.

67 http://www.cfa.harvard.edu/hitran/ (2013) (accessed May 30, 2013).

68 Ballester, J. and García-Armingol, T. (2010) Diagnostic techniques for the monitoring and control of practical flames. *Progress in Energy and Combustion Science*, **36** (4), 375–411.

69 Lackner, M. (2007) Tunable diode laser spectroscopy in the process industries: a review (invited). *Reviews in Chemical Engineering*, **23** (2), 65–147.

70 Shemshad, J., Aminossadati, S.M., and Kizil, M.S. (2012) A review of developments in near infrared methane detection based on tunable diode laser. *Sensors and Actuators B: Chemical*, **171–172**, 77–92.

71 Ruxton, K., Chakraborty, A.L., Johnstone, W., Lengden, M., Stewart, G., and Duffin, K. (2010) Tunable diode laser spectroscopy with wavelength modulation: Elimination of residual amplitude modulation in a phasor decomposition approach. *Sensors and Actuators B: Chemical*, **150** (1), 367–375.

72 Cai, T., Wang, G., Zhang, W., and Gao, X. (2012) Simultaneous measurement of CO and CO_2 at elevated temperatures by diode laser wavelength modulated spectroscopy. *Measurement*, **45** (8), 2089–2095.

73 Seufert, J., Fischer, M., Legge, M., Koeth, J., Werner, R., Kamp, M., and Forchel, A. (2004) DFB laser diodes in the wavelength range from 760 nm to 2.5 µm. *Spectrochimica Acta Part A: Molecular and Biomolecular Spectroscopy*, **60** (14), 3243–3247.

74 Luque, J., Jeffries, J.B., Smith, G.P., Crosley, D.R., and Scherer, J.J. (2001) Combined cavity ringdown absorption and laser-induced fluorescence imaging measurements of CN(B-X) and CH(B-X) in low-pressure CH_4-O_2-N_2 and CH_4-NO-

O_2-N_2 flames. *Combustion and Flame*, **126** (3), 1725–1735.

75 Li, L., Chen, J., Chen, H., Yang, X., Tang, Y., and Zhang, R. (2011) Monitoring optical properties of aerosols with cavity ring-down spectroscopy. *Journal of Aerosol Science*, **42** (4), 277–284.

76 Schulz, C. and Sick, V. (2005) Tracer-LIF diagnostics: quantitative measurement of fuel concentration, temperature and fuel/air ratio in practical combustion systems. *Progress in Energy and Combustion Science*, **31** (1), 75–121.

77 Williams, B., Ewart, P., Wang, X., Stone, R., Ma, H., Walmsley, H., Cracknell, R., Stevens, R., Richardson, D., Fu, H., and Wallace, S. (2010) Quantitative planar laser-induced fluorescence imaging of multi-component fuel/air mixing in a firing gasoline-direct-injection engine: Effects of residual exhaust gas on quantitative PLIF. *Combustion and Flame*, **157** (10), 1866–1878.

78 Orain, M., Grisch, F., Jourdanneau, E., Rossow, B., Guin, C., and Trétout, B. (2009) Simultaneous measurements of equivalence ratio and flame structure in multipoint injectors using PLIF. *Comptes Rendus Mécanique*, **337** (6–7), 373–384.

79 Schulz, C. and Sick, V. (2005) Tracer-LIF diagnostics: quantitative measurement of fuel concentration, temperature and fuel/air ratio in practical combustion systems. *Progress in Energy and Combustion Science*, **31** (1), 75–121.

80 Daily, J.W. (1997) Laser induced fluorescence spectroscopy in flames. *Progress in Energy and Combustion Science*, **23** (2), 133–199.

81 Bougie, B., Ganippa, L.C., van Vliet, A. P., Meerts, W.L., Dam, N.J., and ter Meulen, J.J. (2007) Soot particulate size characterization in a heavy-duty diesel engine for different engine loads by laser-induced incandescence. *Proceedings of the Combustion Institute*, **31** (1), 685–691.

82 Menkiel, B., Donkerbroek, A., Uitz, R., Cracknell, R., and Ganippa, L. (2012) Measurement of in-cylinder soot particles and their distribution in an optical HSDI diesel engine using time resolved laser induced incandescence (TR-LII).

Combustion and Flame, **159** (9), 2985–2998.

83 Ryser, R., Gerber, T., and Dreier, T. (2009) Soot particle sizing during high-pressure Diesel spray combustion via time-resolved laser-induced incandescence. *Combustion and Flame*, **156** (1), 120–129.

84 Kock, B.F., Tribalet, B., Schulz, C., and Roth, P. (2006) Two-color time-resolved LII applied to soot particle sizing in the cylinder of a Diesel engine. *Combustion and Flame*, **147** (1–2), 79–92.

85 Keck, C.M. and Müller, R.H. (2008) Size analysis of submicron particles by laser diffractometry—90% of the published measurements are false. *International Journal of Pharmaceutics*, **355** (1–2), 150–163.

86 Zhao, F.-Q. and Hiroyasu, H. (1993) The applications of laser Rayleigh scattering to combustion diagnostics. *Progress in Energy and Combustion Science*, **19** (6), 447–485.

87 Espey, C., Dec, J.E., Litzinger, T.A., and Santavicca, D.A. (1997) Planar laser Rayleigh scattering for quantitative vapor-fuel imaging in a diesel jet. *Combustion and Flame*, **109** (1–2), 65–78, IN1-IN4, 79–86.

88 Dibble, R.W. and Hollenbach, R.E. (1981) Laser Rayleigh thermometry in turbulent flames. *Symposium (International) on Combustion*, **18** (1), 1489–1499.

89 Tu, Q. and Chang, C. (2012) Diagnostic applications of Raman spectroscopy Nanomedicine: Nanotechnology. *Biology and Medicine*, **8** (5), 545–558.

90 Sullivan, J.A., Dulgheru, P., Atribak, I., Bueno-López, A., and García-García, A. (2011) Attempts at an in situ Raman study of ceria/zirconia catalysts in PM combustion. *Applied Catalysis B: Environmental*, **108–109**, 134–139.

91 Eckbreth, A.C. (1988) Nonlinear Raman spectroscopy for combustion diagnostics. *Journal of Quantitative Spectroscopy and Radiative Transfer*, **40** (3), 369–383.

92 Larkin, P. (2011) *Infrared and Raman Spectroscopy; Principles and Spectral Interpretation*, Elsevier, ISBN: 978-0123869845.

93 Kiefer, J. and Ewart, P. (2011) Laser diagnostics and minor species detection in combustion using resonant four-wave

mixing. *Progress in Energy and Combustion Science*, **37** (5), 525–564.

94 Grisch, F., Bertseva, E., Habiballah, M., Jourdanneau, E., Chaussard, F., Saint-Loup, R., Gabard, T., and Berger, H. (2007) CARS spectroscopy of CH_4 for implication of temperature measurements in supercritical LOX/CH_4 combustion. *Aerospace Science and Technology*, **11** (1), 48–54.

95 Roy, S., Gord, J.R., and Patnaik, A.K. (2010) Recent advances in coherent anti-Stokes Raman scattering spectroscopy: Fundamental developments and applications in reacting flows. *Progress in Energy and Combustion Science*, **36** (2), 280–306.

96 El-Diasty, F. (2011) Coherent anti-Stokes Raman scattering: Spectroscopy and microscopy. *Vibrational Spectroscopy*, **55** (1), 1–37.

97 Boyd, R.W. (2008) *Nonlinear Optics*, 3rd revised edn, Academic Press, ISBN: 978-0123694706.

98 Frederickson, K., Kearney, S.P., Grasser, T., Castaneda, J., Luketa, A., and Hewson, J. (2013) Diagnostic Development for Determining the Joint Temperature/Soot Statistics in Hydrocarbon-Fueled Pool Fires: LDRD, SANDIA REPORT SAND2009-6591, September 2009 http://prod.sandia.gov/techlib/access-control.cgi/2009/096591.pdf (accessed May 30, 2013).

99 Zhang, Z. (2012) *LDA Application Methods: Laser Doppler Anemometry for Fluid Dynamics*, Springer, ISBN: 978-3642264580.

100 Raffel, M., Willert, C.E., Wereley, S.T., and Kompenhans, J. (2007) *Particle Image Velocimetry: A Practical Guide*, 2nd edn, Springer, Berlin Heidelberg, ISBN: 978-3540723073.

101 Soid, S.N. and Zainal, Z.A. (2011) Spray and combustion characterization for internal combustion engines using optical measuring techniques – A review. *Energy*, **36** (2), 724–741.

102 Hernández-Jiménez, F., Sánchez-Delgado, S., Gómez-García, A., and Acosta-Iborra, A. (2011) Comparison between two-fluid model simulations and particle image analysis & velocimetry (PIV) results for a two-dimensional gas–solid fluidized bed. *Chemical Engineering Science*, **66** (17), 3753–3772.

103 Aldén, M., Bood, J., Li, Z., and Richter, M. (2011) Visualization and understanding of combustion processes using spatially and temporally resolved laser diagnostic techniques. *Proceedings of the Combustion Institute*, **33** (1), 69–97.

104 Boxx, I., Heeger, C., Gordon, R., Böhm, B., Aigner, M., Dreizler, A., and Meier, W. (2009) Simultaneous three-component PIV/OH-PLIF measurements of a turbulent lifted, C_3H_8-Argon jet diffusion flame at 1.5 kHz repetition rate. *Proceedings of the Combustion Institute*, **32** (1), 905–912.

105 Black, D.L., McQuay, M.Q., and Bonin, M.P. (1996) Laser-based techniques for particle-size measurement: A review of sizing methods and their industrial applications. *Progress in Energy and Combustion Science*, **22** (3), 267–306.

106 Park, S.H., Youn, I.M., and Lee, C.S. (2011) Influence of ethanol blends on the combustion performance and exhaust emission characteristics of a four-cylinder diesel engine at various engine loads and injection timings. *Fuel*, **90** (2), 748–755.

107 Singh, S., Musculus, M.P.B., and Reitz, R.D. (2009) Mixing and flame structures inferred from OH-PLIF for conventional and low-temperature diesel engine combustion. *Combustion and Flame*, **156** (10), 1898–1908.

108 Ashgriz, N. (2011) *Handbook of Atomization and Sprays: Theory and Applications*, Springer, ISBN: 978-1441972637.

109 Kunze, H.-J. (2009) *Introduction to Plasma Spectroscopy*, Springer, ISBN: 978-3642022326.

110 Docquier, N. and Candel, S. (2002) Combustion control and sensors: a review. *Progress in Energy and Combustion Science*, **28** (2), 107–150.

111 Ballester, J. and García-Armingol, T. (2010) Diagnostic techniques for the monitoring and control of practical flames. *Progress in Energy and Combustion Science*, **36** (4), 375–411.

112 Cheskis, S. and Goldman, A. (2009) Laser diagnostics of trace species in low-pressure flat flame. *Progress in Energy and Combustion Science*, **35** (4), 365–382.

113 Wagner, S., Fisher, B.T., Fleming, J.W., and Ebert, V. (2009) TDLAS-based in situ measurement of absolute acetylene concentrations in laminar 2D diffusion flames. *Proceedings of the Combustion Institute*, **32** (1), 839–846.

114 Lackner, M. (ed.) (2010) *Combustion Synthesis: Novel Routes to Novel Materials*, Bentham Science, eISBN: 978-1-60805-155-7, http://www.bentham.org/ebooks/9781608051557/.

115 Qin, X., Xiao, X., Puri, I.K., and Aggarwal, S.K. (2002) Effect of varying composition on temperature reconstructions obtained from refractive index measurements in flames. *Combustion and Flame*, **128** (1–2), 121–132.

116 Syred, N. (2006) A review of oscillation mechanisms and the role of the precessing vortex core (PVC) in swirl combustion systems. *Progress in Energy and Combustion Science*, **32** (2), 93–161.

117 Beér, J.M. (2000) Combustion technology developments in power generation in response to environmental challenges. *Progress in Energy and Combustion Science*, **26** (4–6), 301–327.

118 http://energy.lbl.gov/aet/combustion/LSC-Info/ (accessed May 30, 2013).

119 Fuest, F., Barlow, R.S., Chen, J.-Y., and Dreizler, A. (2012) Raman/Rayleighscattering and CO-LIF measurements in laminar and turbulent jet flames of dimethyl ether. *Combustion and Flame*, **159** (8), 2533–2562.

120 Hu, S., Pitz, R.W., and Wang, Y. (2009) Extinction and near-extinction instability of non-premixed tubular flames. *Combustion and Flame*, **156** (1), 90–98.

121 Fuse, T., Kobayashi, N., and Hasatani, M. (2005) Combustion characteristics of ethanol in a porous ceramic burner and ignition improved by enhancement of liquid-fuel intrusion in the pore with ultrasonic irradiation. *Experimental Thermal and Fluid Science*, **29** (4), 467–476.

122 Soid, S.N. and Zainal, Z.A. (2011) Spray and combustion characterization for internal combustion engines using optical

measuring techniques – A review. *Energy*, **36** (2), 724–741.

123 http://www1.phc.uni-kiel.de/cms/index.php/de/research-m-gfr/147-shock-tube-investigations-of-high-temperature-react-kin.html (2013) (accessed May 30, 2013).

124 Fang, T., Coverdill, R.E., Lee, C.-fonF., and White, R.A. (2008) Effects of injection angles on combustion processes using multiple injection strategies in an HSDI diesel engine. *Fuel*, **87** (15–16), 3232–3239.

125 http://www.ifk.uni-stuttgart.de/forschung/exp_ein/halbtechnisch/festbett/rofea.en.html (2013) (accessed May 30, 2013).

126 Lackner, M. (ed.) (2009) *Scale-up in Combustion*, ProcessEng Engineering GmbH, Wien, ISBN: 978-3-902655-04-2.

127 Randazzo, M.L. and Sodré, J.R. (2011) Exhaust emissions from a diesel powered vehicle fuelled by soybean biodiesel blends (B3-B20) with ethanol as an additive (B20E2-B20E5). *Fuel*, **90** (1), 98–103.

128 Teichert, H., Fernholz, T., and Ebert, V. (2003) Simultaneous In Situ Measurement of CO, H_2O, and Gas Temperatures in a Full-Sized Coal-Fired Power Plant by Near-Infrared Diode Lasers. *Applied Optics*, **42**, 2043–2051. http://www.opticsinfobase.org/ao/abstract.cfm?URI=ao-42-12-2043

129 Sippula, O., Hokkinen, J., Puustinen, H., Yli-Pirilä, P., and Jokiniemi, J. (2009) Comparison of particle emissions from small heavy fuel oil and wood-fired boilers. *Atmospheric Environment*, **43** (32), 4855–4864.

130 Szentannai, P. (ed.) (2010) *Power Plant Applications of Advanced Control Techniques*, ProcessEng Engineering, Vienna, ISBN: 978-3902655110.

131 http://www.sae.org/mags/tbe/SFTY/6181 (2013).

132 Rogers, D.R. (2010) *Engine Combustion – Pressure Measurement and Analysis*, SAE International, ISBN: 978-0768019636.

133 Cheskis, S. and Goldman, A. (2009) Laser diagnostics of trace species in low-pressure flat flame. *Progress in Energy and Combustion Science*, **35** (4), 365–382.

134 Musculus, M.P.B. and Pickett, L.M. (2005) Diagnostic considerations for optical laser-extinction measurements of soot in high-pressure transient combustion environments. *Combustion and Flame*, **141** (4), 371–391.

135 Nathan, G.J., Kalt, P.A.M., Alwahabi, Z.T., Dally, B.B., Medwell, P.R., and Chan, Q.N. (2012) Recent advances in the measurement of strongly radiating, turbulent reacting flows. *Progress in Energy and Combustion Science*, **38** (1), 41–61.

136 Durst, F. (1990) Optical techniques for fluid flow and heat transfer. *Experimental Thermal and Fluid Science*, **3** (1), 33–51.

137 Zhang, Z. (2010) *LDA Application Methods: Laser Doppler Anemometry for Fluid Dynamics*, Springer, ISBN: 978-3642135132.

138 Warda, H.A., Kassab, S.Z., Elshorbagy, K.A., and Elsaadawy, E.A. (1999) An experimental investigation of the near-field region of a free turbulent coaxial jet using LDA. *Flow Measurement and Instrumentation*, **10** (1), 15–26.

139 Vouitsis, E., Ntziachristos, L., and Samaras, Z. (2003) Particulate matter mass measurements for low emitting diesel powered vehicles: what's next? *Progress in Energy and Combustion Science*, **29** (6), 635–672.

140 Widmann, J.F. and Presser, C. (2002) A benchmark experimental database for multiphase combustion model input and validation. *Combustion and Flame*, **129** (1–2), 47–86.

141 Weldon, V., O'Gorman, J., Pérez-Camacho, J.J., McDonald, D., Hegarty, J., Connolly, J.C., Morris, N.A., Martinelli, R.U., and Abeles, J.H. (1997) Laser diode based oxygen sensing: A comparison of VCSEL and DFB laser diodes emitting in the 762 nm region. *Infrared Physics & Technology*, **38** (6), 325–329.

6
Applications

6.1
Burners

6.1.1
The Evolution of Combustion Processes

The discovery and taming of fire is thought to be one of the most significant steps in human evolution. Fire provided the early man, *homo erectus*, with a more nutritious diet and an expanded range of capabilities which, in turn, contributed towards accelerated brain development. Industrial technology, as we know it today, would not exist without the early discovery of fire.

In the past, the development and exploration of different combustion technologies were mostly guided by the demand for steel and other metals. Industrialization and war efforts required previously unseen amounts of shapeable metals, the production of which would have been impossible without proper firing systems. It is thus no surprise that different types of auxiliary devices, like blowers, preheaters and heat exchangers were developed with the metallurgical application in mind. Heavy industry also required heavy amounts of electricity, thus the appearing power generation technologies followed the development of smelting furnaces.

6.1.2
The Flame

In industrial heating equipment, such as furnaces, combustion can be initiated by supplying sufficient ignition energy to a proper mixture of fuel and oxidant.

The combustion process generates a range of thermal and luminous phenomena, which we collectively refer to as the flame. The geometry of the flame, achieved burn-out, and heat transfer properties are affected by a variety of parameters, such as the mixing and reaction rates, the geometry of the combustion chamber, and the physical state and quality of the fuel and oxidant.

The set-up of a simple industrial burner application is shown in Figure 6.1.

Figure 6.2 illustrates a typical, semi-luminous propane flame.

Combustion: From Basics to Applications, First Edition. Maximilian Lackner, Árpád B. Palotás, and Franz Winter.
© 2013 Wiley-VCH Verlag GmbH & Co. KGaA. Published 2013 by Wiley-VCH Verlag GmbH & Co. KGaA.

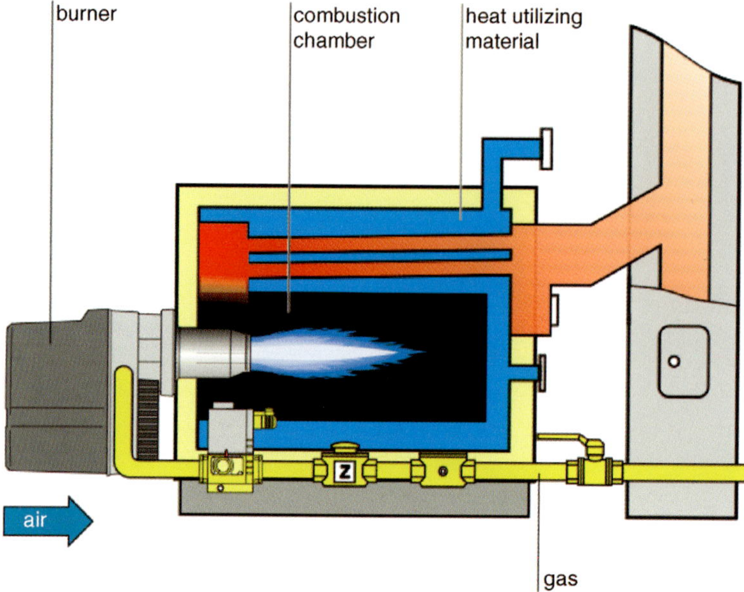

Figure 6.1 Schematic of the combustion system (Courtesy of Weishaupt GmbH).

6.1.3
Fuel Preparation, Pre-Processing

Fuel and oxidant have to be mixed in order to prepare a combustible (ignitable) mixture. The oxidant is rarely pure oxygen – in most practical cases atmospheric air is used. For the mixing process fuels have to be prepared. Depending on the type of

Figure 6.2 Propane flame with slightly luminous tail, indicating some soot formation and corresponding increase in flame emissivity.

fuel, some pre-processing is usually needed in order to produce an efficiently combustible material.

Once produced and refined, *gaseous fuels* do not require much pre-processing apart from the separation of potential liquid and suspended solid contaminants. Gases burn in homogeneous (mostly gas–gas) reactions, which are much less hindered by heterogeneous contamination than the combustion processes of liquid or solid fuels.

Liquid and solid fuels burn in heterogeneous, liquid–gas or solid–gas surface reactions. Heterogeneous reactions can be heavily suppressed by the presence of surface contaminants which, therefore, must be removed before combustion. The typical volume of air needed for combustion is about four times the volume of the liquid or solid fuel to be combusted. In order to achieve complete combustion, a sufficiently large surface area must be available for the heterogeneous reactions. In practical cases, this surface area is achieved by atomization or grinding processes.

Oils are viscous liquids. Their high viscosity makes mechanical atomization fairly inefficient. Fuel oils are, therefore, often pre-heated before atomization, as their viscosities usually tend to decrease with temperature.

Solid fuels are ground or pulverized before combustion, depending on their reaction rates and heating values. In a particulate form, the transportation of the fuel becomes easier as well.

The properties of the fuel affect the geometry of the combustion chamber. Since combustion processes are often designed with a specific desired burn-out rate, slower reactions require longer residence times and, therefore, larger combustion chambers. The physical properties of liquid or solid fuels affect the rate of heterogeneous reactions, therefore, with proper pre-processing the economics of the combustion technology can be improved.

6.1.4
Requirements of a Burner

In industrial practice, there is no commonly accepted, exact definition for burners. There are, however, a number of requirements that a burner must meet.

In most cases, a burner must:

- supply the proper amount of fuel(s) and oxidant(s) for the combustion process,
- propel a jet of fuel and oxidant in order to mix them (except for premixed burners),
- produce the desired flame geometry and heat output,
- ignite a fuel–oxidant mixture reliably and safely,
- produce a stable flame after ignition,
- provide means for controlling the flame.

There are various characterization or classification systems for burners. The most widely accepted characterization is based on

- the types and properties of fuel(s) they burn,
- their nominal thermal power or fuel consumption,
- the pressure of the fuels and the oxidant.

Burners may have some additional, although construction-dependent features:

- the capability to control the flame shape (length), via, for example, swirl parameters,
- the capability to change between fuel-lean or fuel-rich flames,
- the capability of burning more fuels simultaneously or alternately,
- the capability of introducing combustion additives (e.g., water, steam).

Some of the burner's most important characteristics and information must be indicated on the burner body – the exact information is dictated by standards. These labels usually include the following:

- name of the manufacturer,
- type of the burner,
- nominal capacity and/or power consumption
- serial number,
- date of manufacturing.

There are some additional requirements that burners should meet:

- they should be simple to operate, assemble and disassemble,
- there can be little or no control hysteresis,
- long service time is expected,
- minimum maintenance should be required.

6.1.5
Burner Classification by the Fuel Used

Burners can be classified by the fuel used, such as

- gas burners,
- oil burners,
- pulverized coal burners,
- combined (multi-fuel) burners.

The fuel can be specified in more detail, e.g., gas burners can be classified such as:

- natural gas,
- coke oven (chamber) gas,
- blast furnace (throat) gas,
- propane–butane,
- acetylene,
- and so on.

6.1.6
Burner Categories

The two most common burner categories are determined by the way the fuel comes into contact with the oxidant. In this sense, burners can be

- open flame, or direct fired burners, or
- indirect, closed combustion chamber burners (radiant tube furnaces, immersion tubes, radiant elements).

Most of the equipment in practice belongs to the former group (direct combustion). There are, however, various technical reasons why indirect heating might be necessary.

Some burners can operate without additional auxiliary equipment for air transportation. These burners are referred to as atmospheric, injection or entrained-air burners and are typically small-scale. Larger-scale and higher capacity burners often require fans and/or compressors to achieve the desired equivalence ratio in their flames. These are called forced-air or ventilated burners (Figure 6.3).

6.1.6.1 Classification Methods for Gas Burners

Gas burners can be classified based on how they contact the fuel gas with the oxidant:

- premixed natural gas burners with air inlet:
    - atmospheric gas burners (the combustion chamber is operated slightly below atmospheric pressure, air intrusion is not forced)
    - injector burner (air is introduced by the energy of the fuel jet),

- premixed burner with gas–air mixture inlet,

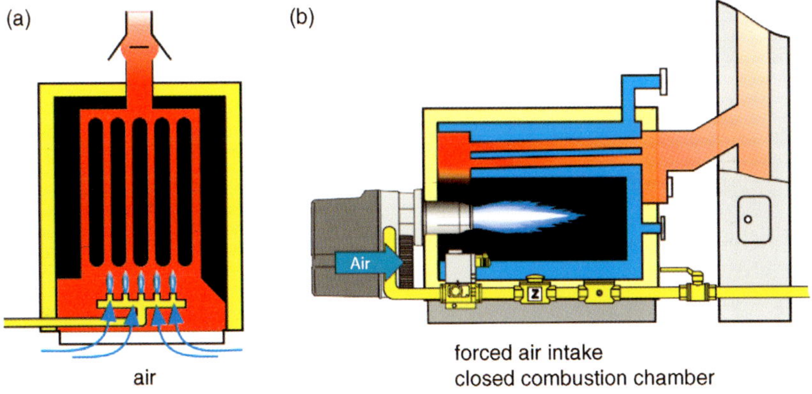

Figure 6.3 Comparison of an atmospheric (a) and a fan (forced air) burner (b) (Courtesy of Max Weishaupt GmbH).

- forced-air burner (with two pipelines), or nozzle-mix gas burners (most gas burners)
    - gas burners with air pipes,
    - block gas burners (with built-in blower system),
    - duo-bloc gas burners (with built-in blower system to each burner),
    - recuperative gas burners (with built-in recuperator),
    - regenerative gas burners (with built-in regenerator),

The classification can be based on the number of steps in which the oxidant is introduced to the fuel stream, for example:

- single-step gas burners,
- gas burners with multi-step gas introduction,
- gas burners with multi-step oxidant introduction.

Burner classifications can be based on flame mixing and flow properties:

- laminar flame gas burners
- turbulent gas burners

Classification can also be based on the flame shape:

- jet-flame burners,
- flat flame burner,
- "flameless" (dark radiant) gas burners.

According to the flow velocity of the burner:

- conventional or low velocity gas burner ($w \leq 30\,\mathrm{m\,s^{-1}}$)
- intermediate velocity or tunnel gas burners ($w \leq 60\,\mathrm{m\,s^{-1}}$)
- high velocity or impulse burners ($w = 100\text{--}120\,\mathrm{m\,s^{-1}}$)

Special burners are designed for different industries, for example,

- the burners of rotary kilns,
- the gas jet burners of glass smelting furnaces,
- the oxy-fuel burners of steelmaking electric arc furnaces and glass furnaces,
- the roof burners of tunnel furnaces,
- the radiant panel burners of the furnaces walls of oil refineries,
- flaring burners,
- channel burners, and so on.

None of these classification methods are exact. A burner can operate at various pressures, temperatures, even with different fuels. The same type of burner can be scaled up to orders of magnitude higher capacities, yet it remains essentially the same design – a capacity-based categorization would place it in a different group. There are burners that can be used at home (domestic), in a church (community) or in a small factory (industrial). This is why categories like those listed above may serve a certain market segment or manufacturer, but a better concept is needed to have a generally meaningful classification system for gas burners.

6.1.6.2 Generalized Classification of Gas Burners

The structure of gas burners defines how and when the fuel mixes with the oxidizer agent (mostly, but not exclusively, atmospheric air), and consequently the operating conditions of the burner. The properties of the flame (speed, width, length, color, temperature) determine the shape of the combustion chamber, as well as – through the dominant heat transfer method – the applicability of the burner in various technologies. The realization of these principles led to the creation of an internationally accepted simple character-coded classification system [26].

The codes are the following:

By the location, where the combustible mixture is created

- inside the burner: I
- outside of the burner: O

Burners with internal mixing have a separate inlet for the fuel and the air. After the internal mixing a combustible mixture is created and it exits through the nozzles as a jet. External mixing means that the air and the fuel do not mix within the burner,

The following characters are assigned if the jet

- is perpendicular to the axis: P
- has rotation or spin: R
- is diverging: D

If a swirl flame is needed, the tangential component can be introduced by

- spinning elements: E
- by holes with angled axes: A
- a cyclone: C

If the combustion happens effectively inside the burner, then it is called a

- tunnel burner: T

Atmospheric burners have no air inlets. When the fuel leaves the nozzles, air is entrained and a combustible mixture forms.

- diffusion burner: K

Based on these letters or codes, most conventional gas burners can be classified into one of the eight basic types:

1) IP burner – Internal mixing, parallel jet. A typical category for injection burners and the simple ventilation burners. The flame has no swirl.
2) OP burner – External mixing burner, where the air leaves the burner through holes with axes parallel to the burner axis. The air flow encloses the fuel jet. Typically long flame burner, no vorticity is generated. Due to the relatively high velocity, intensive heat transfer can be achieved via convection. This type is used in long horizontal rotating kilns, as well as for aluminum melting.
3) IRE burner – Internal mixing burner where the swirl is achieved by the use of rotating/spinning parts.

4) ORE burner – External mixing. Rotating parts give a swirl to the air stream only, while the fuel exits through the nozzle(s) in the center axis.

5) ORA burner – Same as before, but swirl is created by holes whose axes do not intersect.

6) IRC burner – Internal mixing eddy burner. The fuel and the air enter the burner tangentially and produce a short flame with the strongest swirl among the eight basic gas burner types. The rounded burner stone is usually made of fire resistant concrete or ceramic and is an integral part of the burner. The high swirl flame follows the curvature of the burner stone and spreads along the furnace wall or ceiling. Often used as flat flame burners built into the ceiling of large pusher type or walking-beam furnaces. Advantageous when "soft" heat transfer is required via radiation.

7) ID burner – The internally prepared air-fuel mixture exits the burner at the surface of either a cone or a cylinder. A short and wide flame is produced without swirl.

8) OK burner – External mixing, diffusion burner where only the fuel is introduced to the burner. As the fuel leaves the nozzles the jet or spray mixes with air in the surrounding environment.

Figure 6.4 shows a series of sketches for each of these basic burner types with their characteristic flame shapes.

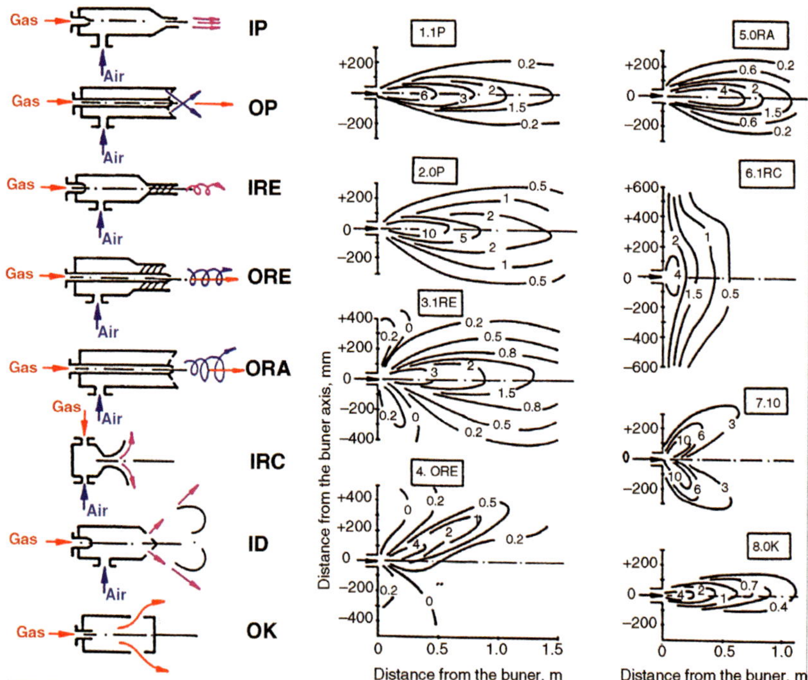

Figure 6.4 Basic gas burner types and their characteristic flame shapes (including velocity data) [26].

Figure 6.5 The ORE type gas burner head capable of producing attached flames. Illustrates the purpose of the air nozzle orientation and the two swirling elements.

Figure 6.5 illustrates a fairly complex system of nozzles and deflection panels for the control of flame shape of an ORA burner.

6.1.7
Burner Control, Automation

The automation of burners can have tremendous advantages. A well-controlled burner is not only safer, but operates much more efficiently. Even a small reduction in fuel consumption can be very beneficial economically in the long term.

The major types of burner controls are shown in Figures 6.6 and 6.7.

6.1.8
Flares

Vast quantities of certain heavier-than-air gases are collected at typical oil wells, refineries and certain chemical plants. These gases are characterized as hazardous waste due to their toxic or explosive properties and thus they cannot be emitted into the atmosphere. These gases are often neutralized by combustion. Flames in which these hazardous gases are combusted are typically referred to as flares. A flare is shown in Figure 6.8.

Flares are typically very large flames (20–30 m long), often with tremendous heat output. Due to the impurities of the fired waste gases, flares are usually sooting flames and are, therefore, highly radiating. For this reason, in most cases flaring burners are mounted on high steel structures that allow the direct emission of flue gases into the atmosphere. The burner heads must be able to withstand extreme weather conditions: the flame must be stable even in rain, snow or high wind

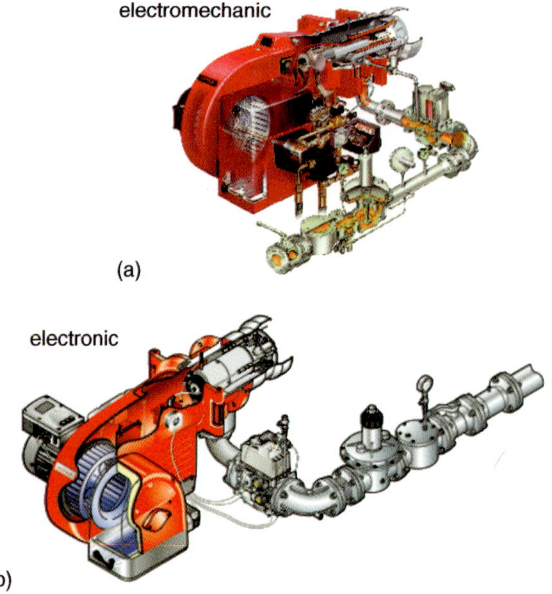

Figure 6.6 Electromechanic (a) and electronic (b) burner control (Courtesy of Max Weishaupt GmbH).

($\sim$120 km h^{-1}). Remotely controlled ignition and pilot flames, flame detection and process control can stabilize the flame. Inert gas shrouds and even water shrouds are used to ensure stable operation.

Traditional flares emit large amounts of soot due to the nature of diffusion flames, in which the oxidant is atmospheric air and mixing is solely provided by diffusion. Water or steam injection can help reduce soot and smoke emission in these flames.

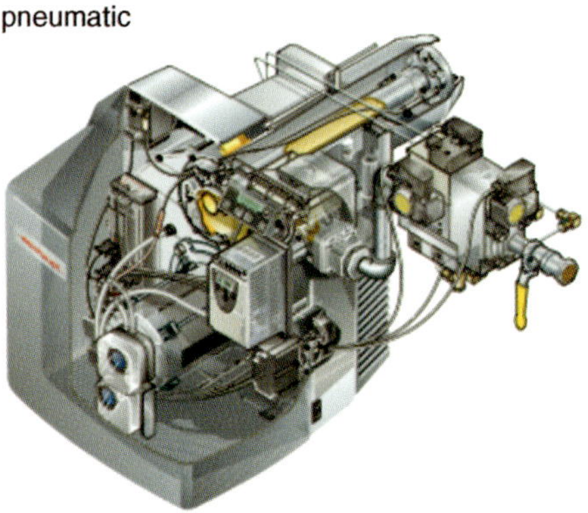

Figure 6.7 Pneumatic burner control (Courtesy of Max Weishaupt GmbH).

Figure 6.8 Industrial flare for the combustion of waste gases.

6.1.9
Categorization of Oil Burners

Oil burners can be classified by the type of air introduction:

- natural flow oil burners (oil refineries)
- forced-air oil burners

    - oil burners with air pipes
    - oil burners built in to the plenum of air
    - block oil burners
    - duo-block oil burners, rotary oil burners.

Based on air and fuel admission to the flame:

- single-step oil burners,

- multi-step oil burners,
- gasifying oil burners.

Based on the luminosity of the oil flame:

- luminous flame oil burners,
- 'cold' flame oil burners.

Finally oil burners can also be classified according to the atomization process of the fuel. Due to the importance of the atomization the process is discussed in more detail.

6.1.10
Atomization of the Fuel

For oil burners the fuel is atomized within the combustion chamber to increase the specific surface area of the fine droplets, which in turn accelerates gas production. In general larger surface areas contribute to increased overall reaction rates.

The couple of hundred micron diameter oil droplets evaporate quickly due to the temperature increased by the radiation, hot combustion air and the recirculating combustion products (flue gas). The oil vapor then reacts with the air similarly to gases, and burns out quickly. The smaller the droplet diameter, the higher the specific surface area. The velocity of burn-out is determined by the droplet size as well as the temperature difference between the gases and the oil. Therefore the quality of atomization can have a substantial effect on the flame shape formed.

Atomization is based on the sheer stress caused by the surface tension and the viscosity of the bulk oil. The atomization requirement to the applied force is to overcome these counterforces. Atomization in most cases occurs as a result of friction on adjacent surfaces, as well as the internal vapor generation in the case of pressure atomization, and the resulting cavitation.

In the case of *pressure atomization*, the motion energy generated drag force becomes larger than the counter force resulting from surface tension as well as the viscosity. The speed increase providing the motion energy results in a quick pressure drop that yields to cavitation. The relative velocity of the ambient air with respect to the gas stream influences the quality of the atomization. The higher it is, the finer is the atomization. Efficient atomization can be achieved with low viscosity oil. For low capacity burners ($\leq$500 kW) the required viscosity is around 6–16 cSt, up to about 5 MW it is 12–22 cSt and for the power plant capacity burners it is around 22–30 cSt.

Twin fluid atomization requires an auxiliary medium to achieve the high velocity needed. It is usually low-pressure air, compressed air, water vapor, or in special cases compressed combustible gas. The viscous fuel hitting the high velocity medium disintegrates due to the viscous force.

Emulsion atomization (an emulsion formed from the mixture of the atomized fuel and the atomizing medium) can achieve relatively homogeneous droplet size. This technique is mostly used for medium pressure and viscosity oils.

Low pressure twin fluid atomization is chosen when low viscosity combustion oil is atomized with adequate pressure (70–120 mbar) 25–40% combustion air as a primary atomization agent and the small pressure mixture is added to the small pressure (0.5–1.0 bar) oil stream. The secondary fraction of the combustion air is added to the atomized stream with a rotation.

For the *rotary atomization* the oil is supplied to the perimeter of a fast rotating (velocity 25–30 m s^{-1} at $n = 5000$–6000 1 min^{-1}) disk or cups internal surface that is surrounded by an increased velocity air stream (at a pressure of 12–25 bar the speed can be up to 90–100 m s^{-1}). The velocities of the air stream and the cup determine the cone angle. This method can be used for the combustion of varying quality, higher viscosity ($\leq$40 cSt) combustion oil that is regularly used for the ship boilers of tankers, purchasing various fuels at a number of ports. It is also common for a one or two flame tube 5–20 t h^{-1} capacity boiler. The burners are compact and similar in their construction to those of the block oil burners.

Figure 6.9 illustrates the most common atomization methods and the corresponding atomizers.

6.1.11
Mixed Fuel and Alternative Burners

Burners that can operate with multiple fuels at the same time are called mixed fuel burners. An example of this type is the multiple gas burners, for example, burners

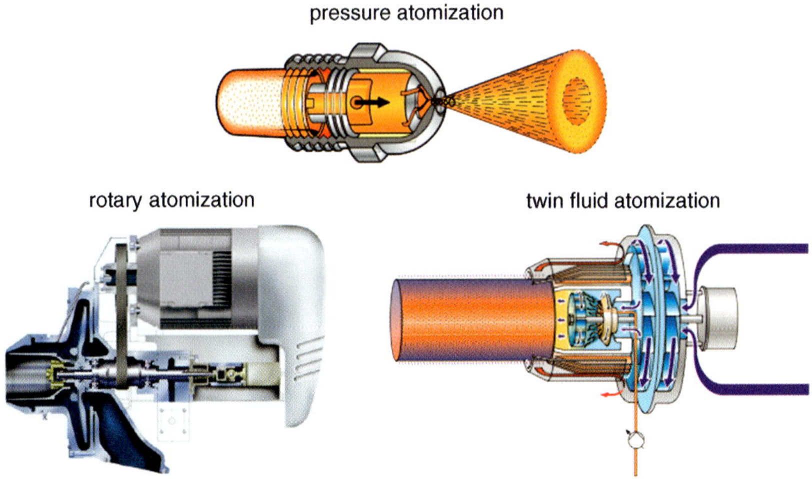

Figure 6.9 Atomizer technologies for liquid fuels (Courtesy of Max Weishaupt GmbH).

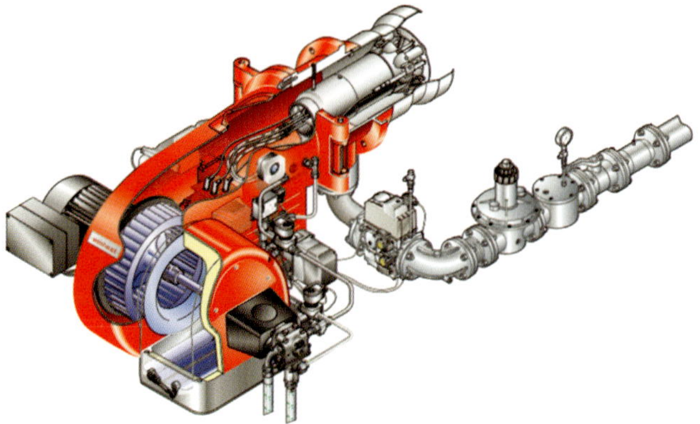

Figure 6.10 Alternating burner that can run both on gas or on oil, depending on fuel availability (Courtesy of Max Weishaupt GmbH).

for low quality waste gases with the option to boost the energy output with natural gas, PB gas, and so on. A special example is a gas burner with oil or coal dust injection to increase the luminosity of the flame. Luminous flames can be characterized by efficient radiative heat transfer while non-luminous (invisible or light blue) natural gas flames are typically non-radiating flames. Here the secondary fuel is introduced to generate soot and not because of its energy content.Technically, though they are mixed fuel burners.

Burners that are capable of switching from one type of fuel (e.g., natural gas) to another (fuel oil) are called alternative burners. One example is shown in Figure 6.10.

Burners have important characteristics, including specific dimensions and geometries, different properties of the produced flame, and so on. These more in-depth properties are rarely listed in catalogs, since they are dependent on many application-specific factors, such as combustor geometry, burner placement, refractory type, and so on.

Large burner manufacturers have dedicated research and/or test centers where the burners can be mounted on a completely controlled testing furnace equipped with all the necessary instrumentation. Testing is often performed for development purposes, but such a facility is also necessary for the handling of customer issues. A very well equipped testing laboratory is shown in Figure 6.11.

6.2
Industrial Boilers

Boilers are fairly common heating equipment in the industry. Their capacity ranges from domestic size (10–20 kW) to large power generation units. The largest boilers can be characterized by their steam producing capacity of several thousand tons/hour. Owing to its physical properties (excellent heat capacity, latent heat of vaporization),

Figure 6.11 Burner testing laboratory (Schwendi, Germany).

water is very well suited for use as a heat transfer medium. Table 6.1 depicts an overview of different heat transfer media comparing their physical property data.

A steam boiler consists of two major parts, a firing and a heat transfer system. According to the design, the heat transfer system can be realized as a fire-tube, or a water tube boiler. Fire-tube boilers generate hot water or steam by a hot flue gas passing through tubes which are surrounded by a water reservoir. Water tube boilers generate steam by a flame heating the tubes filled with water from outside. Fire-tube boilers are limited to a pressure of about 25 bar and a maximum capacity of $20\,\mathrm{t\,h^{-1}}$ steam by design, whereas water-tube boilers are suitable for large capacities.

The class of water-tube boilers can further be subdivided according to the physical concepts of natural circulation, forced circulation, and once-through boilers [1]. Natural circulation boilers require a steam drum for phase separation. This is the reason why natural circulation boilers are limited to subcritical steam conditions since, above the critical point, liquid and gaseous state conditions can no longer be

Table 6.1 Comparison of physical property data of common heat transfer media [20].

Medium	Heat Capacity c_p $[\mathrm{kJ\,kg^{-1}\,K^{-1}}]$	Latent Heat of Vaporization Δh_V $[\mathrm{kJ\,kg^{-1}}]$	Thermal Conductivity $\lambda\,[\mathrm{W\,m^{-1}\,K^{-1}}]$
Water	4.2	2257 (at 100 °C)	0.65
Organic liquids	2.1	—	0.11
Diphenyl/Diphenylether	2.1	≈ 300 (at 256 °C)	0.10
Molten salt	1.55	—	0.3
Liquid metals (Na)	1.3	—	60
Air and CO_2	1.0	—	≈ 0.04
Helium		—	≈ 0.3

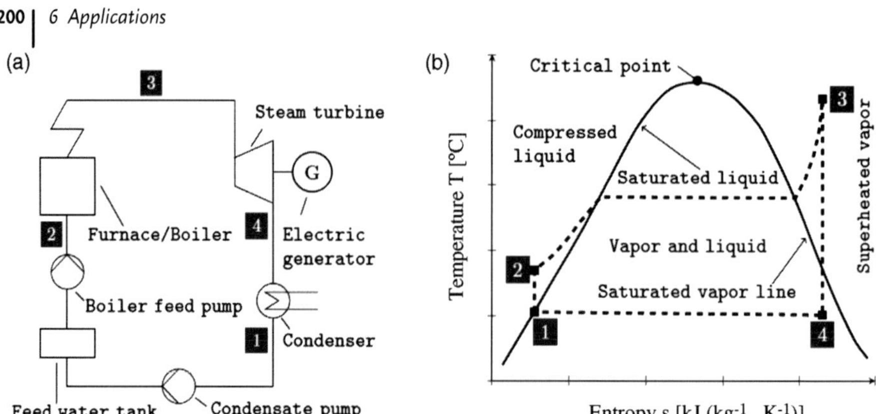

Figure 6.12 (a) Schematic flow diagram of a conventional steam power plant, (b) Rankine cycle with subcritical boiler pressure on *T–s* coordinates.

distinguished. In order to reach higher degrees of efficiency, modern once-through boilers apply supercritical steam parameters.

Figure 6.12 reveals a simplified set-up of a conventional steam power plant.

6.2.1
Combustion Systems for Steam Generation

According to the physical state of the fuel, solid, liquid, or gaseous firing systems can be distinguished. There are different combustion systems for each fuel type.

Gas, mainly natural gas, represents an ideal fuel. There are other less common types of gaseous fuels such as city gas, blast furnace gas, refinery gas, biogas, and landfill gas. One of the biggest advantages of gas-fired furnaces over other combustion systems is the reduced exhaust emissions [2].

Liquid fuels show similar combustion behavior to gas, since they have to be vaporized or atomized prior to ignition. The use of heavy oil requires preheating in order to lower the viscosity of the oil for better atomization. Due to high oil prices liquid fuels have become less important in the area of power generation over the last years.

The utilization of solid fuels is associated with a complex plant design. There are noticeable differences in the complexity for storage, conditioning, and the feeding systems of the fuel and oxidizing agent compared to gas- or oil-fired combustion systems.

A widespread classification of solid firing systems according to the gas–solid contact is shown in Figure 6.13.

In the fixed or packed bed regime the pressure drop per unit height, $\Delta p/L$ is given for a fixed bed of spherical particles by Ergun's equation [3]:

$$\frac{\Delta p}{L} = 150 \frac{(1-\varepsilon)^2}{\varepsilon^3} \underbrace{\frac{\mu U_0}{d_\mathrm{p}^2}}_{\substack{\text{dominant} \\ \text{at Re} < 1}} + 1.75 \frac{(1-\varepsilon)}{\varepsilon^3} \underbrace{\frac{\rho_\mathrm{G} U_0^2}{d_\mathrm{p}}}_{\substack{\text{dominant} \\ \text{at Re} < 1000}} \tag{6.1}$$

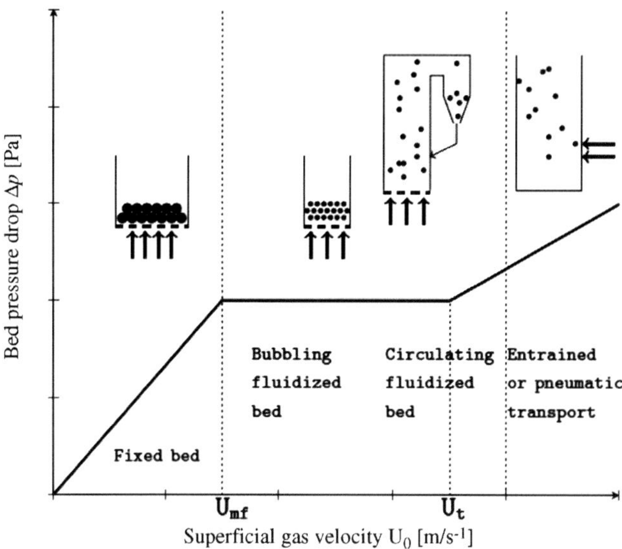

Figure 6.13 Classification of solid firing systems based on pressure drop flow diagram.

The superficial velocity U_0 is defined as the gas flow rate percolating the fixed bed per unit cross section of the bed, ε is the void fraction (porosity) of the bed, μ and ρ_G are the dynamic viscosity and gas density of the fluid, respectively, and d_p represents the uniform diameter of the particles. By increasing the flow rate through the bed, a fixed bed regime is transformed into a fluidized bed by exceeding the minimum fluidization velocity U_{mf}, which is calculated by Eq. (6.2). In this state the suspended particles behave like a liquid. In the range between U_{mf} and U_t the pressure drop remains constant by increasing the superficial velocity up to the point of onset of the particle elutriation.

$$U_{mf} = \frac{\mu}{\rho_G \cdot d_p}\left[\sqrt{33.7^2 + 0.0408 \cdot Ar} - 1\right] \tag{6.2}$$

Ar is the dimensionless Archimedes number, which describes the ratio of gravitational forces to viscous forces. In the fluidized bed regime the pressure drop is equal to the weight of the bed and is given by:

$$\frac{\Delta P}{L} = (1 - \varepsilon_{mf})(\rho_S - \rho_G)g \tag{6.3}$$

ε_{mf} is the void fraction at the fluidization point and g the gravitational acceleration. The terminal velocity U_t of a spherical particle is an equilibrium velocity obtained from a balance of forces of a freely falling particle and is given by:

$$U_t = \sqrt{\frac{4}{3}\frac{\rho_p - \rho_G}{\rho_G}\frac{d_p \cdot g}{C_W}} \tag{6.4}$$

The drag coefficient of a single sphere, C_W is depending upon the range of the Reynolds number and is calculated either by Stokes law, Intermediate law or Newton's law. On further increase of the gas flow rate the particles migrate from the bed and get entrained by the gas flow. Each of the above mentioned flow regimes shows distinct characteristics. The effects on the solid fuel combustion systems are discussed in detail in the following section.

6.2.2
Fixed Bed Combustion

The application of fixed bed (also known as packed bed) combustion systems is limited to a nominal boiler capacity of 0.25 to 150 MW, whereby the latter corresponds to a production of approximately 200 tons steam per hour [4].

Especially low capacity fixed bed firing systems often have advantages over other firing systems, due to low investment costs, simple operability, and high flexibilities according to the fuel [5]. Since pulverized fuel combustion and fluidized bed combustion systems achieve higher efficiencies, the application of fixed bed combustion in firing systems for electricity production (via steam generation) is decreasing. When fuels are difficult or impossible to reduce in size, fixed bed combustion can be a good alternative to the other firing systems [6].

In order to classify fixed bed combustion systems, various possibilities of distinction are useful. First, the system is characterized according to the grate used. Grates can be classified as moving or fixed, and as horizontal or inclined [7].

The direction of flow of the fuel and the flue gas leads to the distinction between co-current, counter-current and cross flow systems. Accomplished designs are traveling grates, vibrating grates, reciprocating grates, and fixed grates.

As an important application of grate firing systems the thermal treatment of municipal waste is sketched. Such combustion systems for waste burning are known as incinerators.

Since the end of the last century a great deal of attention has been given to energy recovery of waste. Waste incineration plants combine the endeavor of energy recovery from waste and the reduction of the volume of the waste. Additional inertization and stabilization processes take place. The strongly fluctuating composition is a typical feature of municipal waste. There are strong differences in the calorific value, moisture content, volatile matter, ash content, bulk density, particle size distribution, and content of heavy metals. Especially, these properties of municipal waste lead to a grate firing system. The efficiency of such combustion devices depends on suitable grate and the combustion chamber design. Figure 6.14 depicts a schematic view of a waste combustion chamber and the flow configuration of the air and the flue gas [2].

Typically, new municipal waste incineration (MWI) plants are designed for an annual capacity between 150 000 and 200 000 tons of waste, depending on the calorific value this results in a thermal capacity of 50 to 80 MW [7]. The generated power is typically used to meet the auxiliary power requirement of the incineration unit. To increase the overall efficiency of the plant hot water or steam can be applied for district heating.

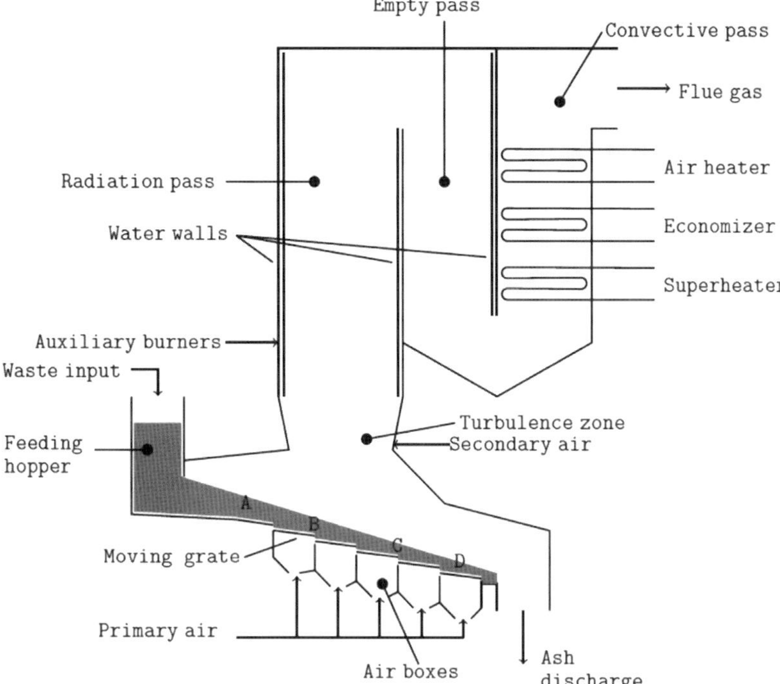

Figure 6.14 Schematic diagram of a firing system of a typical MSW incineration facility.

Unlike conventional steam generation units, waste incineration plants typically operate at moderate steam parameters (usually in the range of 5×10^6 Pa and $400\,^\circ$C [7]) to reduce the risk of corrosion and erosion, caused by the characteristic high chlorine contents of waste used as fuel.

Due to the inhomogeneous character of the municipal waste, a waste pit with a high capacity is needed to achieve a homogenous mixture of the different deliveries (control energy input) to prevent inhomogeneous combustion and operation problems associated therewith. The untreated waste (only a preparation of size reduction of bulky waste is necessary) is fed to the feeding hopper, usually by crane.

A crane loads the feeding hopper. The waste continuously flows in line with the grate, which forms the bottom of the combustion chamber. The grate area extends up to $100\,\mathrm{m}^2$. The entering input stream starts drying by reflected radiation from the hot combustion zone.

By means of the moving single elements of the slight inclined reciprocating grate the waste is transported through the different zones of the combustor (drying (A), pyrolysis and gasification (B), combustion (C), and sintering/burnout (D)). The grate motion ensures sufficient distribution of the primary combustion air, supplied from below, to the layer of the waste. Each zone's air demand is individually adjustable, resulting in an optimized combustion performance. Primary air makes a large share of 60 to 80% of the whole air demand of the combustion process.

Usually, a global excess air ratio of 1.6 to 2.2 is applied to incinerators. In addition, the primary air flow cools the grate. To lower the grate's mechanical strength water cooled grates are sometimes used.

The last zone on the back-end of the grate acts as the burn-out section for the ash, to ensure sufficient combustion efficiency. Residual solids (grate ash) of incombustibles, mostly formed by inorganic constituents, and incompletely burned waste (slag/bottom ash) fall into a water bath at the end of the grate.

The combustion of the formed gaseous compounds takes place above the grate. The injection of secondary air (referred to as over-fire air) as well as the constriction design cause high turbulences and flow conditions similar to a stirred tank, which are required for total burn-out of the flue gas.

High temperatures (at least 850 °C) and a certain residence time (2 s) after the last air injection are set in some countries by law [7] to ensure the breakdown of toxic substances. Additional burners are situated in this part to guarantee the minimum temperatures both in operation and for start-up of the plant.

Furnace walls can be designed as water walls or refractory-lined walls [8]. In the upper region of the furnace, plug flow is desirable for a uniform residence time of the total flue gas. In order to avoid unwanted flue gas leaks and odor emissions, the combustion chamber is designed to operate under slightly sub-atmospheric pressure.

Due to ingredients that may be present in the waste, the incineration process forms and releases many potentially harmful substances.

The raw flue gases include compounds like particulate matter, oxides of nitrogen and sulfur, carbon monoxide, dioxins and furans, metals (lead and mercury), acid gases, volatile chlorinated organic compounds, and polycyclic aromatic compounds [9]. To reduce emissions, incinerators require back-end air pollution control (APC) measures. Typically, these devices consist of units for dedusting (ESP, fabric baghouse filter), removal of acid gases (mainly wet scrubbers), denitrification (SNCR, SCR) and removal of PCDD/PCDF (absorption on activated coke). Due to the biogenic proportion of municipal waste, CO_2 emissions out of a MSW incinerator are CO_2-neutral in part.

The following list gives an overview of some important characteristics of fixed bed combustion systems.

Advantages:

- Highly flexible in terms of fuel selection
- No need for prior sorting
- Thermal destruction of chemical ingredients in municipal waste at high furnace temperatures
- Saving of land for waste disposal due to reduction in amount of waste
- Potential extraction of valuable residues via mineralization
- Waste incineration is a concept of resource economizing and energy recovery.

Disadvantages:

- High firing losses due to high excess air
- Capital and investment costs are relatively high
- Waste incineration requires low steam parameters (due to chlorides), which results in low efficiencies

A very special form of fixed bed combustion is the rotary kiln incineration. As the rotary kiln is in contrast to the usual fixed beds a rotating device, the previous statements are not or only partly valid.

Rotary kilns are used for the incineration of clinical or hazardous wastes, as well in industrial applications (e.g., the cement industry). The input material may be present in the gaseous, liquid, or solid state. Even whole drums and bulk containers can be burned in rotary kilns. The rotary kiln incineration represents a proven technology, due to the high versatility of the input material.

The rotary kiln incineration process, as depicted in Figure 6.15, is a widespread technology in the hazardous waste management industry.

Such systems consist of the rotary kiln itself, a post-combustion chamber, a heat recovery section and a back-end air pollution control (APC) system [8]. The surface shell of the rotary kiln is refractory-lined.

The diameter of a rotating cylinder is 1 to 5 m and has a length up to 20 m [8]. A length-to-diameter ratio between 2 and 10 is common [10]. Both, continuous and batch operation modes are possible, whereby capacities up to $20\,\mathrm{t\,h^{-1}}$ are possible [8]. To ensure high combustion temperatures, in the range 800 to 1600 °C, co-firing of fuel or liquid waste is required occasionally. Due to the prevailing temperature conditions, toxic substances are destroyed.

The slight inclination of the rotating cylinder causes the transport of the solids through the rotary kiln.

The adjustment of the inclination and the rotational speed of the cylinder controls the residence time of solids, which is strongly influenced by the properties of the waste. The residence time ranges from a few seconds to one hour [10].

Combustion residues (slag and metallic objects) are removed at the end of the rotary kiln via an ash chute, and are usually water quenched. Based on the short retention time of the flue gas, a post-combustion chamber is situated downstream. This additional section may or may not be fired. After a complete reaction the flue

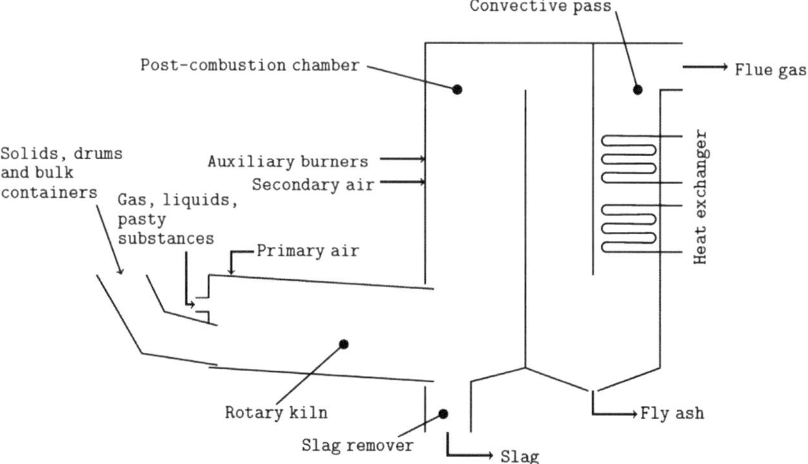

Figure 6.15 Typical diagram of a rotary kiln incineration system.

gas leaves the post-combustion chamber and enters the heat recovery section and the subsequent air pollution control system.

As the rotary kiln does not have any moving parts and no additional pre-treatment is necessary, this technology is advantageous over others. Disadvantages are high capital costs for installation and high maintenance costs. The large amount of excess air needed results in low fuel efficiencies and a low thermal efficiency.

6.2.3
Fluidized Bed Combustion (FBC)

Since the development of fluidized bed technology at the beginning of the twentieth century, the years have been marked by steady progress. The shortage and increasing cost of high-grade fuels and the call for emission reduction from power stations [11] resulted in the introduction of the fluidized bed technique to the power plant sector in the early 1960s.

There is a fundamental difference between the fluidized bed combustion systems and the other technologies of solid fuel combustion. Inside the fluidized bed furnace bed material is situated. These solid particles consist of incombustible inert materials, like sand, ash or sulfur sorbents, of a certain particle distribution. By addition of air, which acts as a fluidizing agent as well as for primary combustion, fluidization takes place. In this process the distributer plate ensures an equal sharing of the air, resulting in a good mixing of the bed material. Fluidized bed combustion systems are high flexible in the choice of fuel. A wide range of fuels, such as coal and its residues, wood residues, sludge, municipal waste, petroleum products, gas, or agriculture waste are used [12]. There are different technologies for fuel addition into the fluidized bed available. Usually, the content of fuel inside the fluidized bulk is very low, typically 1–3% by weight [2]. Based on the excellent heat exchange characteristics inside the suspended bed, a moderate and uniform temperature level is achieved [7].

The selection of the solid fuels is limited by the ash fusion temperature of the fuel, lying below the bed temperature, and high moisture contents. According to the way of suspending the particles inside a fluidized bed, namely bubbling fluidized bed and circulating fluidized bed can be distinguished. The utilization of relatively large particles of bed material and low velocities of air leads to bubbling fluidized beds, which have characteristically very small rates of entrainment. Apart from the circulating fluidized bed combustion, the entrainment of particles in general is a topic of importance. The entrained gas–solid mixture is subsequently separated by a cyclone. The bed material and the unburned fuel are led back to the furnace.

The following lists give an overview of the advantages and disadvantages of fluidized bed combustion technology.

Advantages:

- Excellent multifuel capability without a major penalty in performance
- Enhanced heat transfer
- Reduced NO_X emissions, due to low combustion temperatures
- Possibility of *in situ* desulfurization.

Disadvantages:

- High dust loads in the flue gas
- Increased erosion risks
- Relatively large amounts of solid residues.

Conventionally, fluidized combustion systems are designed to work under atmospheric conditions. In the last years there have been serious efforts to expand the benefits of the fluidized bed combustion technology by introducing a pressurized operation mode, mainly for combined-cycle applications.

This leads to improved efficiency, smaller plant design (for the same output), lower investment costs, and further reduction of pollutant emissions [13]. Some pressurized fluidized combustion systems in the bubbling mode are already in service, whereas the circulating version is still in development.

6.2.3.1 Bubbling Fluidized Bed Combustion (BFBC)

The typical range in power of a bubbling fluidized bed combustion (BFBC) system lies between 3 and 300 MW_{th} [7]. However, at lower capacities fixed bed systems in many cases offer a better economic performance [14].

A BFB boiler typically consists of two major parts, the combustion chamber or furnace and the convective heat exchange section (also referred to as back-pass). The former is further divided into a dense bubbling bed and a lean freeboard section. Additional facilities for feedstock preparation as well as the handling of air, flue gas, and ash complete the plant.

The principle of operation of combustion systems for power generation is illustrated in Figure 6.16.

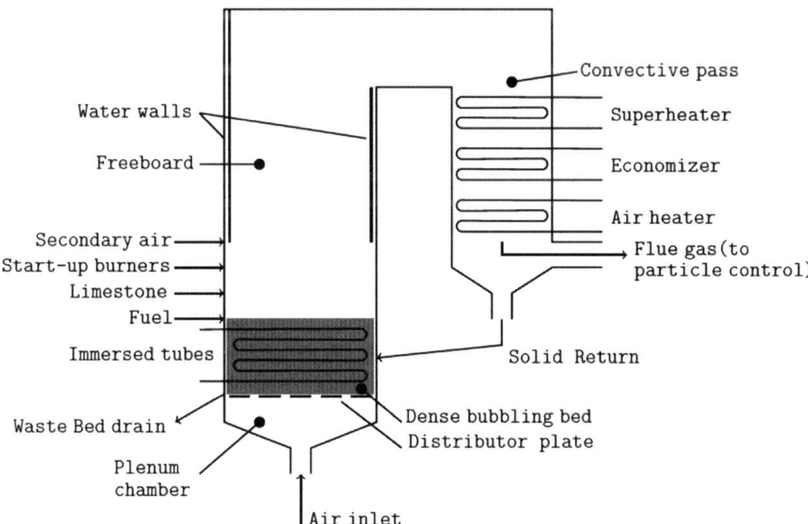

Figure 6.16 Simplified scheme of a typical BFB boiler with immersed heat exchanger.

The fluidization agent, almost exclusively air, which also acts as primary combustion air, is introduced into the bed material. In some cases the air is preheated [12]. The gas velocities are much higher than the minimum fluidizing velocity, typically ranging from 1 to 3 m s^{-1} [7]. Depending on the used flow velocity the mean bed material size lies in the region of 0.5 to 1.5 mm [7]. The fluidizing agent passes the bed material in the shape of bubbles, which occupy 20–50% of the bed volume and leads to an excellent mixing and intense agitation of solid particles [15]. The expanded bed depth is about 0.3 to 1.5 m.

Fuel is conveyed by a pneumatic or mechanical feeding system into the fluidized bed. The feeding of biomass is a major challenge. The share of the fuel in the fluidized bed is very low, a few percent. The particle size of the fuel particles varies, the upper limit lies in the centimeter range. To maintain the low combustion temperatures of 800 to 950 °C [2] dissipation of heat, usually by an in-bed heat exchanger, is necessary.

Furthermore, there is an input of limestone for the *in situ* desulfurization. For a common desulfurization efficiency of 90% a multitude of stoichiometric amounts of sorbent is required. The Ca/S molar ratio is usually in the range of 2.5 to 3.5 [12].

By the above mentioned bubbling fluidized mode the formation of a distinct interface between the dense bed and the lean freeboard regime above is characteristic.

The residence time of the fuel is about 1 min and ensures sufficient combustion. To improve the burn-out of the gasification products and small particles, air staging is usually applied. In this case a secondary air flow of about 10 to 20% of the overall air demand is injected into the freeboard [7]. The optimum of the total excess air lies in the range 10 to 50%. High air excess means improved combustion efficiency, but also increased flue gas volume and higher stack losses [7]. The rate of burn-out mainly depends on the turbulence in this section. The furnace enclosure may consist of water walls or refractory-lined walls.

A further increase in burn-out is achievable by recycling particles. Hot flue gas and fine particulates leave the combustion chamber and enter the second pass. In this convective pass the gas flow is cooled by heat exchangers.

Usually there is no need for post-combustion units (like selective catalytic reduction (SCR) or flue gas desulfurization (FGD)) for the removal of harmful gas emissions for well-performing BFBC boilers. Some plants use a scrubber for the precipitation of mercury and to manage exceptionally low limits for SO_X emissions [12].

Reduction measures for particulates are indispensable. For this purpose particulate collection devices, like an electrostatic precipitator (ESP) or bag-house-filters, are in use.

6.2.3.2 Circulating Fluidized Bed Combustion (CFBC)

Generally, the firing system of a circulating fluidized bed (CFB) is of practical significance for thermal capacities in the power range 30 to 500 MW$_{th}$. Currently, the supercritical power plant Lagisza (Poland), in service since 2009, with a capacity of 400 MW$_e$, is the world's largest CFB unit. However, there are four 550 MW$_e$ ultrasupercritical CFB boilers under construction in Korea with a planned commissioning in 2015. Some manufacturers offer even larger units up to 800 MW$_e$.

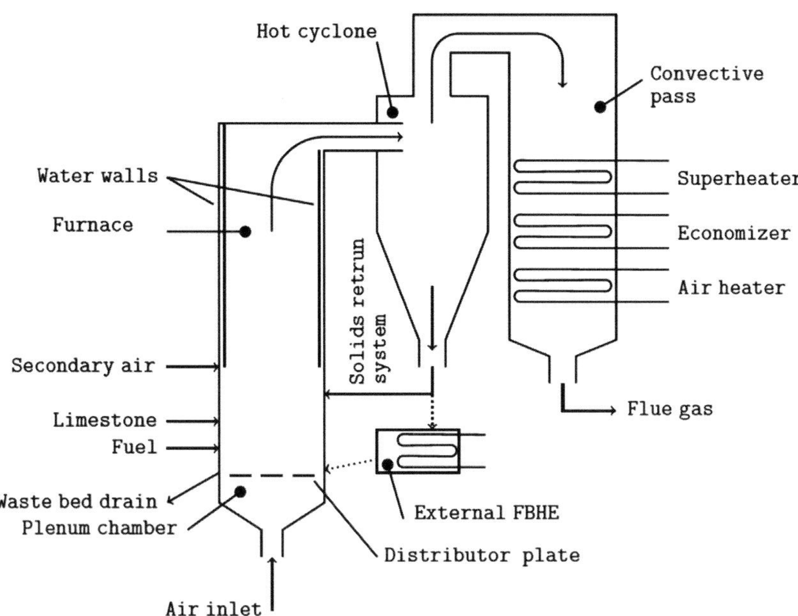

Figure 6.17 Generalized diagram of a CFB boiler with external FBHE.

The CFB technology is a further development of the BFB system. Some disadvantages of the previous technique have been optimized, this includes, among others, smaller bed cross-section areas, cross-mixing in the FB, and in-bed heat exchangers with high erosion hazard.

Figure 6.17 shows a schematic CFB boiler, including the two main sections: CFB loop and convective pass [12]. Additional components like a bed drain and a solid classifier complete the plant.

Unlike the BFB, the bed material is not only situated in a certain low expanded fluidized area. There is a continuous circulation in the CFB loop, which consists of the furnace, gas–solid separation unit, solid recycle device, and in some designs an external fluidized bed heat exchanger (FBHE) [12]. Gas velocities exceeding the particle terminal velocity of the average particles, which results in particle entrainment, reach the particle circulation. Typically, the gas velocities are in the range 5 to $10\,\mathrm{m\,s}^{-1}$ utilizing a bed material of a mean size of 100 to 300 μm [15].

These flow conditions strongly influence the construction. On the basis of the potential high velocity erosion, CFB boilers do not apply in-bed heat exchangers. Unlike BFB boilers, fuel is injected via fewer feeding ports, either into the lower part of the furnace or in the solid recycle system. A distributor plate injects the fluidization agent, exclusively air, from below. This air serves as the primary combustion air and is usually added in sub-stoichiometric amounts. Due to the very abrasive conditions in the lower section of the combustor, the walls are lined with refractory [6].

Sorbent, mostly limestone, is further added to the furnace for direct SO_2 pollution control. Therefore no post-combustion pollution control devices are necessary. Sulfur capture in a CFB boiler requires less sorbent (Ca/S molar ratio of about 2 [12]) in comparison to BFB boilers. The typical low combustion temperature of about 780–900 °C in units with air staging minimizes the formation of thermal NO_X.

The walls in the upper part of the combustor include water walls. The entrained solids enter the gas–solid separation unit, mostly a mechanical cyclone, but an impact separator would also be appropriate. The selected particles are conveyed by the recycle system in the lower part of the combustor. Besides the bed material, unburned fuel particles larger than the cut-size are recycled to achieve the typical high combustion efficiencies of CFB firing systems. Some units have an additional external fluidized bed heat exchanger (FBHE), which is situated in the recycle system. By regulation of this bypass stream, the thermal conditions of the plant can be adapted to the current operation mode. The off-gas and finer solid residues, like fly ash or spent sorbent, leave the cyclone on top. Subsequently, they are discharged to the convective pass. This pass is similar to that of a PC or BFB boiler and usually includes reheater, superheater, economizer, and air-preheater surface [12]. Finally, solids are precipitated by a bag-house or electrostatic precipitator (ESP) further downstream before treated flue gas is emitted via a stack.

6.2.3.3 Dust Firing

The concept of dust firing, or pulverized fuel combustion, represents an opportunity for all types of solid fuels. Coal is preferably used as the fuel, but biogenic fuels, like sewage sludge, wood dust or meat- and bone meal, are also applied.

In the last years co-firing has become increasingly important, especially in pulverized coal systems. Usually there is only a small fraction, 3–5%, of biomass fired in a mixture with fossil fuels [16].

The main requirement of the fuel is a common particle size below 80 μm. Fuels vary substantially in the amount of energy needed for particle size reduction. Generally, the grinding of biomass materials is energy intensive, due to their fibrous nature, therefore the use of biomass is not very common [14]. The capacity of dust firing ranges from 50 to 1300 MW_e [7], whereby no maximum of the range of capacity has been identified so far [2]. Dust firing systems are applied in power generation and for large-scale industrial boilers.

There are different ways of characterizing dust firing systems. Systems are distinguished according to the fuel feeding system, the design of the lower part of the combustion chamber for the purpose of ash removal, and the state of the removal ash. Regarding the dust feeding system there are two main types, bin or storage systems and direct firing systems.

Pulverized coal-fired units are typically classified according to the state of the removed ash. There are typically two different main types, so called dry-bottom and wet-bottom furnaces (which are also known as slag-tap furnaces).

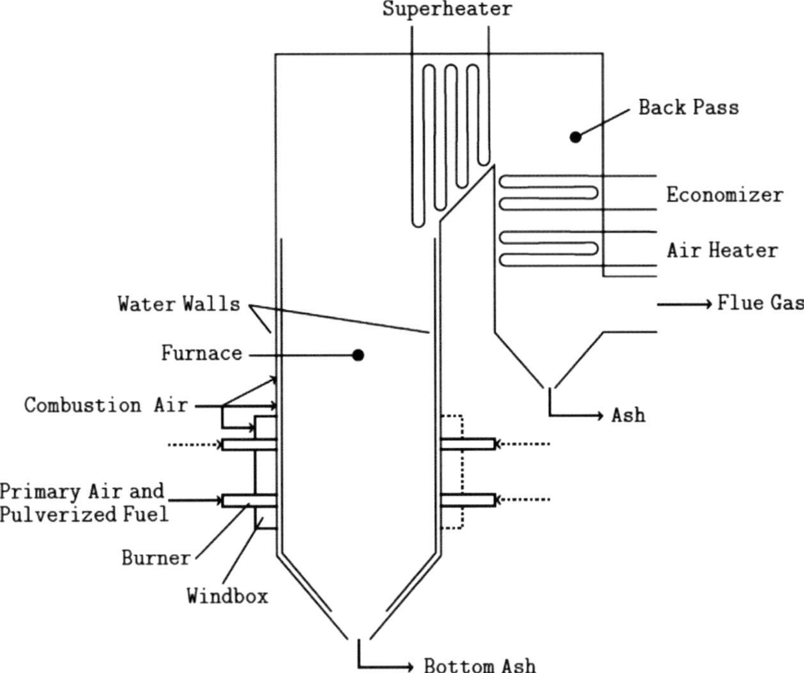

Figure 6.18 General layout of a once-through pulverized coal fired boiler in two-pass construction and single reheat.

Most large-scale pulverized coal-fired power plants utilize subcritical steam parameters (<22.1 MPa and $540\,^\circ$C). Figure 6.18 depicts a simple scheme of a pulverized coal-fired unit.

The usual particle size of pre-crushed coal is in the range of a few centimeters [2], which is not suitable for direct utilization. The pre-treatment typically includes combined milling and drying. Due to the higher moisture content of lignite, the required mills are larger than those for hard coal. The drying process of coal with high water contents (like lignite) is very energy intensive. Varying properties among different coal types led to the development of various plant designs. For lignite with typically high water contents, up to 20% of the energy content of the fuel is used for drying [17]. Therefore major efforts have been made to lower the energy consumption of the drying process by an improvement in heat integration and recovery. The heat for water removal can be supplied by hot air or recirculated hot flue gas. Subsequently, the hot air flow is used for pneumatic transport of the pulverized coal to the burner section and can also serve as primary combustion air. Difficult fuels often require an intermediate storage or bunker system in order to ensure continuous operability of the plant. This concept is rare in modern units [18]. The function of the burners is the intimate mixing of the powdered fuel particles with the primary as well as the secondary air flow. This ideally leads to a stabile ignition and complete combustion with low emissions at low excess air ratios to reduce flue gas losses. Further, an improved combustion also means less pollution problems in the equipment.

The arrangement of the burners at the furnace walls depends on their type and characteristics, [18]. For plant start-up there are start-up burners needed, mostly oil- or gas fired to heat the combustion chamber to a temperature level where the injected dust particles are able to ignite. To enhance particle burn-out and ensure total combustion of the volatile gases, a tertiary-air flow is added in the upper parts of the furnace. After a few seconds the combustion process is finished. A draught fan takes resulting gases and fly ash particles along. There are large differences in the different ash removal systems, which influence design and the ash-split ratio. In wet-bottom furnaces (so called slag-tap furnaces), the flue gas, respectively the flame temperature, has to be 100 to 200 °C above the ash fusion temperature [17]. This is realized by ceramic lining of the melting chamber, which minimizes heat fluxes. In this way up to 60% of the ashes can be removed in the molten state at the bottom of the furnace. In comparison, deliberate heat removal from the combustion chamber results in a lower temperature level. Only a low share, 10 to 15%, [17] of the whole ash content of the fuel is accumulated in sintered or fused form. Dust firing plants with the concept of wet-bottom are usually more cost intensive. That is why wet-bottom furnaces are only used for fuels which are unsuitable for dry-bottom ash removal, like coal with very low contents of volatile matter [17]. The typically high emission level of pulverized coal-fired plants (especially NO_x and SO_x) requires stringent measures for reduction.

The following list gives an overview of the advantages and disadvantages of dust firing systems in comparison to other firing systems.

Advantages:

- No need for a grate
- Combustion similar to gaseous fuels
- Low excess air ratio $\rightarrow$ lower flue gas losses
- Suitable for high-end plants up to 1300 MW_e.

Disadvantages:

- High energy demand for milling and drying
- High investment costs
- High emission level.

6.2.4
Summary of Combustion Technologies for Boilers

Table 6.2 provides an overview of some practically relevant key data of different firing systems, which were presented in detail in the previous sections.

Table 6.2 Comparison of technologies for solid fuel firing systems for steam generation according to [2,15,17–19].

Characteristics	Unit	Fixed Bed	BFBC	CFBC	Pulverized Combustion
Height of bed or fuel burning zone	m	0.06 to 0.45	1 to 2	10 to 30	27 to 45
Superficial velocity	$m\,s^{-1}$	1to 3	0.5 to 3	3 to 9.1	4 to 30
Typical U/Ut ratio	–	≈0.01	≈0.3	≈2	≈40
Excess air	%	20 to 150	10 to 50	20 to 30	10 to 30
Fuel Feed Size (Mean bed material size)	mm	6 to 40	Up to 50 (0.5 to 1.5)	About 5 (about 0.25)	0.02 to 0.1
Overall voidage	–	0.4–0.5	0.5–0.85	0.85–0.99	0.98–0.998
Attrition		Little	Some	Some	Considerable
Agglomeration		Considerable	Some	No problem	No problem
Furnace Temperature	°C	900 to 1400	800 to 900	850 to 950	1200 to 1700
Typical bed-to-surface heat transfer coefficient	$W\,m^{-2}\,K^{-1}$	50–150	200–550	100–200	50–100
Sulfur dioxide capture in furnace	%	None	80–90 (*in situ*)	80–90 (*in situ*)	None
Combustion efficiency	%	85 to 97	95 to 99	96 to 99.5	99–99.5
Unit size		Up to 200 t steam per h	Up to 80 MW_{th}	30 to 500 MW_{th}	Steam flow above 80 $t\,h^{-1}$
Residence time (fuel)		15 to 160 min	≈1 min	infinite (theoretically)	1 to 6 s

6.3
Industrial Technologies

6.3.1
Characteristics of Industrial Heating Installations and Furnaces

Combustion is one of the major technologies across various industries.

In large-scale applications, heat generation can be achieved most economically usually with burners firing hydrocarbon fuels. The combustion process and the heat of combustion are not necessarily contained in an enclosed space, but if they are, the structures that contain them are generally called furnaces. The containment might be necessary for numerous reasons, for example, to utilize the generated heat as fully as possible, to contain flue gases and process byproducts, and to control the amount of oxidant introduced to the flame. An example of open-flame installations is industrial flares.

Combustion technology is applied on a large scale in many industries, including the metallurgical, ceramic and glass industries. Smaller scale applications include welding, soldering, drying, washing, pickling and surface treatment technologies, and so on.

Furnaces are structures that enclose a portion of space by refractory walls. The walls not only provide insulation and reduce the heat loss to the environment, but also help control the amount of oxidant added to the combustion process. Therefore, the containment increases the efficiency of the process and enables higher temperatures and more controlled combustion environments.

In order to achieve high temperature environments economically, the walls must withstand the heat generated or transferred inside or into the contained space. The structure and construction of the walls and the properties and optimal arrangement of different refractory materials can be the topic of a standalone book. There are of course exceptions, furnaces that do not necessarily need refractory lining, for example, cold wall vacuum furnaces, which have heat-resistant steel plate walls or some boilers, which have cooled steel walls [27–29].

6.3.2
Metal Industry

6.3.2.1 Shaft Furnaces

Vertical cylinder-shaped smelting shaft furnaces have been used for hundreds of years as the melting equipment in iron foundries, though they are also used by the heat insulating fiber industry to melt slag and basalt. It is a continuous operation furnace, but it can also be operated with daily or weekly shutdowns.

One of the most common type is the Blast furnace, the heart of metallurgical steel plants. For larger castings the necessary amount of pig iron is gathered by a built-in pre-collector, which is connected with a siphon that assures continuous tapping to the hearth of the cupola furnace.

In order to replace some of the coke for cost efficiency reasons hydrocarbon (either natural gas or oil) is injected into the furnace laboratory (hydrocarbon added firing) through the nozzle above the tuyère level. Experiments have been conducted with pure natural gas fired cupola furnaces for decades for environmental protection reasons (The CO content of the flue gas and ash content are significant). Illustrative examples of former Blast Furnaces are shown in Figure 6.19a and b on a Hungarian and German postal stamp, respectively.

6.3.2.2 Aluminum Melting Furnaces

Traditionally, aluminum smelting furnaces were built with brick regenerators. In later models the regenerator has been replaced by a recuperator. These equipment are equipped with radiative recuperators constructed from chemically resistant materials in order to withstand passing flue gases. The chambers are rectangular-shaped. A typical batch consists of 100–120 t aluminum, which is melted in 4–6 h melting cycles. The latest trend is the utilization of regenerative burners on furnaces having no direct heat recovery on themselves. The regenerators are incorporated into the burner system. Modern furnaces are equipped with tilting mechanisms that allow efficient tapping of the molten metals.

Figure 6.19 Old postal stamps illustrating Blast Furnaces from (a) Hungary, circa 1953 and (b) Germany, circa 1975.

Figure 6.20 Photo of a crucible furnace with molten metal.

6.3.2.3 **Crucible Furnaces**

Crucible furnaces are low-capacity heated vessels used in small-scale metallurgical applications. They can be heated by firing at their refractory lining from the outside or by using electrical wall or submerged heating elements. Such heating elements can be used when melting low melting point non-ferrous metals. For melting steel, inductive electrical heating can be used. Crucible furnaces are also used to produce the molten salt bath of, for example, put furnaces. Figure 6.20 illustrates a crucible furnace with molten metal.

Special crucible furnaces are also used in the glass industry. These furnaces consist of a large primary chamber which contains many smaller crucibles.

Crucibles can be made of metal, silicon carbide, high strength graphite, and traditional thick-walled ceramics. Thick-walled ceramic crucibles are prone to cracking and breaking due to their poor thermal conductivities and low tolerance of heat stress. To mitigate the risk, collector pools are often built next to the crucibles.

6.3.2.4 **Annealing and Heat Treatment Furnaces**

Annealing and heat-treating furnaces are used in the metallurgical industry in order to produce grades that meet certain specifications. They differ from melting furnaces in that they do not produce melts, but rather accurately heat the batch according to a pre-defined program. Both continuously and periodically operating annealing furnaces exist.

Continuous Annealing Furnaces

Pusher-Type Furnaces. In the case of continuous furnaces it is usually the batch that moves through a heated chamber.

In pusher furnaces, the batch travels through the chamber in lines, in which each portion pushes the one in front. The very last portion is pushed by special pushing mechanisms. The heated chamber is usually elongated – the exact length depends on the desired residence time, capacity and temperature profile. The batch usually slides on water-cooled and insulated rails. In some cases, portions of the batch are placed on specially designed support structures, also called sliding shoes. The shoes provide sufficiently wide gaps between the portions and are made of heat-resistant materials.

The flue gas of pusher furnaces usually travels against the direction of batch movement, which allows pre-heating of entrained combustion air and cooling of the batch at the end of the chamber. Multiple-zone pusher furnaces with burners firing both from above and below the batch are not uncommon. Depending on the capacity and temperature program of a pusher furnace, long-flame burners firing at a small angle onto the batch from the roof or many small roof burners can be utilized to achieve the desired material properties. Recently, cross-fired units have appeared with regenerative flue gas utilization. The typical capacity of a pusher furnace can be as high as $500\,\mathrm{t\,h^{-1}}$, with dimensions up to 60×15 m. Such long furnaces have suspended roof structures.

Walking-Beam Furnaces. These furnaces provide alternatives for pusher furnaces. In walking-beam furnaces, the batch is moved by raising and lowering water-cooled beams. The beam structure is constructed in such a way that the beams provide a locomotive motion and move the batch forwards step by step. In other aspects, walking-beam furnaces are similar to pusher furnaces. They can be either top-fired or cross-fired. Their advantages over pusher furnaces include better mixing inside the chamber and the easier evacuation of the batch in case of breakdowns.

Carousel (Annular-Bottom) Furnaces. Carousel furnaces are also called rotating hearth furnaces. Again, they are similar to pusher furnaces or tunnel furnaces. The only difference is the shape of the tunnel – in the case of carousel furnaces, the tunnel has an annular shape and the floor is moved by wheels or rails. The flue gases are in counter flow and the entering batch is pre-heated by them. Rotating hearth furnaces are also used in heavy clay production. These furnaces are similar to annealing carousel furnaces, but in their case the whole chamber rotates, while the batch is stationary.

Tunnel Furnaces and Roller Hearth (Annealing) Furnaces. In these tunnel furnaces the batch is moved on carts that run on rails. This type of furnace is predominantly used in the ceramic industry in the production of bricks, but is also used in the heat treatment of steel.

Tunnel furnaces are high-capacity units featuring highly automatic feeding and extraction. Carts push each other similarly to how units of batch push each other in pusher furnaces. Tunnel furnaces operate continuously, although the carts move in a semi-periodic fashion. The periodicity is caused by the entering and exiting carts through the sealable entry and exit ports. Sealing is provided by heavy doors.

Depending on the products roasted or heat-treated inside cart-tunnel furnaces, the heating mechanism can be direct or indirect. Generally, direct-fired, open-flame cart-tunnel furnaces are higher-capacity units, with dimensions often surpassing 200×6 m. More sensitive batches (fine ceramics) may require indirect heating. Indirectly fired units are usually smaller-scale, with typical dimensions about 30×2 m. The indirect heating is achieved by utilizing radiant heating tubes.

Roller Hearth Furnaces. Using air-cooled rollers to transport the batch in tunnel furnaces allows higher annealing temperatures to be reached. Roller hearth furnaces are widely used in the continuous heat treatment of steel products, but they are also applied in glass annealing.

Variants of the roller hearth furnace include plate conveyor furnaces and chain link conveyor furnaces. Roller hearth furnaces can be inclined. Inclination reduces the energy consumption of the transportation mechanism.

Continuous Strip-Annealing Furnaces. Strip-annealing furnaces are built for the heat treatment of wires, strips and chains. The material is coil-fed into a puller mechanism. Additional coils are fed by welding the ends of the already input wire with the coiled-up wire. The heat-treated product is cut up once again to specific sizes.

Wire patenting furnaces can produce the patenting temperature very accurately by soaking the wire in pre-heated patenting baths. The baths usually consist of molten lead.

The highest capacity strip-annealing furnaces can reach a productivity of 1.4 million $t\,a^{-1}$.

They are used for the heat treatment of carbon steel and soft steel. Strip-annealing furnaces are used in series with rolling mills.

6.3.3
Ceramic Industry

6.3.3.1 Glass Melting Furnaces

Glass melting furnaces are direct-fired furnaces with elongated chamber geometries. The chamber is fired from the side walls, typically with gas burners that produce long, radiating flames. Typically, these furnaces are operated in a side-switching manner, as they are equipped with brick regenerators that pre-heat combustion air by utilizing the waste heat of flue gases. The raw materials (quartz, lime, soda, waste glass) are introduced continuously at one end of the long chamber and extracted at the other end. The operating temperature of the melting zone is 1450–$1550\,°C$. The ductile, but solidified glass is typically pulled out by plier mechanisms on the product end. (Figures 6.21 and 6.22.)

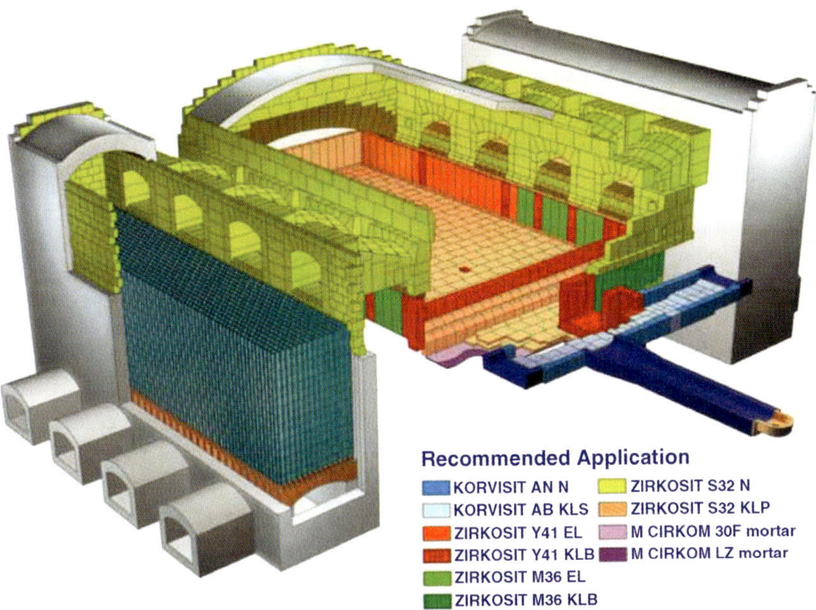

Figure 6.21 Cross fired float glass furnace structure color coded by refractory type (Courtesy of MOTIM).

End fired furnace

1. Regenerator connection
2. Port neck
3. Port arch
4. Burner block
5. Doghouse sidewall
6. Doghouse arch
7. Doghouse corner block
8. Breastwall
9. Tuck stone
10. Peep hole block
11. Shadow wall
12. Melting end paving
13. Melting end sidewall
14. Electrode block
15. Weir
16. Throat
17. Feeder entrance block
18. Working end sidewall
19. Working end paving

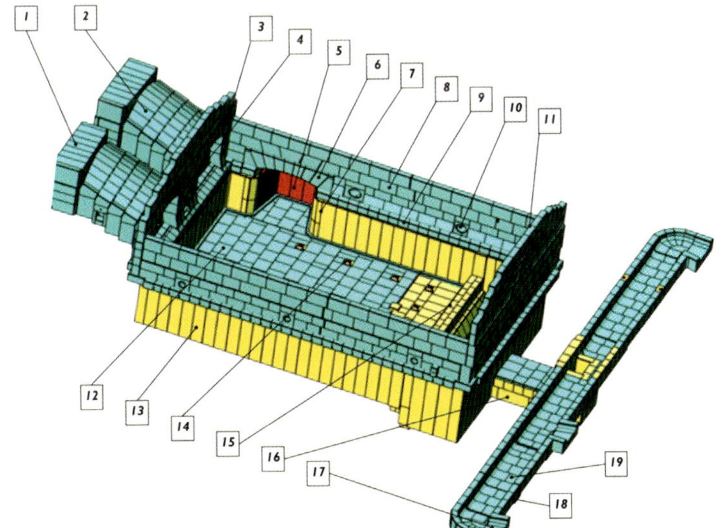

Figure 6.22 End fired furnace for glass making color coded by refractory type (Courtesy of MOTIM).

Cross-fired, high-capacity float glass smelter furnaces utilize the so-called U-jet burners built into the side walls. To increase melting capacity, oxy-fuel burners can also be utilized, which produce higher temperature flames. Auxiliary burners can be built into the roof as well and the vertical temperature of the bath can be homogenized by using heating electrodes built into the chamber floor. These electrodes are made of heat-resistant $MoSi_2$ and they penetrate the hearth bottom from below.

The capacity of the largest glass melting float furnaces reaches $600\,\mathrm{t\,d^{-1}}$.

6.3.4
Furnaces Used in Various Industries

6.3.4.1 Cylindrical Rotary Kilns

Rotary kilns are used for drying, sintering and cement clinker roasting. Their feeds are usually grainy or particulate materials, the motion of which is provided by the rotating inclined cylinder. The length of the cylinder together with its rotation speed and inclination angle control the residence time of the feed. A usual rotary kiln for the cement industry is shown on Figure 6.23.

Rotary kilns are constructed with a refractory lining for sintering and melting processes and without refractories for drying applications. Smaller-scale cylindrical kilns are typically 6–12 m long, but kilns as long as 150 m are routinely used in the cement industry. The inclination angles vary between 1 and 12 °. The kilns rotate at a rate of 0.5–5 rpm.

In direct-fired rotary kilns the burner is typically positioned at the tail end of the cylinder, thus there is a counter-flow of flue gases within the kiln. The heat of the fluidized, spilling grainy feed is transferred towards the entry side by the flowing flue gases, which are subsequently pre-heated.

Burners typically utilized in rotary kilns produce 8–10 m long impulse flames. The burner body itself reaches into the cylinder and is protected from melting by water-cooling or ceramic insulation. These impulse burners are usually mounted on

Figure 6.23 Rotary kiln from the cement industry.

mobile platforms in order to be able to pull them out of the kiln during shutdowns and maintenance.

6.3.4.2 Chamber Furnaces

Direct Fired Kilns   Chamber furnaces are one of the most common furnace types in the insdustry. These furnaces are are usually not operated in a continuous mode. The batch of the furnace stays in the same position throughout the heating process. High temperature forging furnaces are usually of this type.

They are widely used in the industry: the chamber shaped hearth is suitable for heating metals to the temperature of their forming and heat treatment, drying of large casting forms, burning of light-clay products and heat treatment of glass wares. They have many variations.

The square or rectangle-shaped chamber of forging furnaces allow the manipulation of the batch to be heated with one or two upward sliding door(s), either with forging tongs or rotary automatic manipulator. Their firing system and flue gas route can be also very different, depending on the construction.

6.3.4.3 Indirectly Fired Chamber Furnaces

Heat treating furnaces carry out both the heating and cooling of the batch in a continuous or periodical operation schedule. Indirectly fired chamber furnaces are used to treat products that must be protected from oxidizing gases, for example, the flue gas. To ensure the safe separation of the product from the oxidizing gas, protective atmospheres are also used. For specific surface treatment, some furnaces are capable of switching between a protective and active atmosphere (surface carburization, nitridization, etc.). In order to separate the batch from the flue gases, specially designed cases, muffles or retorts are used that contain the flue gas within closed chambers. The majority of heat is transferred from the casing wall to the batch via radiation.

There are many different types of furnaces designed for indirect heating, differing in geometries, capacities and loading methods. They are the following:

- Bell furnaces. The most common type of indirectly heating furnaces. They are designed to heat-treat large amounts (60–150 t) of products. The batch is usually located at the floor level of the facility, supported by a base structure, and the bell-shaped retort is mounted on a crane and positioned and lifted onto the batch (see example on Figure 6.24). Protective gases (N_2, a mixture of N_2 and H_2 or pure H_2) flow between the retort walls and the batch. Bell furnaces carry out specific heat treatment programs and are, therefore, able to heat and cool the batch.
- Elevator kilns. These are similar to bell furnaces, except in the way the batch is loaded into the furnace. Instead of lifting the retort onto the batch, the products are placed on an elevator platform and are lifted into the retort. Elevator kilns are very common in the ceramic industry.
- Pit furnaces. Pit furnaces are cylindrical heating chambers positioned under the floor level of a facility. The batch is hung in the chamber from a bridge crane. The

Figure 6.24 Bell furnace with raised bell and glowing hot batch inside.

chambers have dual-layer walls that contain the radiant heating elements. Heat-resistant steel muffles circulate the protective atmosphere inside the chamber.

- Muffle furnaces. Most commonly used in enameling and the ceramic industry. These are usually low-capacity furnaces that separate the flue gas from the enameled surfaces by utilizing muffles built in between the dual-layer walls.
- Radiant tubes. Radiant tubes are heating elements built into or mounted on furnace walls. Modern radiant tubes can achieve surface temperatures between 900 and 1200 °C. Stationary radiant domes can operate continuously by utilizing carrier rails or carts underneath the dome.
- Immersion furnaces or salt baths. In some cases, the desired surface properties of a batch cannot be achieved by circulating active gases. As in the case of, for example, wire patenting, the batch is immersed in molten salt or liquid metal baths. The baths must ensure a specific, steady temperature. Immersion furnaces can be heated from the outside by firing at their outer casing or from the inside by firing into a submersed crucible.

6.3.5
Heat Treatment Systems, Heat Treatment Furnace Plants

6.3.5.1 Complex Heat Treatment Systems

Chamber furnaces can be constructed in ways that allow multipurpose operation. More complex systems are usually operated periodically. The heated chambers can be split such that in certain zones different heat-treating processes can be carried out simultaneously. Each zone can have its own transportation mechanisms, elevators, atmosphere, sealing, and so on.

Heat treatment of metals can also be done in fluidized bed furnaces. In these furnaces, fluidized grainy material, typically sandy SiO_2 or ZrO_2, transfers the heat to the feed. Through holes in the baseplate, hot flue gases are introduced in high-velocity jets into the chamber, thus both the heating and fluidization are carried out through the distributor hole patterns in the baseplate. Fluidized bed furnaces are also used for the pre-processing of certain ores and the calcination of limestone. In these cases, the feed itself is fluidized.

6.3.5.2 Continuous Heat Treatment Furnace Plants

The continuous furnace plant consists of more furnaces for *annealing, tempering or carburizing heat treatment systems* to increase surface hardness, that ensure the heating at varying temperature and the carrying out of supplementary operations with rollers or conveyors that enable the inner transport, depending on the task. These are widespread in vehicle component part manufacturing.

In the case of a *tempering-carburizing furnace system* the shaped small parts that are stored in heat-resistant steel buckets are transferred on a roller conveyor to a pusher machine in front of the air-lock door of the roller bottom-type furnace with protective gas. The pusher machine pushes the buckets periodically into the radiant tube fired hardening furnace. The material leaving the furnace after the proper soaking goes into the hardening tank that has adjustable heating. From the hardening tank the material, after washing and drying, enters into the radiant tube fired roller bottom-type temper furnace, then after water cooling it leaves the furnace system dried.

In the case of small utility goods, such as screws and hinges, the roller transfer system is replaced by a conveyer, and the roller bottom-type furnaces by steel belt *conveyer type furnaces.*

6.3.6
Petroleum Industry Tube Furnaces

Tube furnaces are part of the basic equipment of the petrochemical and petroleum industry. Their purpose is the heating of liquid.

The feed is circulated in wall tubes that run along the walls of the square-shaped or tubular combustion chamber. The burner can be installed on the roof or on the walls. In case of side burners, the flames are deflected to protect the wall tubes.

Other types of petrochemical tube furnaces carry out the cracking of crudes or intermediate petroleum products. These furnaces utilize higher velocity impulse flames.

6.3.7
Internal Combustion Engines

6.3.7.1 Introduction

Reciprocating internal combustion engines – in short referred to as internal combustion engines – are engines designed to generate work output by intermittently burning mixtures of fuel and air inside them. Internal combustion engines as we know them today were significantly developed in the second half of the nineteenth century by Nicolaus Otto (the four-stroke spark ignition engine) and

Table 6.3 Classification of internal combustion engines with examples.

	Spark Ignition	Compression Ignition
Four-stroke	automotive use, conventional gasoline engines	automotive use, conventional Diesel engines
Two-stroke	lawn mowers, mopeds, small electric power generators and other small engines	ship engines, some locomotives and other large engines

Rudolf Diesel (the four-stroke compression ignition engine) [21]. Early developments date back to the late seventeenth century [22].

Internal combustion engines can be characterized by the number of strokes needed to perform one working cycle as well as by the principle which is used to ignite the fuel–air mixture: Spark ignition engines use a spark generated by a spark plug to ignite a compressed fuel–air mixture, whereas compression ignition engines compress air to increase the temperature, then the fuel is added and autoignites under further compression. Table 6.3 gives an overview of these classifications with examples.

By looking at Figure 6.25 that shows a four-stroke spark ignition engine as an example, we see the most important parts of an engine. The *crankshaft* sits in the *crankcase*, connected to the *piston* via the *connecting-rod*.

6.3.7.2 Four-Stroke Engines

In a four-stroke engine, like the one shown in Figure 6.25, an operating cycle consists of four piston strokes, that is, two revolutions of the crankshaft:

1) During the *intake* or *induction stroke*, the piston moves down from the the top dead center position (TC) to bottom dead center position (BC). The inlet valve is open; the outlet valve is closed; thus drawing in fresh air/fuel mixture into the cylinder.

2) The next stroke is the *compression stroke*. The piston moves up while both valves are closed, reducing the chamber volume to a fraction of its maximum. This fraction is represented by a characteristic figure for all types of internal combustion engines, the compression ratio r_c,

$$r_c = \frac{\text{max.cylinder volume}}{\text{min.cylinder volume}} = \frac{V_d + V_c}{V_c} \tag{6.5}$$

where V_d is the volume swept by the piston and V_c is the clearance volume.

3) The *power stroke* or *working stroke* is characterized by ignition (either by a spark from the spark plug or autoignition) and combustion of the mixture. The pressure and temperature rise sharply, driving the piston down while work is transmitted to the crankshaft. At BC the outlet valve is opened.

4) Finally the *exhaust stroke* removes the majority of burned gas from the cylinder. At the beginning, the pressure difference is high, and then the volume is reduced as the piston moves up again, driving the burned gas out. At TC, the outlet valve

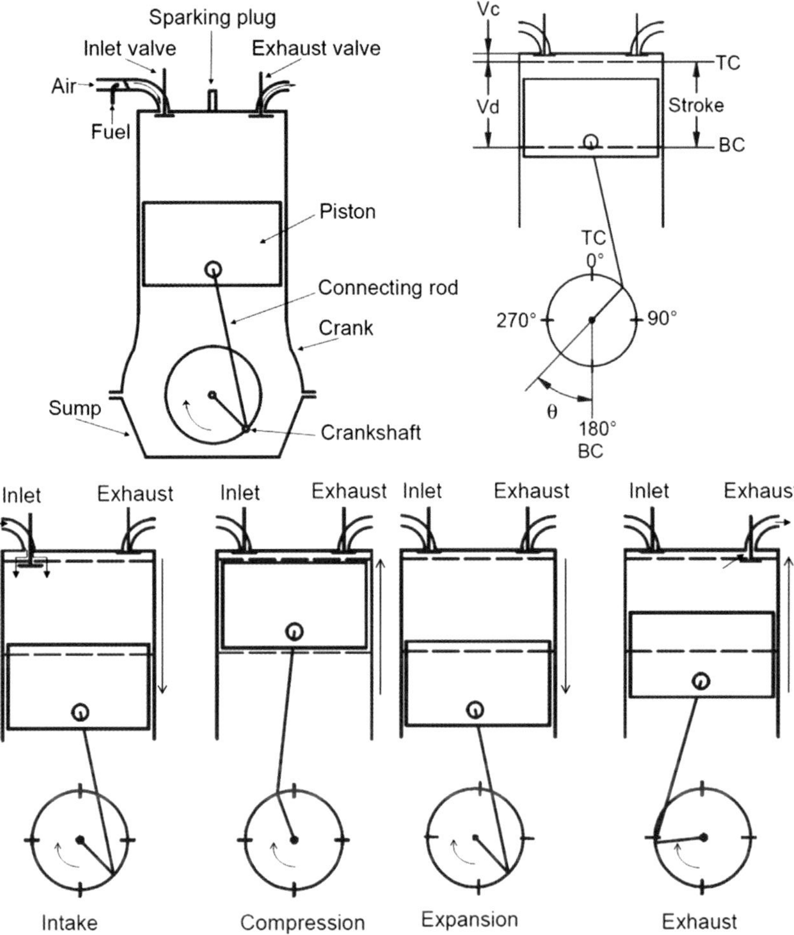

Figure 6.25 A four-stroke spark ignition engine as an example for internal combustion engines.

closes. A small amount of burned gas V_c still remains in the cylinder and will be diluted by the next operating cycle.

6.3.7.3 Two-Stroke Engines

In two-stroke engines, there is one power stroke every crankshaft revolution, resulting in an increase in weight specific power output. The induction and exhaust strokes are omitted by using the lower part of the crankcase as a second chamber for mixture intake.

1) The *compression stroke* compresses the trapped mixture which is ignited near TC. As the piston moves upward, the low pressure in the lower part of the crankcase opens the spring-loaded inlet valve and a fresh charge is drawn in.

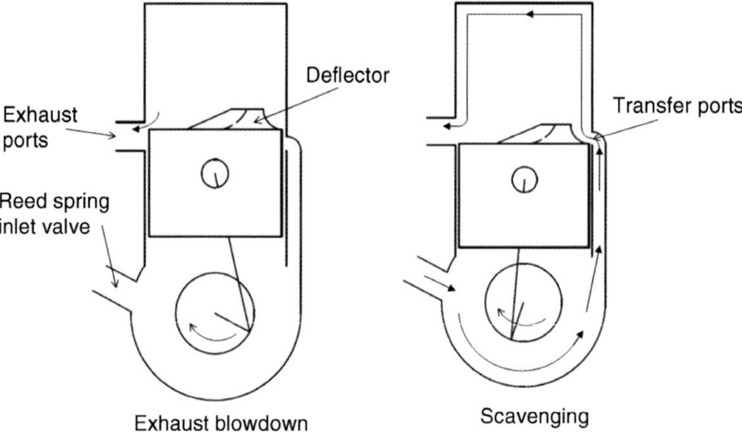

Figure 6.26 A crankcase-scavenged two-stroke spark ignition engine [22].

2) The *power stroke* again forces the piston downward, compressing the fresh charge in the crankcase and opening the exhaust port, thus the pressure in the cylinder decreases. Moving further down, the piston opens the transfer port. Now the compressed fresh charge flows from the crankcase into the cylinder, scavenging the rest of the burned gas. A deflector on top of the piston prevents most of the fresh charge from short-circuiting to the exhaust port.

As a new cycle begins, the ports are closed by the moving piston.

Crankcase-scavenged spark ignition engines (Figure 6.26) are widespread for small engines because of their high power output. Major drawbacks are their high fuel consumption and poor engine emissions caused by short-circuiting of unburned mixture. Nevertheless, efforts are being made to optimize these machines by altering the piston and injection [21]. Other important applications are marine and stationary engines which work with diesel or fuel oil.

6.3.7.4 Spark Ignition Engines

In spark ignition engines, a fuel–air mixture is ignited by a spark drawn from, usually, a single *spark plug* per cylinder shortly before the piston reaches the TC position. This mixture is prepared by a *carburetor*, or in larger and more modern engines by electronically controlled *fuel injection systems*.

Figure 6.27 shows the principle of a fixed jet carburetor. In front of the throttle valve, air is led though a venturi, causing pressure in the venturi to drop. This leads to fuel being drawn into the venturi, where it is broken up to droplets which evaporate building the air–fuel mixture. Fuel overflow is prevented by a float valve. Further improvements are made to optimize the equivalence ratio of the mixture, depending on the position of the throttle valve [21,22].

Ignition timing is important: If the mixture is ignited too early, cylinder peak pressure and temperature rise, increasing the risk of *knock*. Too late ignition will lead to incomplete combustion and too high exhaust gas temperatures, damaging the

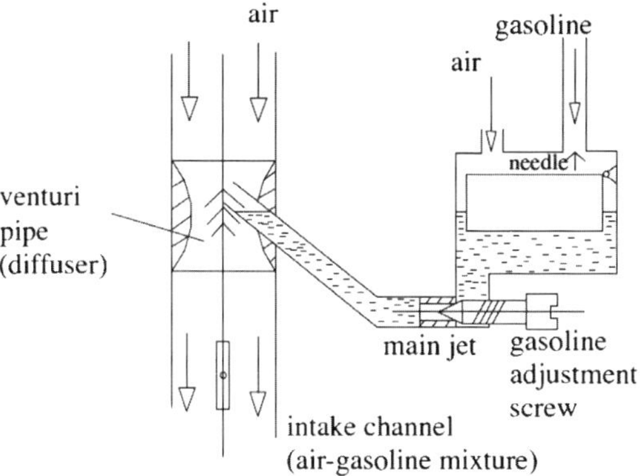

Figure 6.27 Cross-section of a fixed jet carburetor [22].

exhaust valve. An idealized diagram plotting in-cylinder pressure as a function of *crank angle* can be seen in Figure 6.28.

6.3.7.5 Compression Ignition Engines

In compression ignition engines, the fuel is directly injected into the air-filled cylinder as one or more jets of droplets at the end of the induction stroke. This results in a good part-load behavior because only the amount of fuel injected is reduced for part load. At full load, the mass of fuel injected is approximately 5% of the mass of air in the cylinder [22]. Since heat is transferred from the cylinder to the evaporating droplets, compression ignition engines have a minimum cylinder volume of about 400 cm^3 to keep heat loss through the surface acceptable [21]. The maximum speed of compression ignition engines is less than for SI engines of the same size because combustion is slower. On the other hand, cylinder pressures are higher. After injection, there is a short ignition delay period (A–B) where the fuel is vaporized and mixed. Then (B–C) the mixture autoignites and rapid combustion takes place when the piston is at the TC position and during the beginning of the

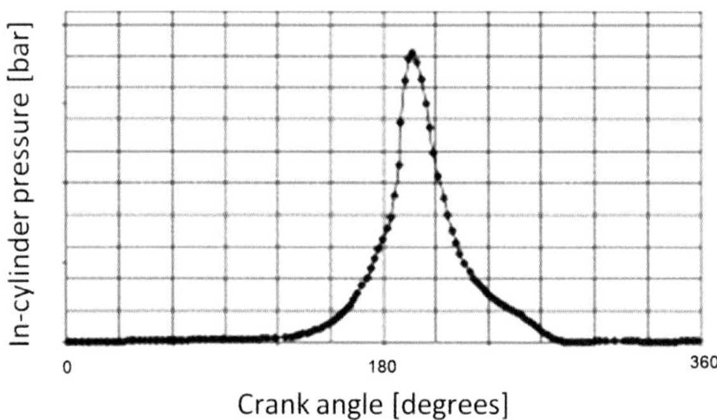

Figure 6.28 Idealized pressure diagram for a spark ignition engine [25]. Reproduced with permission.

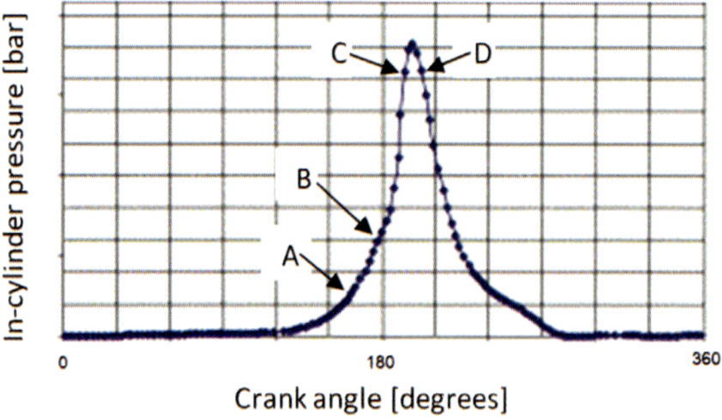

Figure 6.29 Idealized pressure diagram for a compression ignition engine from [25], reproduced with permission.

working stroke, causing a sharp rise in cylinder pressure. A short period of controlled combustion (C–D) follows. Here the combustion is limited by the speed of fresh supply of fuel. After D, the combustion ends diffusively governed until all fuel is burned. (See Figure 6.29)

6.3.7.6 Cycle Analysis and Key Parameters

An important tool for assessing engine efficiency is the comparison with ideal air standard cycles. It is assumed that air, fuel and exhaust gas behave ideally, and combustion processes are reduced to heat transfer for simplicity. These cycles are thermodynamic cycles, whereas real engines run mechanical cycles. Despite these simplifications, the analysis of air standard cycles indicates trends to compare different engine types. The difference between a fired and an unfired Otto cycle can be seen in Figure 6.30. The area enclosed by the cycle corresponds to the work output.

The *rational efficiency* η_R relates the actual work output W of a cycle to a reversible thermodynamic cycle W_{rev}

$$\eta_R = \frac{W}{W_{\text{rev}}} \leq 1 \tag{6.6}$$

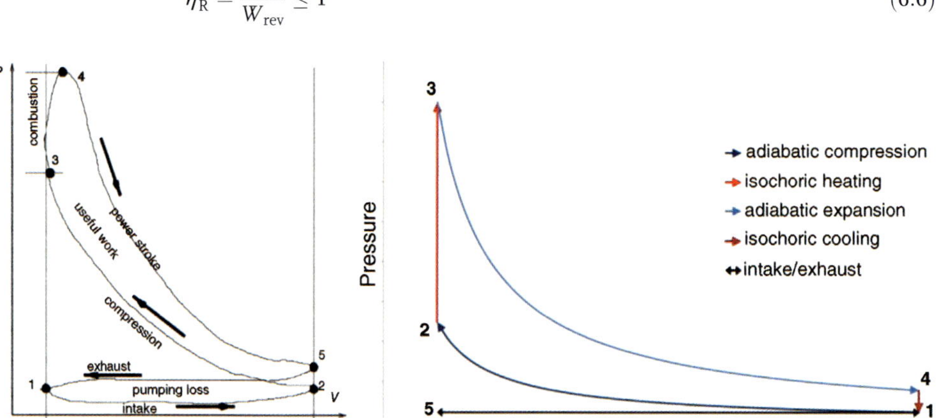

Figure 6.30 Comparison of a fired [23] (Reproduced with permission) and an unfired [24] Otto cycle.

Usually calorific values $CV = -\Delta H_0$ are used instead of W_{rev} for simplicity. They are readily available and easy to measure. This is why the *arbitrary efficiency* η_0, also referred to as *thermal efficiency* or *brake efficiency* η_b, is widely used instead. Typical values for η_0 are around 30%.

$$\eta_0 = \frac{W}{CV} = \frac{W}{-\Delta H_0} \tag{6.7}$$

A demonstrative figure is the *specific fuel consumption* (sfc). This relates the mass of fuel burned to the engine work output generated. When using sfc values, it is indispensable to quote the fuel for which the values were obtained.

The *indicated mean effective pressure* (imep) is independent of engine speed, size and the number of cylinders.

$$\text{imep} = \frac{\text{indicated work output per cylinder per mechanical cycle [N m]}}{\text{swept volume per cylinder [m}^3]} \tag{6.8}$$

Another important figure is the *brake mean effective pressure* (bmep). Despite its name, it is an indicator for the work output which is measured by a brake or a dynamometer.

$$\text{bmep} = \frac{\text{brake work output per cylinder per mechanical cycle [N m]}}{\text{swept volume per cylinder [m}^3]} \tag{6.9}$$

6.3.7.7 Ideal Air Standard Otto Cycle

In the ideal air standard Otto cycle (Figure 6.31), the following four steps are made:

1) Reversible adiabatic compression ($1 \rightarrow 2$) with the compression ratio $r_c = \frac{V_1}{V_2}$
2) Isochoric heating ($2 \rightarrow 3$) by the amount of Q_{23}

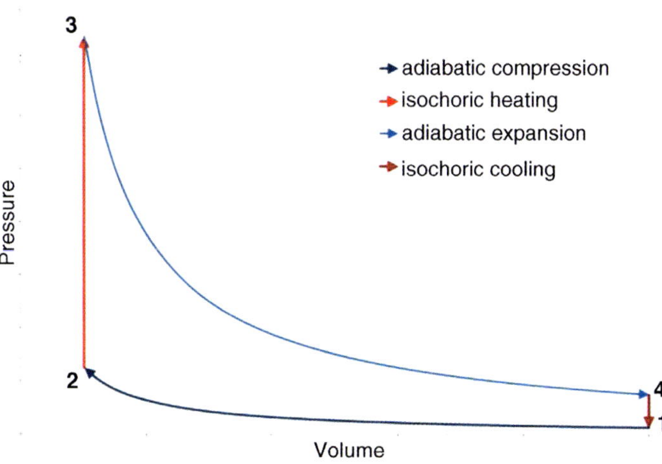

Figure 6.31 Ideal air standard Otto cycle.

3) Reversible adiabatic expansion (3 → 4) with the compression ratio $r_c = \frac{V_4}{V_3} = \frac{V_1}{V_2}$
4) Isochoric cooling (4 → 1) by the amount of Q_{41}

If specific heat capacities c_V are held constant, and $\gamma = \frac{c_p}{c_V}$ is introduced, the efficiency of the Otto cycle can be defined as:

$$\eta_{\text{Otto}} = 1 - r_c^{1-\gamma} \tag{6.10}$$

For the deduction of this equation, refer to [21] or [22]. It is important to see that the cycle efficiency only depends on the compression ratio.

6.3.7.8 Ideal Air Standard Diesel Cycle

In the ideal air standard Diesel cycle (Figure 6.32), the following four steps are made:

1) Reversible adiabatic compression (1 → 2) with the compression ratio $r_c = \frac{V_1}{V_2}$
2) Isobaric heating (2 → 3) by the amount of Q_{23} with the *load* or *cut-off* $\alpha = \frac{V_3}{V_2}$
3) Reversible adiabatic expansion (3 → 4) with the compression ratio $r_c = \frac{V_4}{V_3} = \frac{V_1}{V_2}$
4) Isochoric cooling (4 → 1) by the amount of Q_{41}

If specific heat capacities c_V are held constant again, the efficiency of the Diesel cycle can be defined as:

$$\eta_{\text{Diesel}} = 1 - r_c^{1-\gamma} \cdot \frac{\alpha^{\gamma-1}}{\gamma \cdot (\alpha - 1)} \tag{6.11}$$

For the deduction of the above equation, refer to [21] or [22]. The efficiency of the Diesel cycle depends not only on the compression ratio but also on the load, which deserves closer inspection.

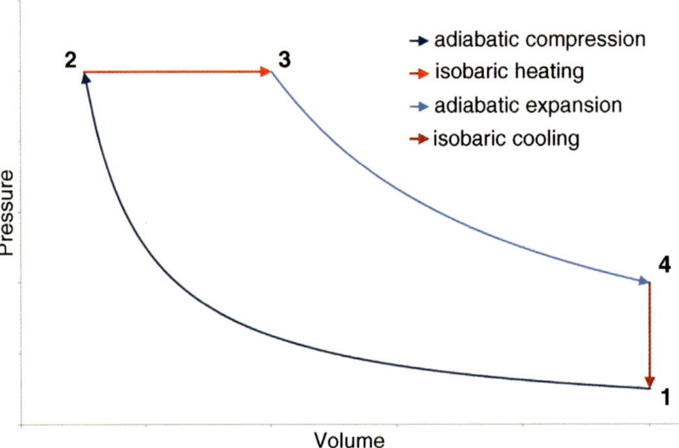

Figure 6.32 Ideal air standard Diesel cycle.

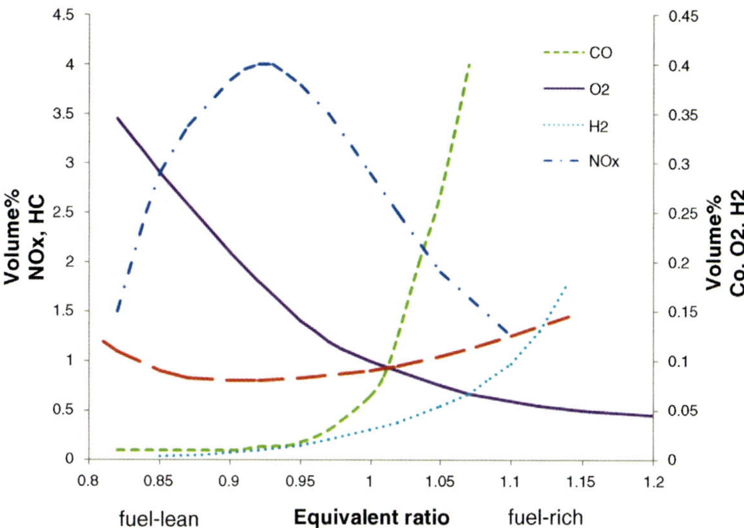

Figure 6.33 Composition of spark ignition engine emissions depending on the equivalence ratio of the mixture.

α has values between 1 and r_c, thus $\frac{\alpha^{\gamma-1}}{\gamma \cdot (\alpha-1)} > 1$ so for a given r_c $\eta_{\text{Diesel}} < \eta_{\text{Otto}}$. If we look at the extreme values for α, we see that

- For $\alpha \to 1 : \eta_{\text{Diesel}} \to \eta_{\text{Otto}}$
- For $\alpha \to r_c : \eta_{\text{Diesel}} \to \max$

6.3.7.9 Engine Emissions

Emissions vary with engine type and load, ignition timing and equivalence ratio. The dependence on the equivalence ratio is shown in Figure 6.33 for spark ignition engines. Here, there are two opposite requirements: For low carbon monoxide (CO) and unburned hydrocarbons (HC) emissions, the engine should operate with lean fuel mixtures, which increases emissions of nitric oxide (NO). The formation of NO is reduced when burning rich fuel mixtures since peak temperatures are lower. On the other hand, this increases the emission of CO and HC. This is why catalytic converters are used to optimize the exhaust gas composition of spark ignition engines.

6.3.8
Gasification and Pyrolysis

6.3.8.1 Introduction to Gasification and Industrial Applications

Gasification is a process in which a solid or liquid, carbon-containing feedstock is reacted with oxygen, air, steam, or a mixture of these gases to produce a "synthesis gas" (syngas) containing hydrogen (H_2) and carbon monoxide (CO). Overall, the objective is to transfer energy contained in a solid or liquid to an energy-containing

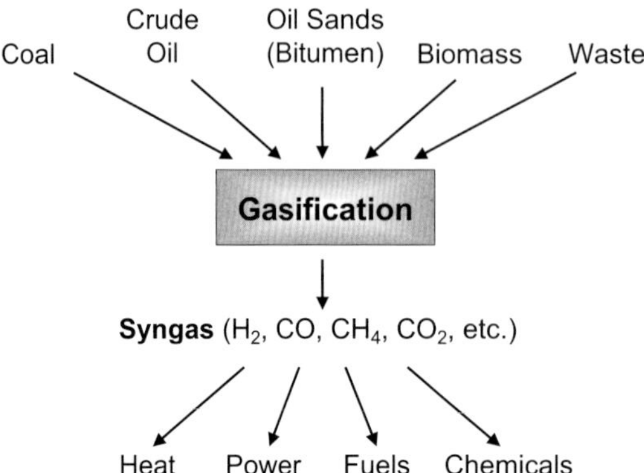

Figure 6.34 Flexibility of gasification.

gas, which can be more easily transported, and which is more suitable for processing in chemical reactors. Typical feedstocks include coal, petroleum coke, heavy oil and other liquid petroleum derivatives, wood, agricultural residues, municipal solid waste, and mixtures of these fuels.

Conversion by gasification is achieved primarily by reaction with oxygen, and in some cases steam. Typically, pure oxygen, rather than air, is used so that the syngas is not diluted with nitrogen from the air. However, some gasifiers are fed with air since it is much less expensive than oxygen. For such systems, the nitrogen-containing product gas is called producer gas. For either type of system, oxygen is input to partly combust the fuel and provide heat for the process, but the amount of O_2 fed is carefully controlled since excessive oxygen results in production of H_2O and CO_2 at the expense of desirable H_2 and CO, thus reducing the energy efficiency of the process. Some gasifiers use steam instead of oxygen as a reacting gas, partly because steam is much less expensive than oxygen. Heat input for such systems is provided indirectly.

As shown in Figure 6.34, syngas has a variety of uses. The most simple application is undoubtedly combusting syngas, either alone or as a supplemental fuel, for generation of heat. Syngas may also be used for generation of electric power, either through combustion in a reciprocating engine coupled to a generator, or by combustion in a gas turbine coupled to a generator. Some of the most efficient and least-polluting coal-fired power plants in the world use gasification in a so-called IGCC (integrated gasification combined cycle) configuration. Finally, syngas may be used as feedstock for production of automotive fuels or chemicals. The hydrogen and carbon monoxide are building blocks for catalytic synthesis of, for example, methanol, dimethyl ether and Fischer–Tropsch fuels.

6.3.8.2 Fuel Conversion During Gasification

The chemical environment in a gasifier is relatively complex, and can vary significantly throughout the system. The gas composition within the reactor is the result of both reactions of the feedstock and homogenous gas-phase reactions.

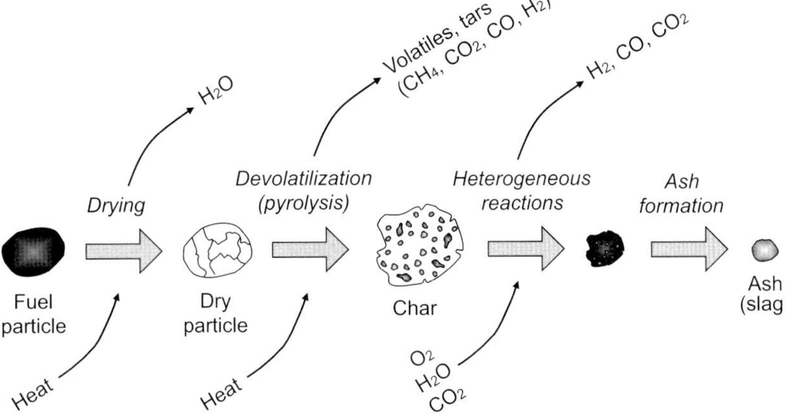

Figure 6.35 Progression of solid fuel conversion during gasification. (Image used with permission from K. Whitty).

Stages of Conversion Gasification is similar to combustion in the regard that fuel particles go through a number of distinct stages as they are being converted. These are depicted schematically for a solid fuel in Figure 6.35.

Drying. Upon entering the hot gasification environment, the fuel temperature increases to the steam saturation temperature and water contained in the particle evaporates. The rate of particle drying is heat transfer-controlled and the time required may range from about a millisecond for micronized particles in a hot entrained-flow gasifier to more than a second for larger particles in fixed or fluidized beds. Water vapor released during drying remains in the gas phase.

Devolatilization. As the particle continues to heat, it is subjected to thermal decomposition and release of volatile components in a process known as devolatilization or pyrolysis. This process is largely the same as the pyrolysis that occurs during combustion and does not involve chemical reaction with other gas-phase components. As with drying, the time required for devolatilization is controlled by the rate of heat transfer to the particle. Devolatilization times are very fuel and environment dependent, but are typically about an order of magnitude longer than the time required for drying.

The gaseous components released during devolatilization also depend on thefuel properties and reactor environment. Generally, they are a mix of light gases, including H_2, CO, CO_2 and CH_4, larger alkanes and alkenes, and so-called "tars." Tars result from partial breakdown of the molecular network that makes up the fuel, resulting in large, typically polyaromatic hydrocarbons with molecular weights of roughly 100–200. Biomass and waste, in particular, produce significant quantities of tars. These components have relatively low vapor pressures, so if they remain unconverted as they travel to cooler regions of the system they can condense on process equipment, forming an odorous, viscous residue that can result in significant operational problems. Tars can be broken down in the gasifier reactor, either thermally or through reaction with gaseous oxidizing species.

Heterogeneous char gasification. The solid char material that remains after devolatilization contains primarily carbon and ash, and the carbon can only be converted by heterogeneous reactions with gas-phase components, which are described in the next section. Char gasification is by far the slowest of the conversion stages, and time associated with char conversion may be 3–4 orders of magnitude longer than devolatilization times. In some low-temperature processes, it may take many hours to convert the char. Char gasification is responsible for much of the production of the desirable syngas components CO and H_2, so it is important that the gasifier be designed to allow sufficient time for these reactions to take place.

For small fuel particles, or at a particular location within larger fuel particles, the three steps presented above occur in succession. Because drying and devolatilization are overall endothermic processes limited by the efficiency of heat transfer, it may take longer for the interior of large particles to dry and devolatilize than the surface of the particle, resulting in composition gradients. Similarly, for relatively reactive, large particles, the rate of char conversion may be limited by the rate at which the reacting gases can diffuse within the particle to contact unconverted carbon, resulting in an unreacted core. In extreme cases, for example with large chips of moist biomass, the exterior of a fuel particle might be fully converted while the center is still wet.

6.3.8.3 Gasification Chemistry and Reactions

The primary chemical reactions responsible for conversion in char are reaction with steam and carbon dioxide:

$$C(s) + H_2O(g) \rightarrow CO(g) + H_2(g) \tag{6.12}$$

$$C(s) + CO_2(g) \rightarrow 2\,CO(g) \tag{6.13}$$

Both of these reactions are endothermic. The heat necessary may come from partial oxidation of the fuel, or it may be transferred to the gasifier indirectly. The intrinsic rate of the carbon–steam reaction is generally 3–4 times faster than the CO_2–carbon reaction. In addition, in gasifiers using oxygen or air as the reacting gas, the char may be converted simply by combustion with oxygen:

$$C(s) + 1/2\,O_2(g) \rightarrow CO(g) \tag{6.14}$$

$$C(s) + O_2(g) \rightarrow CO_2(g) \tag{6.15}$$

Finally, it is also possible for solid carbon to be converted through methanation by reaction with hydrogen:

$$C(s) + 2\,H_2(g) \rightarrow CH_4(g) \tag{6.16}$$

Several homogeneous gas-phase reactions are also important for establishing the composition of the syngas. The most significant of these is the water-gas shift reaction:

$$H_2O(g) + CO(g) \leftrightarrow H_2(g) + CO_2(g) \tag{6.17}$$

This reaction can be promoted by catalysis and is frequently used to adjust the H_2/CO ratio for processes that involve catalytic production of chemicals or fuels from the syngas.

6.3.8.4 Gasification Technologies

Most gasification processes can be classified into one of three technology types: fixed bed, fluidized bed and entrained-flow. All three types may use oxygen, air and/or steam for reaction, and may be operated at high pressure.

Fixed Bed Gasifiers   In a fixed bed gasifier (Figure 6.36a) the reacting gas passes through a slowly-moving bed of solid fuel. The gas typically flows upwards through the bed, although in some technologies the gas flows downwards, which helps minimize elutriation of fines from the bed. Fuel may be fed onto the top of the bed, resulting in a downwards-flowing bed, or it may be fed from underneath the bed with the help of for example, augers. The temperature of fixed bed gasifiers is generally kept below the melting point of the ash, which falls through a grate at the bottom or periphery of the reactor chamber. Many types of fixed-bed gasifiers employ slowly-moving rakes near the floor to help mix the fuel, minimize agglomeration and promote flow of ash through the grate.

Fixed bed gasifiers may operate in co-current or counter-current flow. In a co-current system, the fuel and reacting gas are fed together, either from the top or from the bottom of the reactor. The oxidant is, therefore, exposed to fresh fuel and reacts primarily with volatile material released during the early stages of fuel conversion.

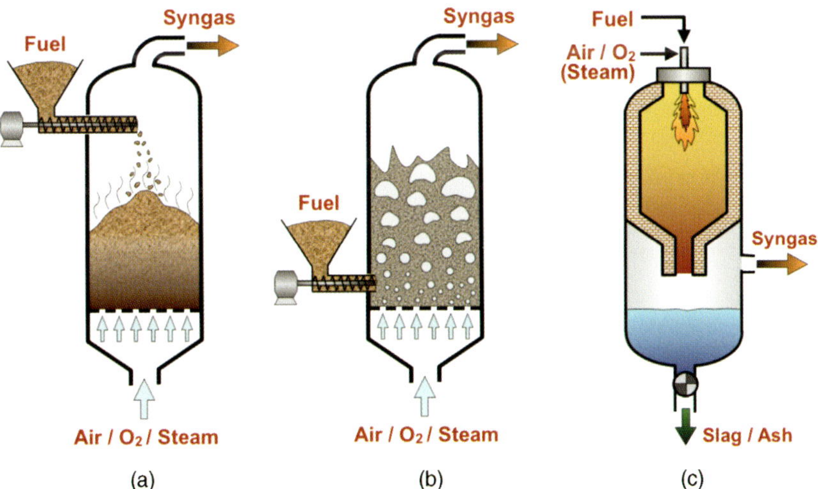

Figure 6.36   Different types of gasifiers. (a) fixed bed, (b) fluidized bed, (c) entrained-flow. (Images used with permission from K. Whitty).

Burn-out of the char further along in the bed must, therefore, occur through reaction with steam or carbon dioxide, and at the relatively low temperatures of fixed bed gasifiers, reaction can be slow, resulting in poor fuel conversion.

A more common configuration is counter-current flow, in which the fuel and oxidant flow in opposite directions. In the system shown in Figure 6.36a, fuel is fed onto the top of the bed and is heated by hot, oxygen-depleted gases flowing from below. As the fuel travels downwards towards hotter regions of the reactor, it slowly pyrolyzes and the resulting pyrolysis gases are swept upwards towards cooler zones of the reactor. Oxygen (or air) enters through a distributor in the bottom of the bed and reacts with fuel char, generating heat that is transported upwards through the bed, providing energy for pyrolysis and drying of the fuel. Because reactive oxygen is available to combust the low-reactivity char, counter-current fixed bed gasifiers tend to achieve better conversion than co-current systems. However, for tar-producing fuels such as biomass and low-rank coals, there is no opportunity for reaction or thermal cracking of the tars since the syngas cools as it flows upwards and out of the reactor. The result is a tar-rich syngas.

Examples of commercial fixed-bed gasifiers include the Lurgi dry bottom gasifier, which currently has the largest number of installed units worldwide, and the Nexterra biomass gasifier.

Fluidized Bed Gasifiers Fluidized bed gasifiers are commonly used for high volatility, reactive fuels, such as biomass and low-rank coal. Fluidized bed gasifiers have also been used for gasification of liquid feedstocks, such as heavy oil and black liquor. Air, oxygen or steam is used both as fluidizing gas and as the gasification reactant and the systems may be configured as low-velocity bubbling bed reactors (Figure 6.36b) or circulating fluidized bed (CFB) reactors. Spouted bed gasifiers have also been operated successfully. Fuel is introduced through, for example, water-cooled augers low in the bed, or in the recirculation leg of a CFB gasifier, to allow maximum residence time for volatiles to be converted.

The bed itself typically comprises particles of diameter 70–300 microns for bubbling and CFB gasifiers, and up to 2 mm for spouted bed reactors. Often naturally occurring materials, such as silica sand or olivine, are used, since they are relatively inexpensive and readily available. Limestone or dolomite may be added both to catalyze gas-phase and tar destruction reactions and to provide *in situ* absorption of hydrogen sulfide and other sulfurous components. Careful attention must be paid to the inorganic chemistry of the fuel and the bed material, so as to avoid compositions that may lower the melting point of the bed as a result of eutectic effects, thereby risking agglomeration and defluidization. This is particularly true of certain biomass fuels and municipal solid waste, which can contain high concentrations of alkali metals and chlorine.

Unlike fixed bed reactors, which have a significant temperature gradient throughout the bed, the temperature in a properly operated fluidized bed gasifier is very stable and uniform, which is desirable for efficient operation. The large mass of hot solids provides good thermal stability, which allows fluidized bed gasifiers to accommodate feedstocks of variable composition, such as municipal solid waste. Although it is generally desirable to operate a fluidized bed gasifier hot to maximize gasification rates, an upper temperature bound is imposed by the need to avoid agglomeration of the bed.

When a fuel particle is introduced into a fluidized bed gasifier, it rapidly dries and devolatilizes, releasing gases that increase the gas velocity in the reactor. The pyrolysis gases may react (combust) with oxygen fed to the system, producing heat, or they may thermally decompose or undergo other gas-phase reactions (possibly promoted by catalytic materials in the bed) to produce lighter gases, including H_2, CO, CH_4 and CO_2. The char resulting from devolatilization remains in the bed and must be converted by heterogeneous reaction with steam, oxygen or CO_2. Eventually conversion and break-up resulting from the turbulent environment in the bed reduce the char/ash size to the point where the gas velocity exceeds the particle's terminal velocity and it is elutriated from the bed.

Several successful fluidized bed gasification technologies have been developed. KBR's TRIG transport gasifier is a fast-fluidized, circulating fluidized bed gasifier that has been successfully operated on coal and various biomass materials. Other circulating fluidized bed systems have been developed by Carbona and Battelle.

Entrained-Flow Gasifiers Entrained-flow gasifiers, also called suspension gasifiers, involve very high temperature reaction of fuel with either pure oxygen or, in some cases, air in reactors that can operate at pressures approaching 100 bar. Typically 30–60% of the oxygen needed for complete combustion is supplied. The partial combustion provides the heat needed for reaction and produces H_2O and CO_2 necessary for the heterogeneous char gasification reactions.

Fuel and the oxidizing gas are introduced together through one or more injectors, whose function is to break up the fuel stream and mix the fuel and gas. Steam may also be introduced to aid atomization and conversion to hydrogen. Solid fuels, such as coal and petroleum coke, are typically pulverized and mixed with water to form a pumpable slurry around 60–65% solids by weight. Dry feed systems may also be used, but it can be challenging to increase the pressure of a solid feed stream to the reactor pressure without introducing considerable amounts of inert gas (e.g., for lock hopper systems).

Entrained-flow gasifiers may be downwards-flowing (e.g., GE, Siemens) or upwards flowing (e.g., ConocoPhillips, Shell). Typically, these gasifiers operate above the melting point of the inorganic components in the fuel, so that the ash melts, forming slag that runs down the walls of the gasifier. A small concentration of fluxing agent may be added to the fuel to reduce the viscosity of the slag and minimize the risk of reactor plugging. Most entrained-flow gasifiers are lined with special high-temperature refractory bricks that can withstand the harsh environment of temperature, pressure, syngas and slag. Some designs, notably those of Siemens and Shell, employ so-called membrane walls that are constructed of pipes that are welded axially to narrow webs of flat metal. The inside surface of the membrane wall may be coated by a conductive protective material, such as silicon carbide. Water or steam flows through the pipes in the membrane wall, so it is cooler than the reaction chamber. The result is that slag adjacent to the wall solidifies, providing a protective layer of frozen slag. Further from the cooled metal wall, the slag is hotter, and at some thickness the temperature is hot enough that the slag is liquid and flows down the layer of frozen slag.

A flame results in the near-injector region as oxygen reacts with the fuel and released volatiles, and represents the highest temperature in the gasifier. For coal gasifiers, the flame temperature can exceed 1500 °C. Beyond the flame region, unreacted fuel char is converted by the heterogeneous reactions presented earlier. Much of the char conversion occurs in suspension, but char that contacts the wall may stick to the molten slag layer there and flow down the wall, thus increasing its residence time and chance for conversion.

Due to the very high operation temperature, entrained-flow gasifiers produce almost no tar. Although residence times are generally only a few seconds, the resulting syngas composition is very near the equilibrium composition. Fuel conversion tends to be relatively high, above 95%. Better conversion could be achieved by increasing the temperature even more, but the rate of refractory degradation would also increase dramatically, and it is not considered worth the cost to pursue the last few percent conversion.

There are many varieties of entrained-flow gasifiers currently in operation. GEs technology (formerly developed by Texaco) involves a single downfired injector and a cylindrical, refractory-lined gasifier chamber. The Siemens gasifier is similar, but uses a membrane wall reactor. The ECUST gasifier, developed in China, is overall downwards-fired, but it uses multiple horizontally-fired injectors along the outer wall near the top of the reactor. ConocoPhillips uses a two-stage refractory-lined updraft gasifier, with fuel and all oxygen introduced in the first stage low in the gasifier, and fuel alone fed to the second stage. The Shell gasifier is also an updraft design, but has a membrane wall and does not have a second fuel feed stage. The authors would like to thank Dr. Kewin J Whitty (University of Utah) for his contribution on Gasification and Pyrolysis.

Abbreviations/Notations

APC	Air pollution control
ESP	Electrostatic precipitator
MWI	Municipal waste incineration
MSW	Municipal solid waste
SCR	Selective catalytic reduction
SNCR	Selective non-catalytic reduction
PCDD	Polychlorinated dibenzodioxin (dioxin)
PCDF	Polychlorinated dibenzofuran
BFB	Bubbling fluidized bed
FGD	Flue gas desulfurization
FB	Fluidized bed
CFBC	Circulating fluidized bed
FBHE	Fluidized bed heat exchanger

Δp Pressure drop [Pa]
L Height [m]
ε Void fraction (porosity) $[-]$
ε_{mf} Void fraction at the fluidization point $[-]$
μ Dynamic viscosity $[\mathrm{kg\,m^{-1}\,s^{-1}}]$
ρ_G Gas density $[\mathrm{kg\,m^{-3}}]$
ρ_P Particle density $[\mathrm{kg\,m^{-3}}]$
g Gravitational acceleration $[\mathrm{m\,s^{-2}}]$
d_P Particle size [m]
U_0 Superficial velocity $[\mathrm{m\,s^{-1}}]$
U_{mf} Minimum fluidization velocity $[\mathrm{m\,s^{-1}}]$
Ar dimensionless Archimedes number $[-]$
c_W Drag coefficient of a single sphere $[-]$

Web Resources

There are many good resources on the internet to learn more about combustion applications. A few are listed here for further the interested reader.

National Fuel Cell Research Center, tutorial "gas turbine"	http://www.nfcrc.uci.edu/ EnergyTutorial/gasturbine.html
International Flame Research Foundation	http://www.ifrf.net/
Animated engines	http://www.animatedengines.com/

References

1 Kessels, J. (2009) *Good Practice for Industrial Coal-Fired Boilers*, IEA Clean Coal Centre, ISBN: 978-92-9029-474-0.

2 Effenberger, H. (2000) *Dampferzeugung*, Springer Verlag, ISBN: 3-540-64175-0.

3 Ergun, S. (1952) Fluid flow through packed columns. *Chemical Engineering Progress*, **48**, 89–94.

4 Grote, K.H. and Feldhusen, J. (2011) *Dubbel – Taschenbuch für Maschinenbau*, Springer Verlag, ISBN: 978-3642173059.

5 Lehmann, H. (2000) *Dampferzeugerpraxis – Grundlagen und Betrieb, Band 4*, Resch Verlag, ISBN: 3-935197-03-9.

6 Buecker, B. (2002) *Basics of Boiler & HRSG design*, PennWell Corporation, ISBN: 978-1-59370-186-4.

7 Lackner, M., Winter, F., and Agarwal, A.K. (2010) *Handbook of Combustion, Vol. 4: Solid Fuels*, Wiley VCH Verlag, ISBN: 978-3-527-32449-1.

8 Rand, T., Haukohl, J., and Marxen, U. (2000) *Municipal Solid Waste Incineration: Requirements for a Successful Project*, World Bank Technical Papers, ISBN: 978-0821346686.

9 National Research Council (2000) *Committee on Health Effects of Waste Inc and D. R. Mattison, Waste Incineration & Public Health*, National Academies Press, ISBN: 978-0309063715.

10 Martin, E.J. and Johnson, J. (1987) *Hazardous Waste Management Engineering*, Van Nostrad Reinhold, ISBN: 0-442-24439-8.

11 Yates, J.G. (1983) *Fundamentals of Fluidized Bed Chemical Processes (Monographs in Chemical Engineering)*, Butterworth-Heinemann Ltd., ISBN: 978-0408709095.

12 Basu, P. (2006) *Combustion and Gasification in Fluidized Beds*, CRC Press, Taylor and Francis Group, ISBN: 987-0-8493-3396-5.

13 Miller, B.G. (2011) *Clean Coal Engineering Technology*, Butterworth-Heinemann, ISBN: 978-1-85617-710-8.

14 van Loo, S. and Koppejan, J. (2010) *Biomass Combustion & Co-Firing*, Earthscan Publishing Ltd., London, ISBN: 978-1-84407-249-1.

15 Miller, B.G. and Tillman, D.A. (2008) *Combustion Engineering Issues for Solid Fuel Systems*, Elsevier, ISBN: 978-0-12-373611-6.

16 Ngô, C. and Natowitz, J.B. (2009) *Our Energy Future- Resources, Alternatives, and the Environment*, Wiley-VCH Verlag GmbH, ISBN: 9780470473788.

17 Strauß, K. (2009) *Kraftwerkstechnik- zur Nutzung Fossiler, Nuklearer und Regenerativer Energiequellen*, Springer Verlag, ISBN: 978-3-642-01430-7.

18 Nag, P.N. (2008) *Power Plant Engineering*, McGraw-Hill Publishing, ISBN: 987-0-07-064815-9.

19 Basu, P. and Fraser, S.A. (1991) *Circulating Fluidized Bed Boilers: Design and Operations*, Butterworth-Heinemann, ISBN: 0-7506-9226-X.

20 Wagner, W. (1990) *Thermische Apparate und Dampferzeuger – Planung und Berechnung*, Vogel Verlag und Druck KG, ISBN: 3-8023-0157-9.

21 Stone, R. (2012) *Introduction to Internal Combustion Engines*, Palgrave Macmillan, ISBN: 978-0-230-57663-6.

22 Heywood, J.B. (1988) *Internal Combustion Engine Fundamentals*, McGraw-Hill, ISBN: 0-07-028637-X.

23 Lackner, M. Winter, F. and Agarwal, A.K. (eds) (2010) *Handbook of Combustion*, Wiley-VCH Verlag GmbH, p. 391, ISBN: 978-3527324491.

24 Sahin, Z., Durgun, O., and Bayram, C. (2011) *Experimental Investigation of Gasoline Fumigation in a Turbocharged IDI Diesel Engine*, vol. 95, Elsevier, pp. 113–121.

25 Perez-Blanco, H. (2004) *Experimental Characterization of Mass, Work and Heat Flows in an Air Cooled, Single Cylinder Engine, Energy Conversion and Management*, vol. 45, Elsevier, p. 13.

26 Biro, A. (2007) *Combustion Technology.* Press of the University of Miskolc.

27 Palotás, Á. (2012) *Industrial Combustion Technology*, e-book University of Miskolc, Hungary.

28 Eddings, E. Combustion in Kutz, Myer (ed): Ch.16. Energy and Power – Mechanical Engineers Handbook 3rd Edition, John Wiley & Sons, Inc. p 575–613.

29 Palotás, Á (2012) Industrial Furnaces in Wopera Á. (ed.) Heating Technologies – Aspects of Environmental and Energy Utilizitaion, e-book, University of Miskolc, Hungary.

7
Safety Issues

7.1
Introduction

Combustion is a desired process, generating heat and energy for electricity, transportation and other purposes. On the other hand, unwanted combustion events, that is, fires and explosions, lead to serious property damage and fatalities throughout the world.

Fires are encountered in forests, private homes, hotels, mines, and small and large industrial installations, such as oil rigs, factories and warehouses. Figure 7.1 shows a destroyed sugar factory after a dust explosion (USA, 2008).

Catastrophic fires and explosions have happened frequently; already back in 1769, a gunpowder explosion in a church in Brescia, Italy, caused many fatalities, and in 1785 a dust explosion in a bakery in Turin, Italy, was recorded. Tragedies from the recent past are a hotel fire in China (2011), a fire in a clothing store in the Philippines (2012), a refinery explosion in Hemel Hempstead in the UK (2005) and the explosion of BP's the oil drilling platform Deep Water Horizon in the gulf of Mexico (2010), spilling 4.9 million barrels (780 million liters) of crude oil into the sea. Fires and explosions can occur in all industries involving combustible materials. Table 7.1 shows the most frequent reasons for fires and explosions in the chemical industry. Risky processes are generally cracking, hydrating, distillation, coking, alkylating, reforming, and storage plus transportation.

Figure 7.2 shows a classification of fires. Liquids and solids can produce a fire reaction, whereas gases and fine solids will react in an explosion.

This chapter deals with how inadvertent acts and conditions leading to accidental combustion reactions can be minimized to increase safety.

7.2
Fundamentals

Essentially, all combustible and explosive materials are fire risks. Fire fighters like to draw the fire triangle (Figure 7.3), variants of which are the fire tetrahedron and the dust explosion pentagon (see later).

Combustion: From Basics to Applications, First Edition. Maximilian Lackner, Árpád B. Palotás, and Franz Winter.
© 2013 Wiley-VCH Verlag GmbH & Co. KGaA. Published 2013 by Wiley-VCH Verlag GmbH & Co. KGaA.

Figure 7.1 Sugar factory after a dust explosion, killing and injuring 50 people [1].

Table 7.1 Statistics on fires and explosions in the chemical industry [2]. Data for 1987/1988.

Reason for fires/explosion	%
Open flames	35.5
Hot work activities	18.0
Electricity	10.5
Hot surfaces	7.2
Smoking	5.7
Friction	5.4
Autoignition (gases)	3.7
Hot particles	3.0
Static electricity	2.8
Others	8.2

The fire triangle lists the necessary conditions for a fire to start and/or to continue. So, for a fire to take place, there needs to be sufficient oxidizer (generally the oxygen in the ambient air), fuel (any gaseous, liquid of solid combustible material), and adequate temperature and/or an ignition source. As soon as any of the three is removed (which fire fighters are doing), the fire will cease. Reduced oxygen environments [19] are a means of reducing the fire risk in server rooms and libraries' and museums' storage facilities, for instance.

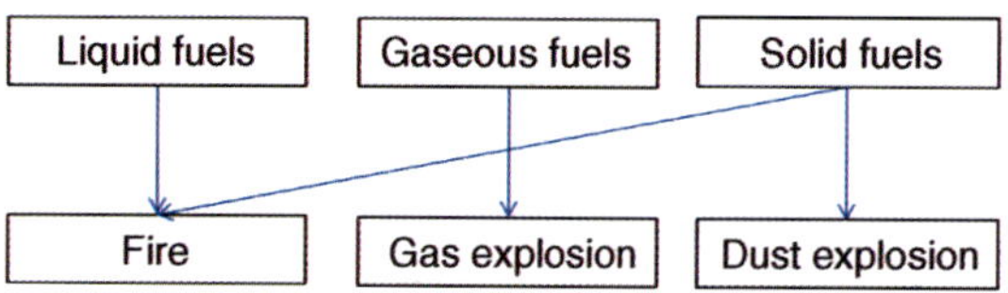

Figure 7.2 Gases and solids can cause an explosion.

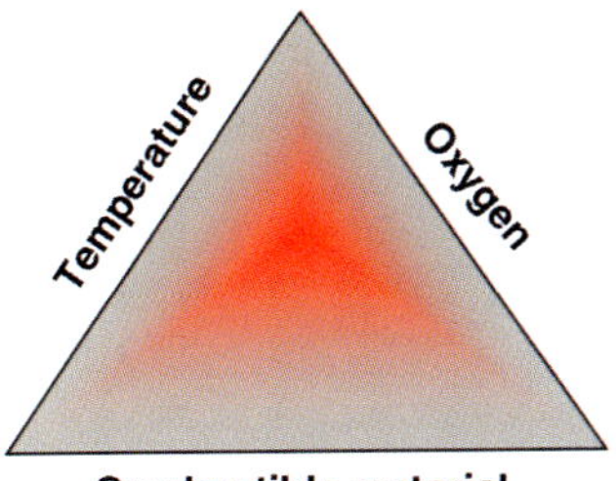

Figure 7.3 Fire triangle. Fuel, oxidizer and ignition source (temperature) are needed simultaneously to start and maintain a fire.

7.3
Fire Classes

In order to facilitate the selection of a suitable fire extinguisher, fires are grouped into classes, for example, by the European Standard "Classification of fires" (EN 2:1992). Table 7.2 presents an overview.

Table 7.2 Fire classes. Note: Other regions of the world have different classes, for example, the US use the NFPA code (NFPA = National Fire Protection Association).

Extinguishing medium	Fire classes					
	A	B	C	D	F	
	Fires with solid fuels	Fires with liquid fuels	Fires with gaseous fuels	Fires with metals	Fires with kitchen fat and oil	Fires in electric installations
Water	+	Fine mist Professional fire fighters only	Fine mist Professional fire fighters only	−	−	Only below 1000 V
Foam	+	+		−	−	
CO$_2$		+	+	−	Vessel cooling Professional fire fighters only	
ABC powder	+	+	+	−	−	
BC powder		+	+	−	−	
D powder	Cost	−	−	+	−	
Fire extinguisher for fat and oil	Cost	Cost	−	−	+	

Figure 7.4 Water is not a suitable fire extinguishing agent for burning oil. As the burning oil has a high temperature, typically 300–350 °C, the water will instantly vaporize and disperse the oil into droplets with a very large surface area, leading to a huge fireball [3].

Note on gaseous fuels: On many occasions, it will be advisable not to extinguish a fire around a gaseous fuel source, because without a (locally confined) fire, the explosive gases are significantly more difficult to control.

Here is an example of what happens if the wrong extinguishing media are used: A class F fire is attacked with water. One can see from Figure 7.4 that instead of extinguishing the fire, a vigorous explosion occurs. This case is particularly relevant for fires in kitchens, for example, with deep fryers (chip pans). Another example where water is absolutely not suitable is fighting burning metals. The high combustion temperature (for instance, aluminum fires generate temperatures in excess of 4000 K) produces H_2 by cleaving the water, and the oxyhydrogen gas will explode. This is most dangerous in confined areas. Metal dusts are increasingly dangerous with decreasing particle size. Moisture can heat up metal dust and lead to auto-ignition.

Fires at home are often underestimated, for example, fires in Christmas trees (see the video in [4]). From ignition to a fire in the entire room, only tens of seconds can pass. In many cases, death is brought about from toxic gases rather than the flames themselves.

7.4
Working Mechanism of Fire Extinguishing Media

ABC powders typically contain ammonium phosphate and ammonium sulfate. They show an anti-catalytic effect. In the case of embers, the ABC powder also produces a shielding sintered layer. The NH_3 further suffocates the fire.

BC powders are generally the "standard" powder extinguishers. They contain sodium bicarbonate ($NaHCO_3$). There are also powder extinguishers that utilize K_2SO_4 and $KHCO_3$.

Fire extinguishers for *class D* fires contain finely ground powder of alkali chlorides, often NaCl. The powder produces a sintered layer of salt over the burning metal, shielding it from further oxygen.

For *class F* (cooking oil fires), the standard class B fire extinguishers are not effective, because the temperatures are higher than those of other combustible liquids. Class F fire extinguishers spray an alkaline solution over the fire, which reacts with the fat/oil in a saponification process to give a non-flammable soap.

The above-mentioned *anti-catalytic fire extinguishing effect* causes radical quenching and chain reaction termination. One can distinguish between homogeneous inhibition and heterogeneous inhibition. In the former, the extinguishing agent forms radicals in the flame which recombine with existing radicals, leading to chain termination. Heterogeneous inhibition is a "wall effect": on the large surface area of the fine powder, radicals are quenched and will stop the chain reaction effectively.

Note: Powder-based fire extinguishers damage electronics. So for household use, *foam*-based extinguishers are recommended. Foam-based fire extinguishers can be in an aspirated (mixed and expanded with air in a branch pipe) or non-aspirated form. They form a frothy "blanket" over the fire, thereby preventing oxygen from reaching the fuel.

7.5
Fire Detectors

As fires tend to develop progressively, early detection and extinguishing is critical. For fire detection, sensors that can measure various signals are available:

- Heat
- Combustion gases (e.g., HCl or CO; CO can be detected using SnO_2 sensors)
- Smoke (optically via light scattering)
- Ionization (^{241}Am)
- UV light

Smoke detectors are low-cost devices. They are a meaningful investment for all kinds of installations.

7.6
Deflagrations and Detonations

One can distinguish between two types of explosion: *deflagration* and *detonation*. Deflagrations are combustion processes that propagate through thermal conductivity at less than supersonic speed. The flame front "eats" itself from one layer of combustible material to the next. Detonations propagate through a shock wave with

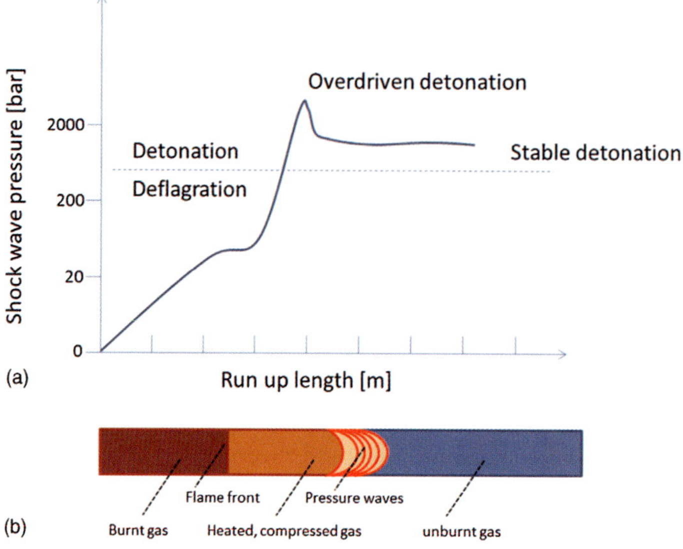

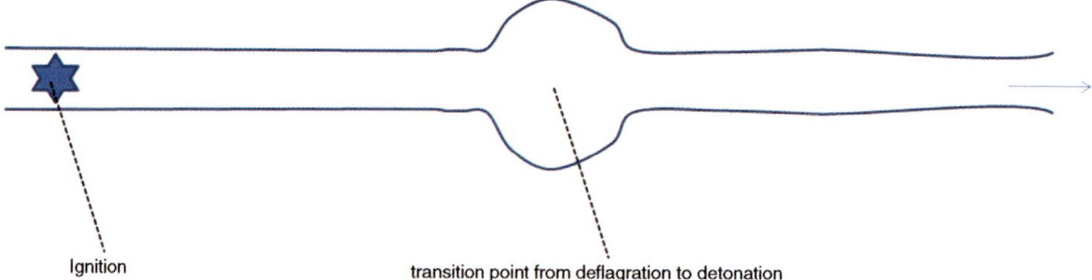

Figure 7.5 Transition of a deflagration into a detonation inside a pipe. Note the logarithmic pressure scale. The run up length is the distance traveled by the flames.

supersonic speed. The shock wave compresses the material ahead, thereby igniting it. Under appropriate conditions, a deflagration can develop into a detonation. This is illustrated below for a pipe that is filled with an explosive gas/air mixture:

In the beginning, the flame front (drawn on the left-hand side of Figure 7.5b) travels through the pipe. The combustion gases have a higher volume than the educts, typically 6–8 fold higher, depending on the involved reaction gases and the temperature increase. Hence the flame front is pushed further at an increasing speed, and the pressure waves can develop into a detonation after a distance which is known as the "run up length". Through pressure wave reflections, bends can shorten the run up length considerably.

Figure 7.6 shows schematically the transition point of a deflagration into a detonation inside a pipe.

Figure 7.6 Schematic illustration of a pipe after the ignition of a contained explosive atmosphere. At the baggy section, the transition from a deflagration to a detonation has occurred [5].

This phenomenon is important to bear in mind for the installation position of flame arrestors. When the distance from a potential ignition source becomes large, or when there are bends, a detonation arrestor is the safer choice. However, detonation arrestors have a larger pressure drop than standard flame arrestors. Flame arrestors are not suitable for gases that contain solids. In that case, a knock-out drum (water seal) might be used.

7.7
Dust Explosions

Fine particles have a large surface/volume ratio, making them very reactive. As the particle size of a solid is reduced, its surface area increases inversely proportionally to d^2 (with d being the diameter). To give an example: A cube of $1 \times 1 \times 1\,cm^3$ that theoretically split into cubes of $1 \times 1 \times 1\,\mu m^3$, increases its surface area by a factor of 10^4 from $6\,\mu m^2$ to $6\,cm^2$, and if further split down to $10 \times 10 \times 10\,nm$ cubelets, increases its surface area from $6\,cm^2$ to $6 \times 10^6\,cm^2 = 600\,m^2$. Dusts of combustible materials become dangerous below $500\,\mu m$ (0.5 mm).

The dust explosion pentagon (Figure 7.7) shows the conditions necessary to bring about a dust explosion.

Bear in mind that even a small fraction of fine particles can become dangerous, as it is possible that it will be enriched in a part of a process plant. A typical value for the lower explosive limit (LEL) of dust is $\sim 30\,g\,m^{-3}$. This is not much: a dust layer on the floor of depth 1 mm, under adverse conditions, can exceed the LEL if made airborne. Hybrid mixtures, that is, combustible dust in the presence of combustible gases, are particularly dangerous. A good database on the properties of dust is "GESTIS DUST EX" [6]. This database lists important combustion and explosion characteristics of more than 4600 dust samples, which can be used as a basis for the safe handling of combustible dusts and for the planning of preventive and protective measures against dust explosions in dust-generating and processing plants. Dust explosions

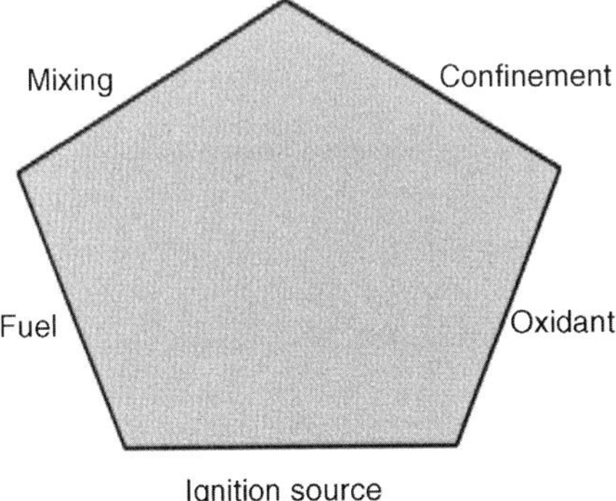

Figure 7.7 The dust explosion pentagon [20]. Confinement is what makes any explosion dangerous.

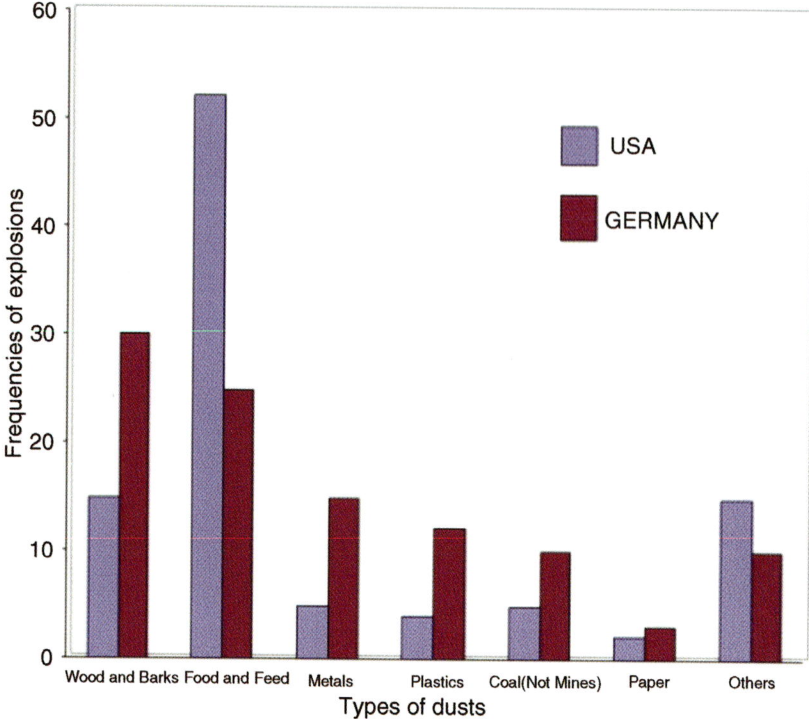

Figure 7.8 Frequencies of dust explosion accidents in Germany and the USA involving different types of dust [20].

can occur in all industries. Storage units (silos and hoppers), pulverizing units, conveying systems, dryers, furnaces, and mixing plants are most often affected [20]. Figure 7.8 shows the dusty materials frequently involved in dust explosions.

Large industrial plants tend to be better protected against the occurrence and the consequences of dust explosions than those of smaller companies, as the high numbers of explosions in wood and food processing plants versus plastics and metals show.

In the case of insufficient housekeeping in a factory, an explosion can lead to a secondary (dust) explosion or even a domino effect, when the blast propels dust deposits into the air, see Figure 7.9.

For details on dust explosions, see [21].

7.8
Legal Framework: Example of ATEX in Europe

The chemical industry which deals with high amounts of flammable substances in costly plants, has developed a wide range of safety tools to minimize process risks. Examples are *HAZOP studies* [7] and *near miss reporting* [8]. The company Dow

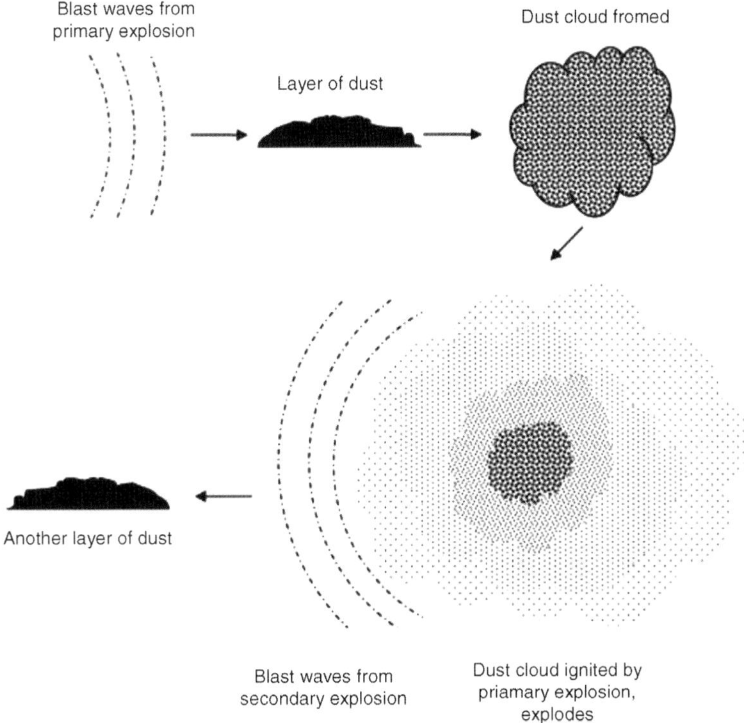

Figure 7.9 Domino effect in dust explosions [20].

started the development in this direction, which is now common practice, leading to a constant decline in work-place related incidents and accidents [9]. Smaller corporations outside the (petro)chemical industry, among them bakeries and repair shops, which have neither the budget for dedicated safety personnel nor the correct safety tools in place, are also at a high risk for fires and explosions. Therefore, legislative authorities have taken some measures with regard to explosion protection. In the EU, for instance, 2 ATEX directives, one for the manufacturers and one for the users of equipment, have been implemented:

- The ATEX 95 *equipment directive* 94/9/EC (Equipment and protective systems intended for use in potentially explosive atmospheres)
- The ATEX 137 *workplace directive* 99/92/EC (Minimum requirements for improving the safety and health protection of workers potentially at risk from explosive atmospheres)

ATEX is derived from the French for "*ex*plosive *at*mospheres". At its core is a zoning concept (area classification), see Table 7.3.

This concept is based on likelihoods and persistence of explosive atmospheres being present. If the user (i.e., operator of an industrial installation) wants to "downgrade" an area to a less critical zone, he has to implement proper measures.

Table 7.3 Zoning concept in ATEX for explosive gas/air and dust/air mixtures.

Zone		Description	Keyword for the presence of explosive atmospheres	Equipment Category
(gases and vapors)	(dusts)			
Zone 0	Zone 20	An area where an explosive atmosphere is present continuously or for long periods	Often	Category 1
Zone 1	Zone 21	An area where an explosive atmosphere is likely to occur in normal operation	Occasionally	Category 2
Zone 2	Zone 22	An area where an explosive atmosphere is not likely to occur in normal operation	Selom	Category 3
No zone	No zone	An area without risk for explosive atmosphere	Never	All (standard) equipment, no precautions necessary

Organizational measures (e.g., housekeeping) are significantly less secure than technical ones, such as N_2 blanketing or a process modification. Figures 7.10 and 7.11 make the zoning concept clear.

Figure 7.10 shows an example where a combustible dust is used. An explosive dust/air atmosphere can form. In this example, organic powders are dumped from bags into a bag slitting station connected to a fabric filter. The second example, Figure 7.11, highlights a situation where explosive gas/air mixtures can form: a petrol station. Note the different zones 20, 21 and 22, and 0, 1 and 2, respectively.

The operator of equipment has to produce an EPD (explosion protection document) which contains the findings of a risk assessment of any work activity involving flammable/or explosive atmospheres, plus the defined mitigating actions.

Further information on zoning is provided in the ATEX directive, the norm EN 60079 or for example, on official safety websites like here in the UK [11].

Classified areas (i.e., those with an assigned zone 0, 1, 2, 20, 21 or 22) must be protected from effective ignition sources. One such measure is that only suitable equipment is used. For instance, zones 0 and 20 require equipment (electrical and non-electrical) that is marked as "category 1", whereas zones 1 and 21 require "category 2" equipment, which is slightly less protected/demanding. Since "category 1" equipment is significantly more expensive to invest in and to maintain than "standard" equipment without (highest) protection, plant operators typically do not by default use such equipment in all areas of their factories where not necessary.

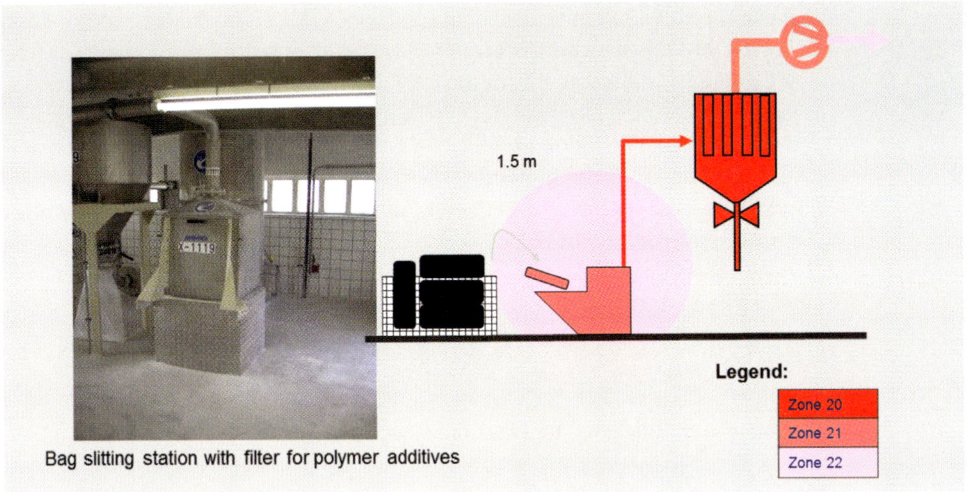

Figure 7.10 Example of a zoning concept for a bag dumping station with adjacent filter.

Effective ignition sources as defined by the European ATEX directive are:

- Lightning strikes
- Open flames (e.g., cigarettes and welding activities)
- Mechanically generated impact sparks (e.g., a hammer hitting a steel surface)
- Mechanically generated friction sparks (e.g., inside rotating equipment)
- Electric sparks (e.g., bad electrical connections, faulty pressure transmitters)
- High surface temperature (e.g., faulty equipment, mechanical friction in a bearing, high pressure steam pipes)
- Electrostatic discharges (e.g., produced by the flow of a non-conductive liquid through a filter screen without grounding) [22]
- Radiation
- Adiabatic compression (e.g., from compressed air systems)

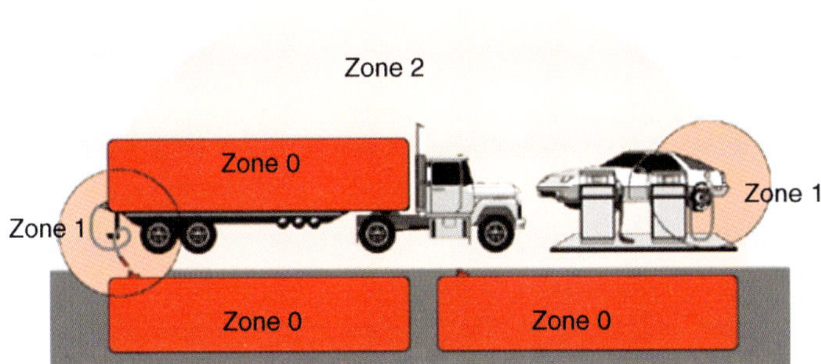

Figure 7.11 Example of a zoning concept for a petrol station [10].

The minimum ignition energy for gases (zones 0, 1, 2) is lower than that for dusts (zones 20, 21, 22), see also Chapter 2, Sections 2.2.8 and 2.2.9.

ATEX is only valid in the EU. In the USA, the classification of hazardous areas is based on the National Electrical Code (NEC), while in Canada the Canadian Electric Code (CEC) has to be followed. Also, the NFPA (National Fire Protection Agency) codes are widely followed in North America and beyond.

In Europe, Asia and Australia, there is also a tendency to follow the recommendations of the International Electrotechnical Commission (IEC).

Other bodies such as OSHA, NIST and CSB offer valuable information, see the weblinks in Section 7.14.

7.9
Preventing and Mitigating the Effect of Explosions in Industry

Once a plant operator has made his zoning plan (EPD), he can think about appropriate measures of explosion protection. There are several possibilities:

- Primary explosion protection: prevention of the formation of an explosive atmosphere
- Secondary explosion protection: prevention of ignition in explosive areas
- Constructive explosion protection: minimizing the effects of an explosion to a harmless level, see also Table 7.4.

Further information can be found in the document "NFPA 68: Standard on Explosion Protection by Deflagration Venting" [13].

For process section isolation, flame arrestors may be used. There are deflagration arrestors and detonation arrestors. Both are only suitable for pure gases, where

Table 7.4 Comparison of explosion protection design measures [12].

	Containment	Explosion venting	Explosion suppression
Pressure resistance	7–10 bar	Without relief pipe: up to 2 bar; With relief pipe: up to 4 bar	ST1 up to 0.5 bar; ST2 and ST3: up to 1.0 bar
Location	Independent	Dependent	Independent
Limits of application	Products which decompose spontaneously	Toxic product and products which decompose spontaneously	Products which decompose spontaneously, metal dust hazard
Environmentally friendly	Yes	No (flame, pressure, and product)	yes
Loss of material	+++++	+++++	++
Maintenance requirements	++++	++++	+++++

Figure 7.12 A commercially available spark detection and suppression system [14].

clogging of the screens can reliably be avoided (due to the short MSEG),[1] they have a relatively large pressure drop. A water seal drum can be used for "dirty" gases, for example, as a knockout drum upstream of a flare. Also, several vendors offer spark detection and suppression systems (spark extinguishing systems), which can be installed for example, upstream of a filter, see Figure 7.12. An alternative are fast-closing gates.

7.10
Aspects of Preventive Fire Protection

Most organic building materials are combustible. Fire protection aims at delaying combustion of the material so that people have more time to flee and fire fighters more time to act. In this section, several techniques for fire protection are shown. Generally, flame retardant additives are added to materials to be protected, for example, textiles aboard planes or polymer casings of electrical equipment.

7.10.1
Flame Retardants Containing Phosphorus

Phosphorus itself is highly combustible (compare a match head). However, phosphorus-containing compounds can be used for flame retardancy purposes, as they will cover the material under a dense layer of phosphorus compounds,

---

1) Maximum experimental safe gap.

keeping fresh oxygen away:

$$(\mathrm{NH_4PO_3})_n \rightarrow\rightarrow (> 250°C) \rightarrow\rightarrow (\mathrm{HPO_3})_n \tag{7.1}$$

$$\mathrm{P_{red}} + \mathrm{O_2} \rightarrow\rightarrow \mathrm{P_4O_{10}} \rightarrow\rightarrow (\mathrm{HPO_3})_n \tag{7.2}$$

Next, the polyphosphoric acid will produce a layer of protection by charring, for example, by water cleavage:

$$(\mathrm{HPO_3})_n + \mathrm{C}_x(\mathrm{H_2O})_m \rightarrow\rightarrow (\text{``C''})_x + (\mathrm{HPO_3})_n \bullet m\mathrm{H_2O} \tag{7.3}$$

7.10.2
Flame Retardants Based on Hydroxides of Al and Mg

The working principle is that the disintegration consumes energy and liberates water:

$$\mathrm{Al(OH)_3} + \text{energy} \rightarrow\rightarrow \mathrm{Al_2O_3} + \mathrm{H_2O} \tag{7.4}$$

$$\mathrm{Mg(OH)_2} + \text{energy} \rightarrow\rightarrow \mathrm{MgO} + \mathrm{H_2O} \tag{7.5}$$

The result is a cooling of the material to be protected, typically polymers, and a dilution of the combustion gases. A disadvantage of this system is that high loadings of several 10% are necessary, resulting in a considerable deterioration of the polymers' mechanical properties.

7.10.3
Organic, Halogen-Containing Compounds

The working principle is interruption of the radical chain mechanism of combustion in the gas phase.

1) Liberation of halogen radicals ($\mathrm{X^\bullet = Cl^\bullet, Br^\bullet}$) from the flame retardant additive (R–X)

$$R - X \rightarrow \mathrm{R^\bullet} + \mathrm{X^\bullet} \tag{7.6}$$

2) Formation of hydrogen halides (HX)

$$\mathrm{RH} + \mathrm{X^\bullet} \rightarrow \mathrm{HX} + \mathrm{R^\bullet} \tag{7.7}$$

3) Neutralization of the energetic radicals

$$\mathrm{HX} + \mathrm{H^\bullet} \rightarrow \mathrm{H_2} + \mathrm{X^\bullet} \tag{7.8}$$

$$\mathrm{HX} + \mathrm{OH^\bullet} \rightarrow \mathrm{H_2O} + \mathrm{X^\bullet} \tag{7.9}$$

7.10.4
Intumescence-Based Systems

An *intumescent* is a material that swells upon heat exposure, thereby increasing its volume. Intumescence-based systems are used for passive fire protection. The voluminous layer which is formed under fire/heat protects the unburnt material.

There are various systems, for example, soft charring and hard charring materials (the latter use sodium silicates and graphite). Generally speaking, an intumescence-system is composed of three materials:

1) Carbon-donor (e.g., polyalcohols, starch, pentaerythrite)
2) Acid donor (e.g., ammonium polyphosphate)
3) Blowing agent (e.g., melamin)

As the temperature increases, the following processes occur:

1) Softening of the binders/polymer
2) Release of an inorganic acid, for example, ammoniumpolyphosphate

$$(NH_4PO_3)_n \rightarrow\rightarrow (> 250°C, -n\,NH_3) \rightarrow\rightarrow (HPO_3)_n \tag{7.10}$$

3) Charring, for example, of polyalcohols

$$(HPO_3)_n + C_x(H_2O)_m \rightarrow (\text{``C''})_x + (HPO_3)_n \bullet mH_2O \tag{7.11}$$

4) Gas formation by the blowing agent, for example, melamin

$$C_3H_6N_6 \rightarrow\rightarrow (\text{heat}) \rightarrow\rightarrow NH_3 \rightarrow\rightarrow N_2 + H_2O \tag{7.12}$$

5) Foaming of the mixture

7.11
Fire Suppression by Oxygen Reduction

Hypoxic air technology for active fire prevention can be used inside buildings, such as warehouses and server rooms: The concentration of oxygen in the air is reduced from 21% in standard air by a few percent, which is still safe for humans who do not perform heavy work, but does not allow a fire to develop (no protection against pyrolysis, though). Other applications for hypoxic air are the preservation of artifacts and historic objects from degradation (oxidation) and the preservation of food from deterioration. Membrane technology can be used to separate air in an energy-efficient way. For details, see [19].

7.12
Safety by Process Design

The avoidance of fires and explosions across different industries was discussed above. Safety is also a key element in the design and operation of a combustor. Good industry practice, such as preventive maintenance, regular operator and engineer training, HAZOP studies [7] and near miss reporting [8] should be applied. Boilers involve (superheated) steam, and all combustors deal with high temperatures. Insulation of hot surfaces is one aspect not to be neglected. The amount of combustible material in the vicinity of a combustor should also be kept to a minimum, and offices (also temporary ones) should be placed at a distance that is safe in the case of a serious malfunction. There are industry-specific hazard codes that should be followed, for example:

- NFPA 85: Boiler and Combustion Systems Hazards Code (NFPA) [24]
- NFPA 86: Standard for Ovens and Furnaces (NFPA) [25]
- CSD-1: Controls and Safety Devices for Automatically Fired Boilers (ASME) [26]

7.13
Other Important Terms Related to Fire Safety

7.13.1
Flashover

When a fire occurs in a closed room, the available oxygen is soon consumed. Hot materials, for example, smouldering furniture, on the floor then keep decomposing and continue to produce combustible gases (e.g., CO) that move to the ceiling. The heat causes further pyrolysis in the entire room and further production of gases. At a certain point in time, the temperature increase reaches the autoignition temperature of the gases, which is around 500 °C, and an instantaneous combustion reaction takes place. This phenomenon in called "*flashover*". Firefighters fear flashovers, as they are very vigorous.

Two special types of flashovers are the "lean flashover" and the "rich flashover". The "lean flashover" is also called *rollover*. This is a quasi-simultaneous ignition of the hot combustible gases under the ceiling, where the air/fuel ratio is at the lower end (i.e., lean region) of the flammability. By contrast, a "rich flashover" occurs when the flammable gases are ignited at the upper region of the flammability range (i.e., rich). Often, a fire in a room has ceased due to missing oxygen. When a door or window is then opened, fresh oxygen is introduced, leading to an instantaneous ignition of the gases. This event is also termed "*backdraft*".

7.13.2
"Loss-of-containment" and Fires

In the process industries, loss-of-containment (LOC) of combustible materials can lead to fires. A study in the UK has revealed that 69% of LOC incidents can be categorized as either being the result of an incorrect action by an operator (37%) or equipment or plant failure (32%). Equipment failure was caused in one third of cases by inadequate maintenance and in another third by inadequate design. Interestingly, 56% of incidents occurred during normal operating mode [23].

7.13.3
Flammable Substance Release

Fires in industry can be classified by their phenomenology [15]:

- Pool fire
- Jet fire
- Boiling liquid expanding vapor explosion (BLEVE)
- Unconfined vapor cloud explosion (UVCE)
- Confined vapor cloud explosion (CVCE)

7.13.3.1 Boiling Liquid Expanding Vapor Explosion (BLEVE) and Vapor Cloud Explosion (VCE)

Fires involving liquefied gases are particularly dangerous. A BLEVE occurs in four stages (see Figure 7.13)

1) A fire around a liquefied gas container breaks out and starts heating the contents.
2) The fire leads to partial failure of the container, the egressing combustible fluid further accelerates the heating.
3) The pressurized liquid starts to boil, excessive pressure develops.

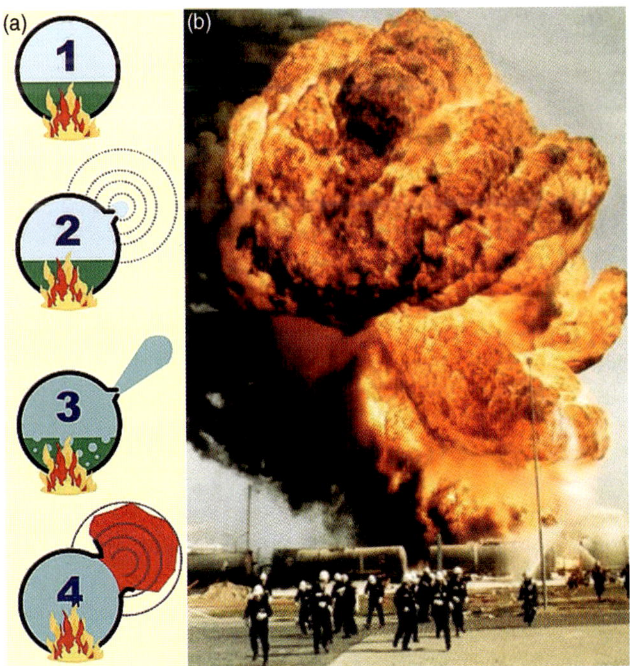

Figure 7.13 A BLEVE. (a) Scheme [16]; (b) Photo of an actual BLEVE in South America [17].

Table 7.5 Burning index.

Rating	Reaction	Example
1	No ignition	Sodium chloride
2	Short ignition and fast extinction	Tartaric acid
3	Local burning or smoldering with expansion to a small area	Lactose
4	Smoldering without emitting sparks or slow decomposition without flames	Tobacco
5	Burning with flames or emitting sparks	Sulfur
6	Deflagration or fast decomposition without flames	Black powder

4) The weakened container can no longer withstand the internal pressure, the container ruptures. As the pressure drops, the liquid is instantly vaporized. The gas cloud mixes with ambient air and ignites, resulting in a severe explosion

A BLEVE can have fatal consequences for fire fighters and the area around.

7.13.4
Burning Index (Danger Class, Rating)

The burning index (Table 7.5) is a value that describes in which kind of reaction a fire spreads out in a dust layer after ignition. It is determined according to the method VDI 2263 ("Dust fires and dust explosions").

7.13.5
K Value (K_G, K_{ST})

The K value is a characteristic of explosive mixtures. For flammable gases and vapors, the K_G value is used, and for dusts the K_{ST} value. It is calculated according to the cubic law from the pressure rise per unit time of an explosion (maximum pressure rise rate):

$$(\mathrm{d}p/\mathrm{d}t)_{\max} V^{(1/3)} = \text{const.} = K \text{ value} \tag{7.13}$$

V is the vessel volume.

7.13.6
Dust Explosion Class

Based on their K_{ST} values, explosive dusts are grouped into three categories, see Table 7.6.

The dust explosion classes are a simplification for engineering purposes. Depending on its particle size, a dust of one material can fall into different ST classes. The data for a given dust can be obtained from GESTIS DUST EX [6] or, best, by measurement. Designing explosion protection for class ST3 dusts is difficult and costly. For details on dust explosion classes, see VDI 2263 [18].

Table 7.6 Dust explosion class (ST1, ST2 and ST3).

K_{ST} Value [bar m s^{-1}]	Dust Explosion Class	Example (see also GESTIS DUST EX)
0–200	ST1	Milk powder, 100 µm
200–300	ST2	Barley grain dust, 50 µm
>300	ST3	Al powder, 40 µm

7.13.7
Explosion Pressure

The explosion pressure, measured in bar, is the maximum pressure that an explosive mixture can produce in a defined, closed container upon ignition (deflagration). For gases and vapors, the standard is EN 13673, and for dusts EN 14034. Turbulence does not strongly influence the result. Note that in long tubes, a deflagration can transition into a detonation, which is characterized by significantly higher pressures. The explosion pressure depends on the initial pressure, temperature and mixing ratio. For dusts, the particle size also plays a role (small particles react more violently). The explosion pressure, the particle size also plays a role (small particles react more violently). The explosion pressure more specifically the maximum pressure and the maximum pressure rise, are necessary for the design of explosion protection and mitigation measures, such as bursting disks.

7.13.8
Limiting Oxygen Concentration (LOC)

The limiting oxygen concentration (LOC) is the maximum content of oxygen in a mixture of a combustible gas with air and inert gas, at which the mixture will just ignite (explode). Below this limit adding any amount of combustible would not form an explosive mixture. Common standards are VDI 2263, ISO 14034 and ISO 14756.

Web Resources

There are good resources on the internet to learn more about fire prevention and accident investigation. A few are listed below for the interested reader:

GESTIS DUST EX	http://www.dguv.de/ifa/en/gestis/expl/index.jsp
US National Institute of Standards and Technology	http://www.nist.gov/fire/
US National Fire Protection Association	http://www.nfpa.org
US Chemical Safety Board	http://www.csb.gov/
US Occupational Safety and Health Association	http://www.osha.gov/dts/shib/shib073105.html
UK Fire Protection Association	https://www.thefpa.co.uk/fpa_home/

References

1 U.S. Chemical Safety and Hazard Investigation Board, Investigation Report "Sugar Dust Explosion and Fire (14 killed, 36 injured), Report No. 2008-05-I-GA, September (2009) http://www.csb.gov/assets/document/Imperial_Sugar_Report_Final_updated.pdf (accessed June 1, 2013).

2 Babrauskas, V. (2003) *Ignition Handbook: Principles and Applications to Fire Safety Engineering, Fire Investigation, Risk Management and Forensic Science*, Fire Science Pub, 978-0972811132.

3 http://www.feuerwehr-glashuette.de/fettexplosion.html (accessed June 1, 2013).

4 http://fire.nist.gov/tree_fire.htm (2013).

5 http://www.safetynet.de/Publications/articles/CMRNov99.pdf (accessed June 1, 2013).

6 GESTIS DUST EX, IFA (Institut für Arbeitsschutz der Deutschen Gesetzlichen Unfallversicherung), http://www.dguv.de/ifa/en/gestis/expl/index.jsp (accessed June 1, 2013).

7 Tyler, B., Crawley, F., and Preston, M. (2008) *HAZOP: Guide to Best Practice*, 2nd edn, Institution of Chemical Engineers (IChemE), 978-0852955253.

8 McKinnon, R.C. (2012) *Safety Management: Near Miss Identification, Recognition, and Investigation*, CRC Press, 978-1439879467.

9 Brandt, E.N. (1997) *Growth Company: Dow Chemical's First Century*, Michigan State University Press, 978-0870134265.

10 Bartec GmbH (2013) http://www.bartec.de.

11 http://www.hse.gov.uk/fireandexplosion/zoning.pdf (accessed June 1, 2013).

12 (2002) *Ullmann's Encyclopedia of Industrial Chemistry*, 6th edn, Wiley-VCH Verlag GmbH, 978-3527303854.

13 http://www.nfpa.org/aboutthecodes (accessed June 1, 2013).

14 http://www.grecon.de/grecon_product_bs_7_en.php (accessed June 1, 2013).

15 Center for Chemical Process Safety (2013) http://www.aiche.org/CCPS/index.aspx (accessed June 1, 2013).

16 http://www.blaulicht.at/grafik-downloads.html (accessed June 1, 2013).

17 http://1.bp.blogspot.com/-jlNNT4isBgA/TbCFpYpYTFI/AAAAAAAAA5c/P9fqNgTrB7M/s1600/bleve_10_C0D0443F-0978-09AC-82A33391DCD10134.jpg (accessed June 1, 2013).

18 VDI 2263, Staubbrände und Staubexplosionen, Gefahren – Beurteilungen – Schutzmaßnahmen – Bestimmung des Staubungsverhaltens von Schüttgütern (Dust fires and dust explosions – Hazards – assessment – protective measures – Determination of dustiness of bulk materials, VDI (Verein der Ingenieure), May (2008) http://www.vdi.de/uploads/tx_vdirili/pdf/1417412.pdf (accessed June 1, 2013).

19 Xin, Y. and Khan, M.M. (2007) Flammability of combustible materials in reduced oxygen environment. *Fire Safety Journal*, **42** (8), 536–547.

20 Abbasi, T. and Abbasi, S.A. (2007) Dust explosions–Cases, causes, consequences, and control. *Journal of Hazardous Materials*, **140** (1–2), 7–44.

21 Eckhoff, R.K. (2003) *Dust Explosions in the Process Industries*, 3rd edn, Elsevier, 978-0-7506-7602-1.

22 Glor, M. (2003) Ignition hazard due to static electricity in particulate processes. *Powder Technology*, **135–136**, 223–233.

23 Haywood, B. (2012) Loss of containment (LOC) data. HSE Safety Information Posts, www.safteng.net, http://www.safteng.net/images/stories/PDF/loss_of_containment_hse_report.pdf (accessed June 1, 2013).

24 NFPA 85: Boiler and Combustion Systems Hazards Code, NFPA (2011) http://www.nfpa.org/aboutthecodes/AboutTheCodes.asp?DocNum=85.

25 NFPA 86: Standard for Ovens and Furnaces, NFPA (2011) http://www.nfpa.org/aboutthecodes/AboutTheCodes.asp?DocNum=86 (accessed June 1, 2013).

26 CSD-1: Controls and Safety Devices for Automatically Fired Boilers, ASME 9780791834107 (2012) http://www.asme.org/.

Index

Combustion: From Basics to Applications, First Edition. Maximilian Lackner, Árpád B. Palotás, and Franz Winter.
© 2013 Wiley-VCH Verlag GmbH & Co. KGaA. Published 2013 by Wiley-VCH Verlag GmbH & Co. KGaA.